BIOCHEMISTRY

BIOCHEMISTRY

L. VEERAKUMARI
Reader in Zoology,
Pachaiyappa's College
Chennai 600 030

MJP PUBLISHERS
Chennai 600 005

Impressions: 2005, 2007, 2010

Cataloguing-in-Publication Data

Veerakumari. L (1957 –)
Biochemistry/ L. Veerakumari. –
Chennai : MJP Publishers, 2004.
xii, 380p. ; 21 cm.
Includes Glossary and Index.
ISBN 978-81-8094-004-0 (pbk.)
1. Biochemistry 2. Metabolism.
I. Title
574.19 VEE MJP 003

ISBN 978-81-8094-004-0

Printed and bound in India

MJP PUBLISHERS
47, Nallathambi Street
Triplicane
Chennai 600 005

Publisher : J.C. Pillai
Managing Editor : C. Sajeesh Kumar
Marketing Manager : S.Y. Sekar
Project Editor : P. Parvath Radha
Acquisitions Editor : C. Janarthanan
Editorial Team : N. Yamuna Devi, Lissy John, M. Gnanasoundari, N. Thilagavathi
CIP Data : Prof. K. Hariharan, Librarian RKM Vivekananda College, Chennai.

To
my teachers and students

PREFACE

This book explains the basic concepts in biochemistry and is primarily intended for beginners in the subject. Knowledge of biochemistry is mandatory for students of zoology, botany, microbiology, biotechnology, bioinformatics, agriculture, home science, veterinary science and health science. Students of biology possess inadequate knowledge of biochemistry and have difficulty in understanding its basic principles. Efforts have been put in to present the subject in a simple and lucid way.

Carbohydrates, lipids, proteins, water and vitamins are the essential nutrient molecules for living organisms. Living cells are self-regulatory chemical engines. Myriads of enzyme-catalysed chemical reactions take place in living cells to release chemical energy which is used to perform various biological functions. Hormones play an important role in chemical coordination. The chapters presented in this book are oriented around these facts. The biological significance and chemistry of water, carbohydrates, lipids, proteins and their metabolic pathways and energetics are explained in the first eight chapters. The structure and fuction of nucleic acids, the biochemistry of enzymes and high-energy compounds are discussed in the next three chapters. Following this are two chapters dealing with the chemistry and physiological functions of vitamins and hormones. The next two chapters explain biological detoxification methods and, chemistry and mode of action of antibiotics. The last chapter construes the basic principles and applications of some important biochemical techniques. Additional information has been provided in the appendices and glossary.

I thank my parents, colleagues and freinds for their constant encouragement and support. I would like to thank my husband Mr. S. Rameshbabu who has all along been a powerful guiding force, encouraging me thoughout the period of my career. I am indebted to all the distinguished authors of related books which provided voluminous information for the preparation of this book. Grateful acknowledgement is made to the publisher for providing technical assistance and for the patience and care with which this book has been produced in an excellent manner.

Valuable suggestions and comments are most welcome.

L. Veerakumari

CONTENTS

1

INTRODUCTION

Life is an intricate mixture of chemicals. Biochemistry deals with the chemistry of living organisms. The study of biochemistry shows how the collections of inanimate molecules that constitute living organisms interact to maintain and perpetuate life, animated solely by the chemical laws that govern the nonliving universe. Biochemical research has revealed that all organisms are remarkably alike at the cellular and chemical levels. Biochemistry describes in molecular terms the structures, mechanisms and chemical processes shared by all organisms and provides organising principles that underlie life in all of its diverse forms.

The findings that all biological macromolecules in all organisms are made from common building-block molecules has provided strong evidence that modern organisms have descended from a single primordial cell line whose fundamental chemistry would be recognisable even today. Furthermore, several billion years of adaptive selection have refined cellular systems to take maximum advantage of the chemical and physical properties of these molecular raw materials for carrying out the basic energy-transforming and self-replicating features of a living cell. The simple organic compounds from which living organisms are constructed are called 'biomolecules'.

1.1 ATOMS AND MOLECULES

An element is composed of atoms, all of the same type. An **atom** is defined as the smallest unit of an element, which participates in the chemical combinations with other elements. Each atom consists of a nucleus containing neutrons which have no electrical charge, and a fixed number of protons, each of which has a unit positive charge. The nucleus is surrounded by electrons, each having unit negative charge equal to that of the proton, and arranged in layers or shells. The number of protons in the atomic nucleus is called *atomic number*; the total number of protons and neutrons in the atomic nucleus is the *atomic weight* of that particular atom.

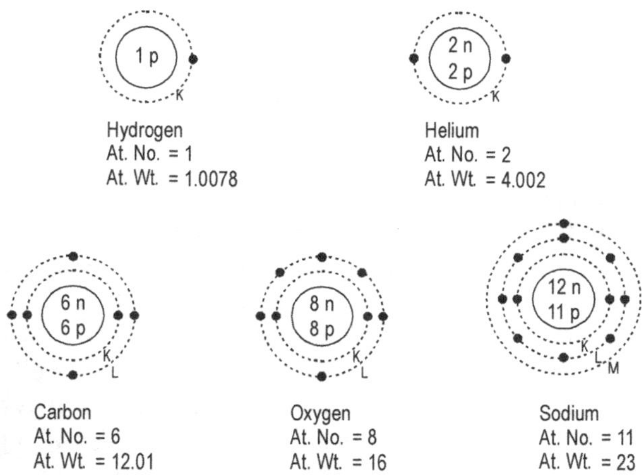

Fig. 1.1 Electronic configuration of some atoms

Hydrogen has one electron, one proton and no neutron. Hence, the outer shell contains one electron. Helium atom has two protons, two neutrons. The outer shell contains two electrons. Carbon atom has 6 electrons, 6 protons and 6 neutrons. The electrons are arranged in two shells. The first shell contains two electrons and the outer shell contains four electrons. Oxygen atom contains 8 electrons, 8 protons and 8 neutrons. The electrons are arranged in two shells. The first shell contains two electrons and the outer shell contains 6 electrons. Sodium atom contains 11 electrons, 11 protons and 12 neutrons. The electrons are arranged in three shells. The inner shell contains two electrons, the middle shell contains 8 electrons and the outer shell contains one electron (Figure 1.1).

Atoms of different elements combine with each other usually in fixed ratios, to give compounds. Most biomolecules are composed of carbon. Carbon, like hydrogen, oxygen and nitrogen, is capable of forming *covalent bonds*. Two atoms with unpaired electrons in their outer shells can form covalent bonds with each other by sharing electron pairs. Atoms participating in covalent bonding tend to fill their outer shell. The hydrogen atom needs one electron, oxygen needs two, nitrogen three, and the carbon atom four to fill their respective outer shells. Thus, a carbon atom can share four electron pairs with four hydrogen atoms to form the compound, methane (CH_4), in which each of the shared electron pairs is a single bond (Figure 1.2).

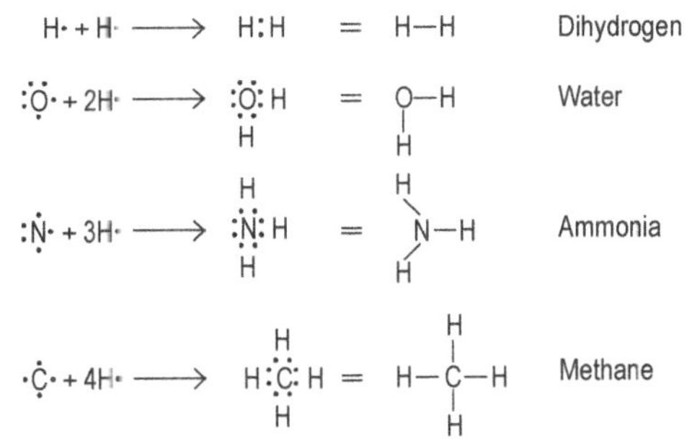

Reaction	Electron-dot structure	Bond structure	Compound
H· + H· ⟶	H:H	= H—H	Dihydrogen
:Ö· + 2H· ⟶	:Ö:H, H	= O—H, H	Water
:N· + 3H· ⟶	H, :N:H, H	= H, N—H, H	Ammonia
·C· + 4H· ⟶	H, H:C:H, H	= H, H—C—H, H	Methane

Fig. 1.2 Covalent bonding

Carbon can form a single bond with oxygen and nitrogen atoms. But most significant in biology is the ability of carbon atoms to share electron pairs with each other to form very stable carbon–carbon single bonds.

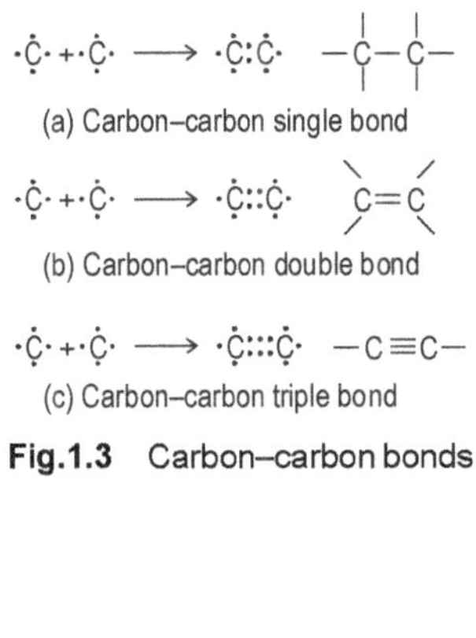

·C· + ·C· ⟶ ·C:C· —C—C—

(a) Carbon–carbon single bond

·C· + ·C· ⟶ ·C::C· C=C

(b) Carbon–carbon double bond

·C· + ·C· ⟶ ·C:::C· —C≡C—

(c) Carbon–carbon triple bond

Fig.1.3 Carbon–carbon bonds

Each carbon atom can form single bond with one, two, three or four other carbon atoms. Two carbon atoms also can share two or three electron pairs with each other, thus forming carbon-carbon double or triple bonds (Figure 1.3).

There are other types of bonding besides covalent bonding. *Electrovalent* or *ionic bonding* occurs when entities of opposite charge are held together by electrostatic attraction. This occurs in crystals and in large molecules where they contribute to cohesion between chains.

$$\vdash NH_3^{+\,-}OOC \dashv$$

Coordinate linkages are a type of covalency but both electrons in the bond are donated by one atom. For example, in certain compounds nitrogen and iron are linked by coordination.

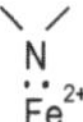

Hydrogen bonds occur when hydrogen is covalently bonded to a more electronegative element. If electrons are drawn away from the hydrogen atom, it acquires positive charge and so can attract another electronegative atom. Prominent electronegative elements are O, and N.

$$N{-}\overset{\delta^+}{H}\cdots\overset{\delta^-}{O}{-}C$$

In a hydrogen bond, a hydrogen atom serves as a bridge between two electronegative atoms, holding one by a covalent bond and the other by electrostatic forces. This electrostatic bond is about one-tenth the strength of a normal covalent bond.

Hydrophobic interactions bring about the association of nonpolar groups with each other in aqueous systems because of the tendency of the surrounding water molecules to seek their most stable state.

Van der Waals interactions are weak interatomic attractions. When two uncharged atoms are brought very close together, their surrounding electron clouds influence each other. Random variations in the positions of the electrons around one nucleus may create a transient electric dipole, which induces a transient, opposite electric dipole in the nearby atom. The two dipoles weakly attract each other, bringing the two nuclei closer. These weak attractions are called van der Waals interactions. As the two nuclei draw closer together, their electron clouds begin to repel each other. At the point when the van der Waals attraction exactly balances this repulsive force, the nuclei are said to be in van der Waals contact. Each atom has a characteristic van der Waals radius, a measure of how close that atom will allow another to approach.

1.2 FUNCTIONAL GROUPS

Nearly all organic biomolecules can be regarded as derivatives of hydrocarbons—compounds of carbon and hydrogen in which the backbone consists of carbon atoms joined by covalent bonds and the other bonds of carbon are shared with hydrogen atoms. The backbone of such hydrocarbons is very stable, because carbon–carbon single and double bonds share their electron pairs equally. One or more hydrogen atoms of hydrocarbons may be replaced by different kinds of functional groups to yield different families of organic compounds. Typical families of organic compounds and their characteristic functional groups are alcohols which have one or more hydroxyl groups, amines which have amino groups, ketones which have carbonyl groups and acids which have carboxyl groups. Several other common functional groups like methyl, ethyl, disulphide, phospho, guanido, phenyl and imidazole are also important in biomolecules. Some of the functional groups prominent in biological materials are given in Table1.1.

1.3 ORIGIN OF LIFE

Biomolecules first arose by chemical evolution. The first scientific account of the origin of life was given by A. I. Oparin in his book, *The Origin of Life*, published in 1936. Oparin based his contentions on the theory of the fiery origin of planets. This fiery origin of planet (earth), postulated by Sir James Jeans states that a mass of the sun's atmosphere was torn away by another star with the result that various fragments spread out in different directions and later on condensed into the respective planets. At that time, temperature was extremely high, and served as the basis for the primary origin of organic compounds on earth. Gradually, as the universe cooled, various saturated and unsaturated hydrocarbons were formed.

Table 1.1 Some functional groups

Structural representation		Example	
$-OH$	Alcohol, Hydroxyl	C_2H_5OH	Ethyl alchohol
$-CHO$ or $-\overset{O}{\overset{\Vert}{C}}-H$	Aldehyde	CH_3-CHO	Acetaldehyde
$>C=O$	Ketone, carbonyl	$CH_3-CO-CH_3$	Acetone
$-C-O-C-$	Ether	CH_3-O-CH_3	Diethyl ether
$-C-S-C-$	Thioether	$H_3C-S-CH_2-CH_2-\underset{}{\overset{NH_2}{\overset{\vert}{C}}}HCOOH$	Methionine
$-OH$	Phenol	$HOOC-\overset{NH_2}{\overset{\vert}{C}}H-CH_2-$ $-OH$	Tyrosine
$-NH_2$	Amino	$HOOC-CH_2-NH_2$	Glycine
$-COOH$	Carboxyl	CH_3-COOH	Acetic acid
$-\overset{O}{\overset{\Vert}{C}}-NH_2$ or $-CONH_2$	Amide	$CONH_2$ N	Nicotinamide
$-SH$	Thiol, Sulphydryl	$HS-H_2C-\overset{NH_2}{\overset{\vert}{C}}H-COOH$	Cysteine

Table 1.1 Some functional groups (contd.)

Structural representation		Example	
$-S-S-$	Disulphide	$HOOC-CH(NH_2)-CH_2-S-S-H_2C-CH(NH_2)-COOH$	Cystine
$-C(=O)-O-$ or $-CO-O-$	Ester	$R-O-CO-R'$	Waxes
$-C(=O)-S-$ or $-CO-S-$	Thioester	$CH_3-CO-S-CoA$	Acetyl CoA

The saturated and unsaturated hydrocarbons combined and recombined in various ways through the processes of condensation, polymerisation and oxido-reduction forming a specialised mixture of water called hot dilute soup by Haldane. This water contained the alcohols, aldehydes, acetates, acids and various amines and amino acids. The amino acids polymerised resulting in the formation of polypeptides—the first stage of protein synthesis.

During various mechanisms of polymerisation and condensation, a particular phenomenon of coacervation took place, which according to Oparin, led to the origin of living matter. Because of mixing of two different colloidal solutions, some microscopic droplets got separated from them and these were called as coacervates. Oparin considered these coacervates as the sole living molecules, which gave rise to life. Internal reactions in these coacervates took place like that of catalysis, etc. with the result, selective complexity specified for particular organisms evolved. At this stage, the enzymes which developed in these coacervates, moulded them into successful droplets. The primitive organisms were successful coacervates. The energy for these coacervates to carry on their activities was provided by the adenylic system.

The most abundant elements in the living organisms are hydrogen, oxygen, carbon and nitrogen. Simple low molecular weight molecules like CO_2, N_2 and H_2O are taken up by the living organisms from the atmosphere and are synthesised into the simple building blocks like glucose, nucleotides, amino acids and fatty acids. From these, the large macromolecules like polysaccharides, nucleic acids, proteins and lipids are formed. Nucleic acids and proteins are called the 'informational macromolecules'. Each nucleic acid and each protein has got a specific nucleotide or amino acid sequence, which contains the information about its functional role in the cell. On the other hand, polysaccharides and lipids have a relatively simpler structure and do not have any function in carrying information. They are the energy sources for living organisms. Carbohydrates, proteins and fat may further combine to form supramolecular systems. The components of supramolecular systems are held together, not by covalent bonds, but by relatively weak, non-covalent forces like hydrogen bonds, hydrophobic interactions and van der Waals interactions. Though the individual bonds are weak, their very large numbers give sufficient stability to the supramolecular complexes. The supramolecular systems are further assembled together into cell organelles like nuclei, mitochondria, chloroplasts, and others. As evolution progressed further, different kinds of organisms ultimately began to interact with each other, to exchange nutrients and energy, thus forming increasingly complex ecological systems.

2

WATER

Water is the major component of the protoplasm and constitutes more than 60% of the total protoplasmic mass. However, it varies in different tissues of the same animal and the same tissues of different animals. For example, the enamel of tooth has only 2% water, whereas, in nervous tissue it ranges up to 80%. Water is the most effective and universal solvent and furnishes a medium in which most biochemical reactions occur.

2.1 BIOLOGICAL IMPORTANCE OF WATER

Water possesses certain peculiar and unique properties very vital for the maintenance of life activities. It is the universal solvent effecting many chemical reactions and facilitating the breakdown of loosely associated molecules into their components which are the electrically charged particles called ions. Being a small molecule, water can pass through the membranes easily and help in ionic regulation. Even very thin layers of water can absorb large quantities of ultraviolet rays and indirectly protect the tissues from these rays. The unique biological properties of water are a reflection of its exceptional physicochemical characteristics.

2.2 PROPERTIES OF WATER

2.2.1 CHEMICAL NATURE OF WATER

Water is formed from hydrogen and oxygen atoms with two covalent bonds, holding the two hydrogen to one oxygen and forming an angle of 104.5°. The length of each bond is 0.99 A° (Figure 2.1).

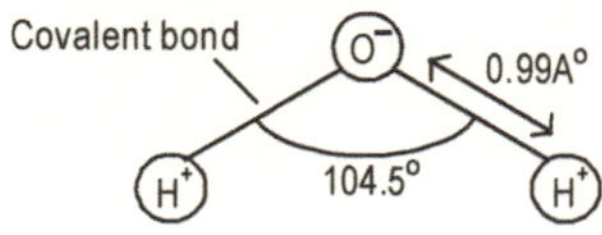

Fig. 2.1 The dipole nature of water molecule

2.2.2 DIPOLE NATURE OF WATER

Water is dipole, a particle with two charges of opposite sign. In a molecule of water the hydrogen atoms are positively charged, whereas the oxygen atom is negatively charged and the configuration of the bond angle makes the electric charge on the molecule unevenly distributed. Therefore, it tends to orient itself as a dipole in an electric field.

2.2.3 WATER AS A SOLVENT

The unequal distribution of charges on a water molecule enables it to form hydrogen bonds by electrostatic attraction (Figure 2.2). However, the bonds are weak as compared to covalent bonds. This property enables water to act as an excellent solvent for salts that are completely dissociated into ions even in the solid, crystalline state.

Water has a strong dielectric constant, i.e. the ability to reduce the constancy of an external electric field, along with a permanent dipole moment. The dipole nature of all water molecules in a solution add and produce a field of opposite charges. This resulting field opposes original external electric field by appropriate orientation. Thus, in an electrolytic solution the pairs of oppositely charged ions have the electrostatic attraction between them greatly reduced by the high dielectric constant of water. After electrolytic dissociation, each ion can behave as a single molecule due to reduction in the forces binding them together.

```
                     H                 H
                     |               /
Hydroxyl group   R—C—O—H·····O
                     |               \
                     H                 H

                                 O
                               /   \
                         O·······H     H
                        /
Carboxyl group   R—C
                        \          H
                         OH·····O
                                   H

                     H     H····· O—H
                     |   /          \H
Amino group      R—C—N
                     |   \          H
                     H     H····· O—H

                       H····· O—H
                      /         \H
Aldehyde         R—C            O
                      \\       /   \
                        O····· H     H

                                 O
                               /   \
Ketone           R—C=O·····H     H
```

Fig. 2.2 The organic groups with which water forms hydrogen bonds and enhances solubility

The charged character of water molecule permits it to form oriented layers around macromolecules such as proteins that contain many charged groups (Figure 2.3). This action not only aids in the separation of solution of proteins but also forms a protective field around them that cuts down the forces of their charged groups.

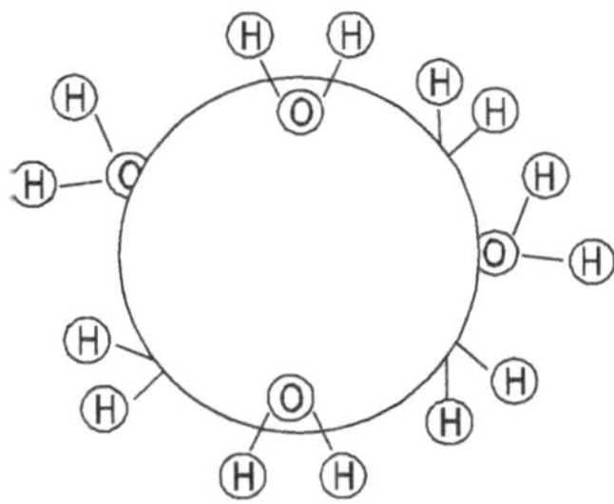

Fig. 2.3 Orientation of water molecules around macromolecules

2.2.4 HIGH SURFACE TENSION

The force with which the surface molecules are pulled towards the interior is known as surface tension. It is a measure of the energy required to expand a surface. Water has a high surface tension because its molecules are more closely bound together at the interfaces between water and other media (Figure 2.4). Hence, dissolved substances tend to be concentrated at the surface. Under such condition, chemical reactions between various dissolved substances occur more rapidly. This effect is much like the action of inorganic catalysts in promoting chemical reactions by holding reactants together in a reactive configuration. This property is of great biological significance and it enables the formation and maintenance of lipoprotein membranes of cell organelles in proper functional configuration.

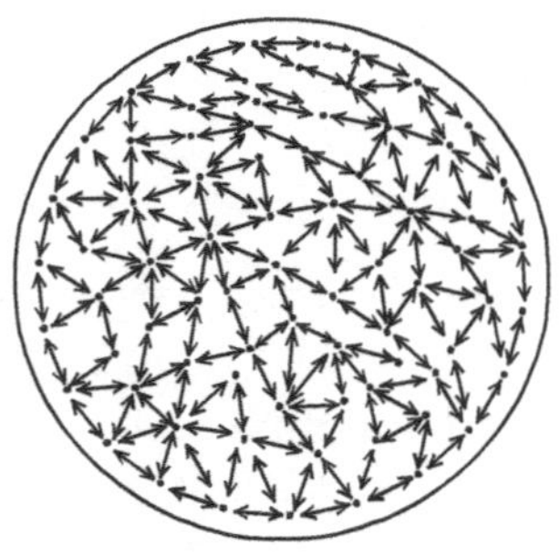

Fig. 2.4 Surface Tension

2.2.5 LOW VISCOSITY

Viscosity is a measurement of the ease or difficulty of flow of a liquid. Water has low viscosity and this property serves as an efficient vehicle for the transport of substances inside a cell as well as through animal body. Low viscosity also enables the movement associated with changes in the form that generally occur during muscle contraction, amoeboid movements, etc.

2.2.6 THERMAL PROPERTIES OF WATER

Water plays an important role in the maintenance of body temperature. Latent heat is the quantity of heat required to change the state of a substance from one form to another without change in the temperature. This property prevents a cell or organism from freezing. High specific heat, high heat of vaporisation and high heat of conductivity are the three physical properties that enable water to act as a temperature regulator.

High specific heat It is the quantity of heat in calories required to raise the temperature of one gram of substance through one degree celsius. The higher the specific heat of a substance, the lower is the change in temperature, when heat is gained by it. Thus, during metabolic reactions, as the temperature rises, the hydrogen bonds of water molecules gradually break up by the absorption of heat energy. The higher specific heat enables heat gain without considerable rise in temperature, thus aiding thermal stabilisation of the cells.

High heat of vaporisation It is the amount of heat absorbed / released when a substance is transformed from liquid state into gaseous state. The high heat of vaporisation accounts for evaporative cooling that avoids overheating. Thus, a large amount of heat can be dissipated by vaporisation of water. This principle operates in the process of sweating, a physiological mechanism employed in thermoregulation, whenever the body temperature rises above tolerable range.

High heat of conductivity This property enables prompt and equal distribution of heat all over the body and thus prevents localised damages which can be caused by overheating.

2.3 ELECTROLYTIC DISSOCIATION OF WATER

Water can be considered to be a weak electrolyte. It can dissociate as follows:

$$H_2O \rightleftharpoons H^+ + OH^-$$

The equilibrium constant for the dissociation of water is given by the equation:

$$K_{eq} = \frac{(H^+)(OH^-)}{(H_2O)} \quad \text{(or)} \quad K_{eq}(H_2O) = (H^+)(OH^-)$$

At 25°, K_{eq} is found to be 1.8×10^{-16}.

In pure water, the molecular concentration of water (H_2O) is 1000/18 i.e. 55.5 moles/litre.

Therefore, $(H^+) \times (OH^-) = 55.5 \times 1.8 \times 10^{-16} = 1.01 \times 10^{-14}$

This product $(H^+) \times (OH^-)$ is called the ionic product of water, abbreviated as K_w.

Therefore, $K_w = 1.01 \times 10^{-14}$ at 25°.

In pure water equal number of hydrogen and hydroxyl ions exist i.e. $(H^+) = (OH^-)$

Therefore $(H^+) \times (OH^-) = (H^+) \times (H^+) = (H^+)^2 = 1.01 \times 10^{-14}$

For ease of representation of (H^+) in numbers rather than as fractions, *Sorenson* developed the concept of pH. pH is defined as the logarithm of the reciprocal of hydrogen ion concentration.

$$pH = \log \frac{1}{(H^+)} = -\log (H^+)$$

For pure water, since $(H^+) = 1.0 \times 10^{-7}$, pH will be 7.0.

2.4 ACIDS AND BASES

According to the theory of Bronsted, an acid is a proton donor and a base is a proton acceptor.

$$\underset{\text{(Acetic acid)}}{CH_3COOH} \longrightarrow \underset{\text{(Acetate)}}{CH_3COO^-} + H^+$$

Acetic acid is a proton donor and hence an acid. Acetate (CH_3COO^-) is a proton acceptor and hence a base. Acetic acid and acetate together constitute a conjugate acid–base pair. In dilute aqueous solutions, an acid will dissociate to give a proton, which is taken up by a water molecule to form a hydronium ion, H_3O^+.

$$HA + H_2O \longrightarrow H_3O^+ + A^-$$

Acids which have only a slight tendency to give up protons, are weak acids (e.g. acetic acid) and acids which give up their protons readily to water, are strong acids (e.g. HCl).

2.4.1 DISSOCIATION OF STRONG ELECTROLYTES

Strong acids, bases and their salts are called strong electrolytes. They dissociate almost completely in water.

$$HCl \longrightarrow H^+ + Cl^-$$
$$NaOH \longrightarrow Na^+ + OH^-$$
$$NaCl \longrightarrow Na^+ + Cl^-$$

2.4.2 DISSOCIATION OF WEAK ELECTROLYTES

Weak acids, weak bases and their salts are called weak electrolytes. They dissociate only slightly in solution.

Taking acetic acid as an example,

$$CH_3COOH \longrightarrow CH_3COO^- + H^+$$

The equilibrium constant for this acid is 1.8×10^{-5}

i.e. $$\frac{(H^+)\,(CH_3COO^-)}{(CH_3COOH)} = 1.8 \times 10^{-5}$$

The pH of a 1.0 molar solution of acetic acid can be calculated from the above equation to be 2.38.

2.5 HENDERSON–HASSELBALCH EQUATION

It is important for understanding the buffer action and acid–base balance in the blood. The quantitative relationship between pH, the buffering action of a mixture of weak acid with its conjugate base, and the pk' of the weak acid is given by the Henderson–Hasselbalch equation.

The tendency of a weak acid (HA) to lose a proton (H^+) and form its conjugate base (A^-) is defined by the equilibrium constant K_a for the reversible reaction,

$$HA \rightleftharpoons H^+ + A^-$$

The equilibrium constant $K_a = \dfrac{(H^+)\,(A^-)}{HA}$

$$K_a \times HA = (H^+)\,(A^-)$$

$$H^+ = K_a \frac{HA}{A^-}$$

Taking negative logarithm on both sides, we have,

$$-\log (H^+) = -\log K_a - \log \frac{HA}{A^-}$$

But, $-\log (H^+) = pH$

$-\log K_a$ is defined as pK_a and

$$-\log \frac{HA}{(A^-)} = \frac{\log (A^-)}{\log HA}$$

Substituting in the above equation,

$$pH = pK_a + \log \frac{(A^-)}{HA}$$

or

$$pH = pK_a \frac{(A^-)}{HA}$$

pK_a is the dissociation constant of the weak acid HA; (HA) and (A^-) are the molar concentrations of the weak acid and its conjugate base respectively.

2.6 BUFFERS

A buffer solution is one which resists a change in pH when an acid or base is added to it. It is usually made up of a mixture of weak acid and its conjugate base (salt of the weak acid), e.g. acetic acid and sodium acetate. Suppose an alkali is added to such a mixture, it will react with the weak acid to form its salt.

$$NaOH + CH_3COOH \longrightarrow CH_3COONa + H_2O$$

If an acid is added, this will be taken up by the base, sodium acetate.

$$HCl + CH_3COONa \longrightarrow CH_3COOH + NaCl$$

In either case, there is no increase in either H^+ or OH^-. The relative concentrations of the acetate and acetic acid are altered slightly.

2.6.1 BUFFER SYSTEMS OF BLOOD

The different buffer systems of blood which play a significant role in the regulation of acid–base balance are the following:

i. Bicarbonate buffer system $H_2CO_3/BHCO_3$
ii. Hemoglobin buffer system HHb/BHb
iii. Plasma protein buffer system H.Protein/B.Protein
iv. Phosphate buffer system BH_2PO_4/B_2HPO_4

B indicates the bases Na^+, K^+, Ca^{++}, or Mg^{++} and H indicates acid.

i. Bicarbonate buffer system The chief acid present in the blood is carbonic acid (H_2CO_3), which is formed by the combination of CO_2 with H_2O. Carbonic acid dissociates to yield H^+ ions as indicated below:

$$H_2O + CO_2 \longrightarrow H_2CO_3 \longrightarrow H^+ + HCO_3^-$$

The only alkali present as such in blood is the bicarbonate HCO_3. It dissociates in water as follows:

$$NaHCO_3 + H_2O \longrightarrow H_2CO_3 + Na^+ + OH^-$$

The presence and the dissociation of bicarbonate will increase the OH^- ion concentration of the solution, thereby increasing its alkalinity.

At normal pH of blood (7.4) the ratio of H_2CO_3 /$BHCO_3$ is 20:1. This ratio is calculated from the Henderson–Hasselbalch equation as follows:

The H^+ ion concentration of a weak acid solution HA and its salt BA is represented as

$$(H^+) = K\frac{BA}{HA}$$

where K = dissociation constant
H^+ = hydrogen ion concentration
HA = concentration of the acid
BA = concentration of the salt

In logarithmic form, the reaction is written as:

$$pH = pK + \log\frac{BA}{HA}$$

pK for H_2CO_3 is 6.1.

To keep pH at 7.4, the ratios of these acids to their salts must be kept constant. Substituting in the Henderson–Hasselbalch equation for H_2CO_3,

$$7.4 = 6.1 + \log \frac{BHCO_3\,(base)}{H_2CO_3\,(acid)}$$

$$7.4 - 6.1 = \log \frac{BHCO_3}{H_2CO_3}$$

$$1.3 = \log \frac{BHCO_3}{H_2CO_3}$$

Since antilog of 1.3 = 20,

$$\frac{BHCO_3}{H_2CO_3} = \frac{20}{1}$$

As long as this ratio is maintained, the pH of the blood will be normal, but when this ratio is altered, the acid–base balance of the blood will be disturbed.

ii. Hemoglobin buffer system The buffering capacity of Hb is attributed to the effect of the imidazole groups, which are constituents of the histidine components of the globins of Hb in the erythrocytes. If the Hb is more oxygenated, the imidazole groups become more acidic and therefore more dissociated and consequently the buffering capacity is also increased. Oxygenated Hb (HbO_2) is a stronger acid than reduced Hb. When Hb is less oxygenated or contains no O_2 as in the case of reduced Hb (HHb), the imidazole groups become less acidic and therefore less dissociated and consequently the buffering capacity is also decreased.

iii. Plasma protein buffer system CO_2 is buffered only to a lesser extent by the plasma protein buffer system H.Protein/B.Protein and this occurs as shown below:

$$CO_2 + H_2O \rightleftharpoons H_2CO_3$$

$$H_2CO_3 + B.Protein \rightleftharpoons BHCO_3 + H^+.Protein$$

H^+.Protein is a weaker acid compared to H_2CO_3.

iv. Phosphate buffer system Phosphates, which are present in low concentration in the blood, are concentrated by the kidneys and constitute the principal buffers in the urine. NaH_2PO_4 and Na_2HPO_4, which are excreted in varying proportions depending upon the reaction of blood, constitute the phosphate buffer system.

2.7 ACID–BASE BALANCE

The pH of blood ranges from 7.35 to 7.45. The metabolic processes of the body form mostly acidic substances like carbonic acid, sulphuric acid, phosphoric acid, hydrochloric acid, hydroxy butyric acid and uric acid. Acids and bases formed in the cells reach the lungs and kidneys and get eliminated. Disturbances in acid–base balance are known as *acidosis* and *alkalosis*, which are mainly due to respiratory abnormalities or metabolic disorders.

Respiratory acidosis Diseases like pneumonia, emphysema, congestive heart failure and asthma may interfere with the elimination of CO_2 by the lung, thereby increasing the level of H_2CO_3 in blood.

Respiratory alkalosis This is due to fall in the level of H_2CO_3 in blood. This occurs due to the elimination of excessive amount of CO_2 and this situation is common in hyperventilation which occurs at high altitudes or due to hysteria and salicylate poisoning.

Metabolic acidosis In this condition the level of bicarbonate ions in the blood is decreased. This can occur in diabetes mellitus, renal failure, severe diarrhoea and vomiting.

Metabolic alkalosis In metabolic alkalosis, bicarbonate level in the blood is increased due to the consumption of large quantities of alkalies in the treatment of ulcer; vomiting due to high intestinal obstruction is also a common cause for metabolic alkalosis.

2.7.1 REGULATORY MECHANISMS

The pH of blood remains constant between 7.3 and 7.5, despite the constant synthesis of acid and base substances. The important regulatory mechanisms are: (i) respiratory mechanism and (ii) renal mechanism.

i. Respiratory mechanism The respiratory mechanism plays an important role in the regulation of acid–base balance. The volatile acid H_2CO_3 is effectively eliminated by the lungs in the form of CO_2 and H_2O. It has been found that 60% of the buffering capacity of blood is due to Hb and another 25% is contributed by the phosphate buffer system in the red blood cells. Thus 85% of buffering action takes place inside the RBC. But most of the buffered CO_2 is carried as bicarbonate in the plasma. The process by which the buffered CO_2 appears in the plasma and circulates as $NaHCO_3$ is described by Hamberger's phenomenon (Figure 2.5).

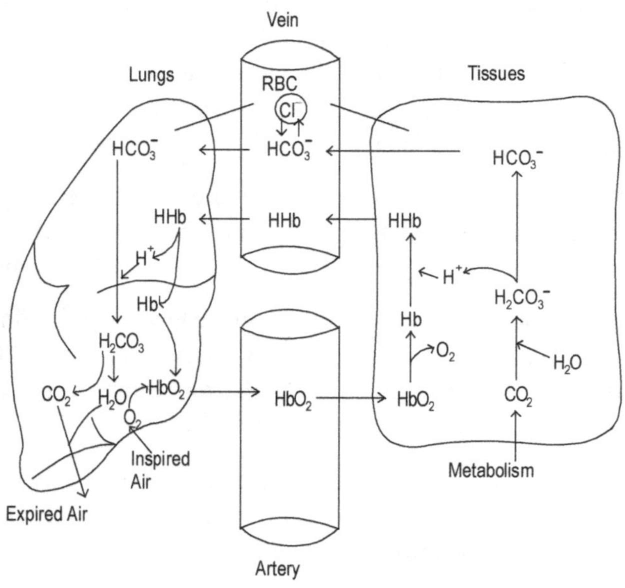

Fig. 2.5 Buffering action of Hemoglobin

In the tissues when O_2 tension is reduced, HbO_2 dissociates and releases O_2.

$$HbO_2 \longrightarrow Hb + O_2$$

Simultaneously, CO_2 produced during metabolism enters the blood, where it is hydrated to form H_2CO_3 which immediately dissociates to form H^+ and HCO^-_3

$$H_2O + CO_2 \longrightarrow H_2CO_3 \longrightarrow H^+ + HCO_3^-$$

Hb binds with H^+ ion forming hemoglobinic acid or reduced Hb (HHb).

$$Hb + H^+ \rightleftharpoons HHb$$

The HCO_3^- ions formed in the erythrocyte enter the plasma. This disturbs the ionic equilibrium. To restore the equilibrium equal amount of Cl^- ions enter the cell. This process is called as chloride shift or Hamberger's phenomenon. When the blood returns to the lungs, HCO_3^- enters the cell and Cl^- moves out into the plasma. HHb combines with O_2 to form HbO_2 and H^+ is liberated.

$$HHb + O_2 \rightleftharpoons HbO_2 + H^+$$

The newly formed H^+ ions react with the HCO_3^- to form H_2CO_3; this dissociates immediately to form water and CO_2, which is continuously eliminated in the expired air.

$$HCO_3^- + H^+ \longrightarrow H_2CO_3 \longrightarrow H_2O + CO_2$$

Increase in pCO_2 and H^+ ion concentration (H_2CO_3) stimulate the respiratory centre to increase the rate of respiration. The H_2CO_3 is converted to H_2O and CO_2, and CO_2 is breathed out allowing the pCO_2 and H^+ to return to normal level. Decrease in pCO_2 and H^+ ion concentration decreases the rate of respiration and conserves CO_2 and H^+ ion concentration. Thus, $H_2CO_3/BHCO_3$ level is maintained in the plasma.

ii. Renal mechanism Kidneys play an important role in maintaining the plasma bicarbonate level. Lungs can eliminate the volatile acid H_2CO_3 effectively while the non-volatile acids like phosphoric, sulphuric and hydrochloric acids are buffered by the plasma bicarbonate system. The plasma bicarbonate neutralises the acidic compounds, which enter the blood. Hence, the bicarbonate content of the plasma is referred to as the 'alkali reserve' of the plasma. Kidneys maintain alkali reserve by reabsorbing, secreting and excreting the acidic or basic substances. The pH of urine is highly variable fluctuating from 4.8 to 8.0. Under normal conditions, the pH of urine is slightly acidic (6.0).

The most important buffer system of urine is phosphate buffer system (B_2HPO_4/BH_2PO_4). At the pH 7.4, this ratio is 5. When there is accumulation of acid (H^+) in the plasma, the acid is removed by the kidney by increasing the amount of BH_2PO_4 so that at a urinary pH of 4.8, the ratio may become 1/99. At an alkaline pH of the urine (8.0), the BH_2PO_4 is greatly increased so that the ratio becomes 95/1.

Non-volatile acids such as lactic acid, ketone bodies, sulphuric acid and phosphoric acid are eliminated by the kidney. In glomerular filtrate, the acids are filtered off as salts of Na^+ and K^+. However, the Na^+ is reabsorbed by the tubule and replaced by H^+ or NH_4^+ secreted by the epithelial cells lining the tubule.

Formation of H^+ ion CO_2 formed during metabolism combines with H_2O and forms H_2CO_3 which immediately dissociate to H^+ and HCO_3^- in the presence of the enzyme carbonic anhydrase.

$$CO_2 + H_2O \xrightarrow{\text{Carbonic anhydrase}} H_2CO_3 \xrightarrow{\text{Carbonic anhydrase}} H^+ + HCO_3^-$$

The HCO_3^- is reabsorbed by the blood along with Na^+ while H^+ is secreted in the urine.

Formation of NH_4 Glutamine is converted to glutamic acid and NH_3 in the presence of the enzyme glutaminase.

$$\text{Glutamine} \xrightarrow{\text{Glutaminase}} \text{Glutamic acid} + NH_3$$

NH_3 combines with H^+(derived from H_2CO_3) to form NH_4^+, which will be secreted in the urine (Figure 2.6).

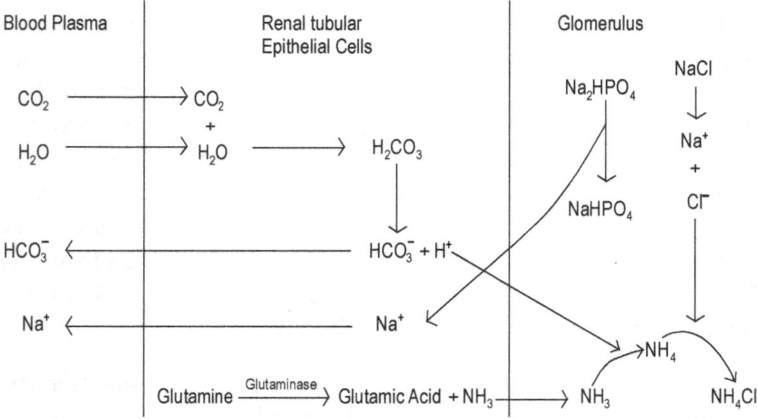

Fig. 2.6 Formation of ammonium salts in kidney

2.8 WATER BALANCE

Human body contains about 73% water, which remains distributed in extracellular and intracellular compartments. The extracellular water constitutes about 30% and the intracellular water constitutes about 50% of the total body weight. The extracellular fluids include plasma, interstitial fluid, cerebrospinal fluid (CSF), lymph, pericardial fluid, and aqueous and vitreous humours. The daily input of water through different sources varies from 2.1–2.5 litre and whereas the daily output of water by different channels varies from 2.0–2.5 litre. The total output through the different channels is comprised of urinary loss (1.1–1.5 litre), perspiration (0.4 litre), faeces (0.2 litre) and respiratory loss (0.4 litre). The values change even for normal persons depending upon seasons, and on the extent and type of work done by the individual.

The intake of water must be balanced by an equal output. Deprivation of the body water is known as dehydration. Loss of 20 per cent of the body water results in death. Such a condition develops by profuse sweating, severe diarrhoea and vomiting. Deficient intake of water also results in dehydration. The plasma becomes concentrated first, followed by the concentration of other extracellular compartments and finally the intracellular compartments. Intracellular potassium comes out along with water and simultaneously it is compensated by entry of sodium ions into the cells. Concentration of extracellular fluids results in increased osmotic pressure. This tends to draw out intracellular water. Dehydration of the intracellular compartments most likely stimulates the thirst centre, situated in the third ventricle, through osmoreceptors and sensory nerves of mouth and pharynx. These nerves respond to dryness of mouth and pharynx. Excess intake of water without proportionate output causes accumulation of water in the body and results in water intoxication. The person feels nausea and muscle cramps. Excess of water in different compartments lowers the osmotic pressure and suppresses the release of antidiuretic hormone (ADH), which in turn lowers the reabsorption of water from the distal convoluted tubule in the kidney, thereby increasing the volume of urine.

2.9 ELECTROLYTE BALANCE

The different body fluids contain a definite proportion of the cations and anions. Not much is known regarding the electrolyte composition of the different intracellular fluids. The important cations present in the body fluids are Na^+, K^+, $Ca^{++}$and Mg^{++} ions, and the anions are Cl^-, HCO_3^-, HPO_4^-, SO_4^-, organic acids and proteins.

Table 2.1 Electrolyte composition of different body fluids (concentration in meq./ltre)

	Extracellular fluids							
Ions	**Plasma**	**Gastric juice**	**Pancreatic juice**	**Bile**	**Intestinal secretion**	**Sweat**	**CSF**	**Intra-cellular fluids**
Cations								
Na^+	142	60	130	145	100	50	146	10
K^+	5	8	5	5	5	2	3.5-4	148
Ca^{++}	5	-	-	-	-	-	3.0	2
Mg^{++}	3	-	-	-	-	-	-	40
Total	**155**	**68**	**135**	**150**	**105**	**52**	**153**	**200**
Anions								
Cl^-	103	90	75	95	100	50	125	-
HCO_3	27	-	110	35	30	-	25	8
SO_4^-	1							
HPO_4^-	2							
Organic acids	6							136 (phos-phates, others)
Protein	16						15-40 mg/ 100ml	56
Total	**155**	**90**	**185**	**130**	**130**	**50**	**165-190**	**200**

By quantitative estimation, it has been found that the total concentration of cations in milliequivalents equals that of anions in blood plasma and intracellular fluids (muscle cell fluid). In gastric, pancreatic and intestinal secretions, the total cations are less than anions whereas in bile, the total cations are more than total anions. In sweat, the concentrations are almost equal.

It is apparent from Table 2.1 that the extracellular fluids have Na^+ as the principal cation and Cl^- and HCO_3^- as the principal anions. The intracellular fluids have K^+ and Mg^{++} as the principal cations, phosphates, and proteins as the principal anions. Intracellular fluids contain only very small amounts of Na^+, Cl^- and HCO_3^-.

2.10 REGULATORY MECHANISMS

i. ADH secretion The antidiuretic hormone (ADH or vasopressin) is released from the neurohypophysis and acts on the distal convoluted tubules of the kidney to cause increased water reabsorption and diminished urine volume (antidiuresis). The stimulus for its release is an increased electrolyte content i.e. increased osmotic pressure of the plasma which stimulates osmoreceptor in the diencephalon, which pass on the stimuli to paraventricular and supraoptic nuclei to produce more ADH. The ADH is passed on to neurohypophysis and released from there. The retention of water enables plasma to be diluted and brings down the osmotic pressure to normal levels. The release of ADH is suppressed when the plasma electrolyte concentration and osmotic pressure are lower than normal. Hence, tubular reabsorption of water is diminished, diuresis occurs and plasma becomes concentrated.

ii. Mineralocorticoids Deoxycortisone and aldosterone particularly exert a potent effect on the mineral metabolism, specifically of sodium and potassium. When plasma osmotic pressure is lowered, the adrenal cortex is stimulated to secrete the mineralocorticoids, which act on the distal convoluted tubules of the kidney to increase the reabsorption of sodium salts, thus restoring the osmotic pressure. The secretion of the hormones is suppressed when there is an elevated osmotic pressure, thus allowing more of sodium salts to be excreted by the kidney to bring down the osmotic pressure.

iii. Mechanism of thirst A 'thirst centre' located in the third ventricle regulates the amount of water consumption. A deficient intake of water leads to concentration of the body fluids (with respect to the solutes) and a rise in their osmotic pressure. This tends to draw out water from the intracellular compartment. This dehydration of the cells seems to be the main stimulus for the thirst mechanism through osmoreceptors as well as sensory nerves of the mouth and pharynx (glossopharyngeal and vagus), which respond to the dryness of the mouth and pharynx. When there is more water in the fluid compartments, there is lowering of the osmotic pressure of the fluids. This causes a suppression of the production or release of the antidiuretic hormone by the posterior pituitary and a decrease in the reabsorption of water by the distal convoluted tubule and thus promotes diuresis and loss of body water.

iv. Volume receptors An increase in the volume of the circulating blood plasma itself may stimulate the mechanism for elimination of more fluid by the kidney. These mechanisms operate through pressure receptors (baroreceptors) and volume receptors (stretch receptors).

REVIEW QUESTIONS

1. Enumerate the properties of water.
2. Explain the concept of pH.
3. Derive Henderson-Hasselbalch equation.
4. Explain the role of buffer systems of blood in the regulation of acid–base balance.
5. Explain Hambergers phenomenon.
6. Write a brief note on:
   i. Acid–base balance
   ii. Electrolyte balance
   iii. Water balance
7. Define:
   i. pH
   ii. Acid and base
   iii. Buffer
   iv. Acidosis
   v. Alkalosis

3

CARBOHYDRATES

Carbohydrates are compounds normally characterised by having carbon, hydrogen and oxygen. The basic units of the carbohydrates are monosaccharides, and glucose is the most important monosaccharide. Carbohydrates are chemically described as polyhydric alcohols having potentially active aldehyde and ketone groups.

3.1 BIOLOGICAL SIGNIFICANCE

i. Carbohydrates supply the major portion of energy required by living cells.

ii. Certain products of carbohydrate metabolism act as catalysts to promote oxidation of foodstuffs.

iii. Certain carbohydrates can be used as the starting material for the biosynthesis of compounds such as fatty acids and amino acids.

3.2 OCCURRENCE

Carbohydrates are widely distributed in nature in plants and animals. The most important carbohydrate found in plants is starch. It occurs abundantly in the roots, tubers, leaves, vegetables and grains. The carbohydrate found in animals is glycogen. It forms the storage form of carbohydrates in animals and is found abundantly in the liver and muscles. Both starch and glycogen are polysaccharides having high molecular weight. Carbohydrates having low molecular weight are also found in nature. They come under the groups monosaccharides and oligosaccharides, which are white crystalline substances, sweet to taste. They are generally known as sugars and are consumed with food.

Sucrose, which is largely taken with food, is obtained from sugarcane. It is also present in the nectar of flowers and in fruits. Glycosides, which are derivatives of carbohydrates, are found in certain plants. Certain glycosides are used in the treatment of heart diseases. Glucose and fructose, which are simple sugars, are widely distributed in plants. Glucose is the sugar of blood and other body fluids. Lactose is found in milk.

3.3 CHEMICAL CHARACTERISTICS OF CARBOHYDRATES

Chemically, carbohydrates contain carbon, hydrogen and oxygen. All simple sugars contain a potential aldehyde or ketone group. All compound sugars, which are made up of simple sugar molecules, also contain the potential aldehyde group either in the free form or in the combined form. The importance of potential aldehyde or ketone group is that they are associated with reducing properties.

3.4 CLASSIFICATION OF CARBOHYDRATES

Carbohydrates are classified into four major groups:

1. Monosaccharides Monosaccharides contain only one molecule of sugar and they cannot be broken into simpler substances by hydrolysis. e.g. glucose, fructose, etc.

2. Oligosaccharides Oligosaccharides yield 2 to 10 monosaccharides on hydrolysis. e.g. raffinose, stachylose and verbacose.

3. Disaccharides Disaccharides ($C_{12}H_{22}O_{11}$) yield two molecules of monosaccharides on hydrolysis. e.g. sucrose, lactose and maltose.

4. Polysaccharides Polysaccharides $(C_6H_{10}O_5)_n$ yield more than 10 molecules of monosaccharides on hydrolysis. e.g. starch, glycogen, cellulose, etc.

3.4.1 MONOSACCHARIDES

Monosaccharides are further classified according to the number of carbon atoms.

i. Diose It contains two carbon atoms. e.g. glycolaldehyde ($CH_2OH.CHO$). It has an aldehyde group and therefore it is referred to as aldose.

ii. Triose Trioses contain 3 carbon atoms ($C_3H_6O_3$). They are considered as true carbohydrates because they contain polyhydroxylic groups.

```
H—C=O             CH2OH
  |                 |
H—C—OH             C=O
  |                 |
  CH2OH            CH2OH
```

D-Glyceraldehyde   Dihydroxyacetone

Glyceraldehyde which can be considered as the simplest monosaccharide, contains an aldehyde group. Dihydroxyacetone is a ketotriose because it is a triose and contains a ketone group.

iii. Tetroses They contain four carbon atoms ($C_4H_8O_4$). e.g. erythrose, threose and erythrulose.

Erythrose and threose are aldotetroses while erythrulose is a ketotetrose.

```
     CHO           CHO          CH2OH
      |             |             |
  HO—C—H        H—C—OH           C=O
      |             |             |
   H—C—OH       H—C—OH        H—C—OH
      |             |             |
     CH2OH        CH2OH         CH2OH
```

D-Threose   D-Erythrose   D-Erythrulose

iv. Pentoses They contain five carbon atoms ($C_5H_{10}O_5$). e.g. ribose, deoxyribose, xylulose and arabinose.

Pentoses are physiologically important because ribose and deoxyribose are constituents of nucleic acids. Xylulose is a metabolite of glucuronic acid and is excreted in the urine of individuals suffering from *pentosuria*, which is an inborn error of pentose metabolism. Arabinose is found in plants.

```
     CHO              CHO              CH2OH
     |                |                |
 H—C—OH               CH2              C=O
     |                |                |
 H—C—OH          H—C—OH          OH—C—H
     |                |                |
 H—C—OH          H—C—OH           H—C—OH
     |                |                |
     CH2OH            CH2OH            CH2OH
  D-Ribose        D-Deoxyribose     D-Xylulose
```

v. Hexoses They are physiologically important compounds. They contain six carbon atoms ($C_6H_{12}O_6$). e.g. glucose, fructose, galactose, mannose, etc.

Glucose, galactose and mannose are aldohexoses each having an aldehyde group. Fructose is a ketohexose having a ketone group. Glucose and fructose occur freely in plants and fruits. Galactose is a component of lactose, which is a disaccharide present in milk. Polysaccharides containing mannose are found in ivory nuts, orchid tubers and yeast.

```
     CHO              CHO              CHO              CH2OH
     |                |                |                |
  H—C—OH          HO—C—H           H—C—OH              C=O
     |                |                |                |
 HO—C—H           HO—C—H          HO—C—H           HO—C—H
     |                |                |                |
  H—C—OH           H—C—OH         HO—C—H            H—C—OH
     |                |                |                |
  H—C—OH           H—C—OH          H—C—OH           H—C—OH
     |                |                |                |
     CH2OH            CH2OH            CH2OH            CH2OH
  D-Glucose        D-Mannose       D-Galactose      D-Fructose
```

vi. Heptoses These contain seven carbon atoms. Sedoheptulose is a heptose, which is formed as an intermediate in the oxidative shunt pathway of carbohydrate metabolism.

CH_2OH
|
C=O
|
HO—C—H
|
H—C—OH
|
H—C—OH
|
H—C—OH
|
CH_2OH

Sedoheptulose

Spatial configuration of sugar—stereoisomerism Many of the monosaccharides contain the same number of atoms and the same types of groups in their structures. But they are entirely different substances. The spatial arrangement of certain constituent groups like hydrogen and hydroxyl in each one of the hexoses varies. This phenomenon is called *stereoisomerism* or space isomerism and the sugars are called *stereoisomers*.

D-Glucose	D-Mannose	D-Galactose	D-Allose
1 CHO	CHO	CHO	CHO
H—2C—OH	HO—C—H	H—C—OH	H—C—OH
HO—3C—H	HO—C—H	HO—C—H	H—C—OH
H—4C—OH	H—C—OH	HO—C—H	H—C—OH
H—5C—OH	H—C—OH	H—C—OH	H—C—OH
6 CH_2OH	CH_2OH	CH_2OH	CH_2OH

Two sterioisomers, which differ from one another only in the configuration around one specific carbon, are called *epimers* of each other. Thus, D-glucose and D-mannose are epimers with respect to carbon atom 2; D-glucose and D-allose are epimers with respect to carbon atom 3 and D-glucose and D-galactose are epimers with respect to carbon atom 4.

Asymmetry of carbon atoms and optical activity Carbohydrates exhibit an important property called the optical activity, which is closely related to the asymmetry of the carbon atoms contained in them. A carbon atom is said to be asymmetric when it bears four different atoms or groups of atoms at the four valency bonds. Any compound having asymmetric carbon atoms is optically active and exists in more than one form or as *isomers*. The isomers are identical in configuration but differ only in spatial configuration. Isomers are said to exhibit optical isomerism, which involves optical rotation. Optical rotation to the left is levorotation and is expressed by a minus (–) sign and rotation to the right is dextrorotation and is expressed by a plus (+) sign. This phenomenon exhibited by asymmetric compounds is called optical isomerism.

D-Glyceraldehyde	L-Glyceraldehyde	D-Glucose	L-Glucose
		CHO	CHO
		H—C—OH	HO—C—H
		HO—C—H	H—C—OH
CHO	CHO	H—C—OH	HO—C—H
H—C—OH	HO—C—H	H—C—OH	HO—C—H
CH_2OH	CH_2OH	CH_2OH	CH_2OH

The geometric representation of the dextrorotatory form is customarily written with the secondary alcoholic hydroxyl group of the carbon atom just adjacent to the terminal primary alcoholic hydroxyl group extending to the right; such compounds are said to belong to 'D' series and are represented by the letter 'D' before their names. On the other hand the geometrical representation of the levorotatory form is customarily written with the secondary alcoholic hydroxyl group of the carbon atom just adjacent to the terminal primary alcoholic group extending to the left. Such compounds are said to belong to 'L' series and are represented by the letter 'L' before their names.

Pyranoses and furanoses Haworth protection formula is commonly used to show the ring form of a monosaccharide. Haworth proposed a scheme by which all sugars forming six-membered rings are to be called pyranoses, as pyran possesses the same ring composed of six members of which five are carbon and one is oxygen. Sugars in pyranose form appear to be most stable.

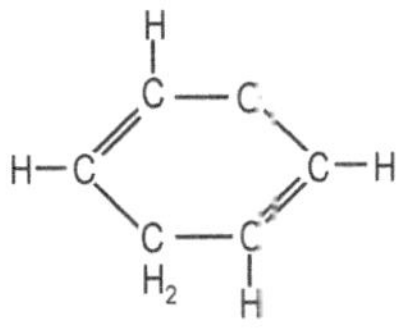

Pyran

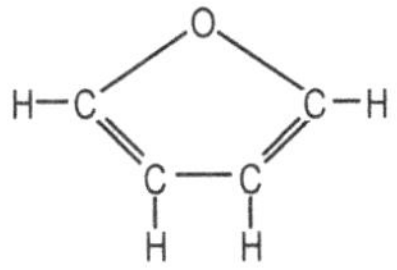

Furan

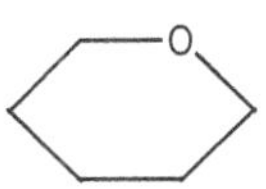

Pyranose ring

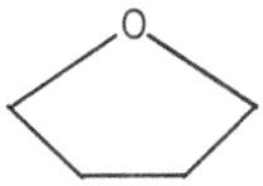

Furanose ring

Haworth designated that sugars having five-membered rings are to be called furanoses, because furan possesses the same ring composed of four carbons and one oxygen. Furanoses are unstable.

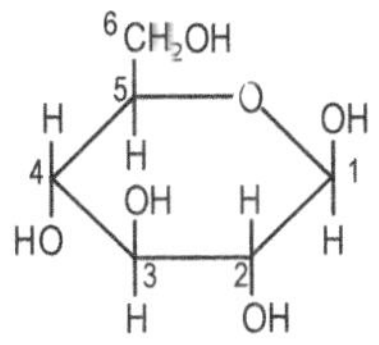

D-Glucopyranose

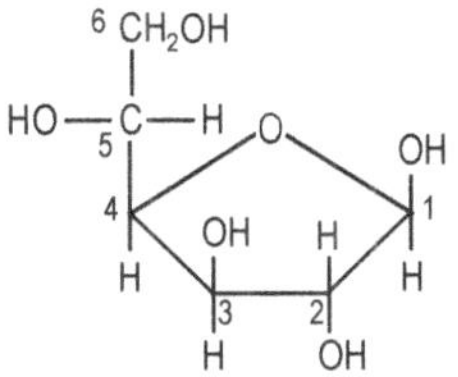

D-Glucofuranose

3.4.2 OLIGOSACCHARIDES

The oligosaccharides consist of 2 to 10 monosaccharides in their molecules. In oligosaccharide sugar the monomers or monosaccharide units are linked with each other by the *glycosidic linkages*.

3.4.3 DISACCHARIDES

The disaccharides are sugars containing two molecules of monosaccharides. These are water-soluble sugars formed by the union of two molecules of monosaccharides by the removal of one molecule of water. In this process of condensation, the –OH group of one monosaccharide is joined to the 'H' of the OH group of another monosaccharide to form one water, and remaining 'O' forms a glycoside bond between the two sugar molecules. The common disaccharides are maltose, sucrose and lactose.

Maltose Maltose does not occur free in nature. It is formed by condensation of two molecules of glucopyranose. It is not so sweet as sucrose. It reduces copper solutions and forms osazone.

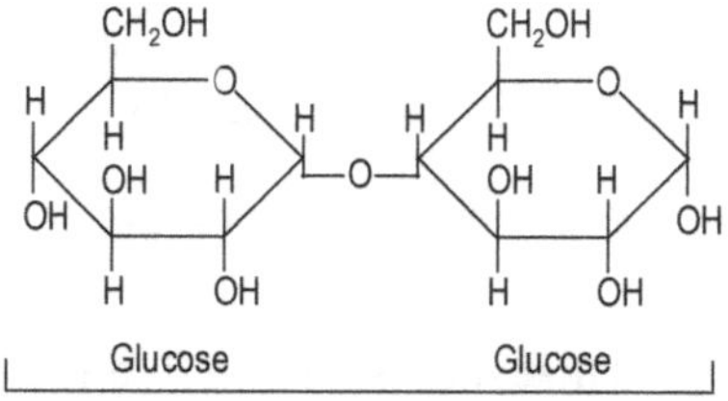

Glucose Glucose

Maltose

Sucrose It is found in greatest abundance in certain plants such as sugarcane, beets, carrots and some sweet fruits. It is sweet, crystalline and freely soluble in water. It is neither an aldehyde nor a ketone as it does not reduce copper sulphate solutions nor does it form osazone. It is formed by the condensation of fructofuranose and glucopyranose.

Glucose

Fructose

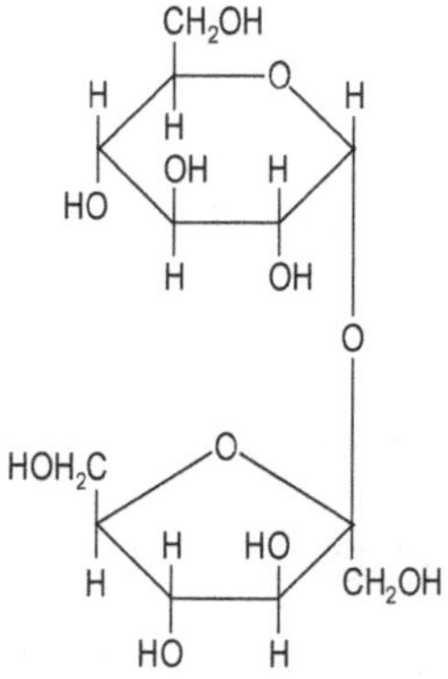

Sucrose

Lactose It is the sugar found in milk. It is formed by the condensation of glucose and galactose. It forms white gritty crystals on concentration of the milk whey. It is not so sweet as others. It is not found in plants. It also reduces copper solutions and forms osazone, indicating the presence of an aldehyde group in the molecule.

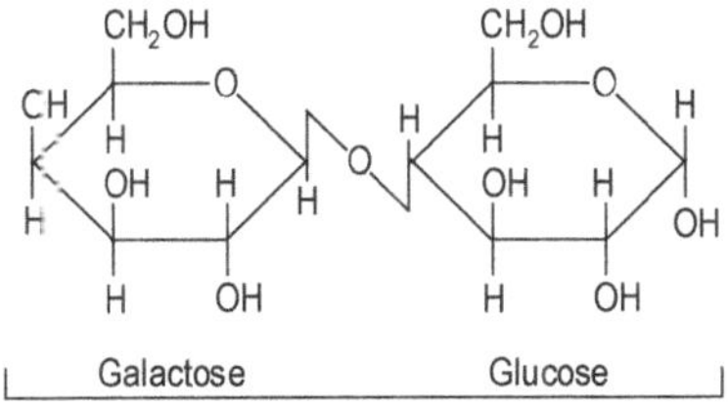

Lactose

Trisaccharides These carbohydrates contain three monomers in their molecules. e.g. raffinose, ratinose.

Tetrasaccharides These contain four monomers in their molecules. e.g. stachylose

Pentasaccharides These contain five monomers in their molecules. e.g. verbacose.

3.4.4 POLYSACCHARIDES

Polysaccharides are made up of tens to many thousands of monosaccharide monomers in their macromolecules. Polysaccharides are insoluble. They may be of two types—homopolysaccharides and heteropolysaccharides.

Homopolysaccharides The homopolysaccharides contain the same type of monomers in their molecules. Some of the important homopolysaccharides are the following:

Starch It is a complex substance, formed by the condensation of amylose and amylopectin, which is a branched polysaccharide with shorter chains. It is found in abundance in plants, seeds, fruits and tubers.

Amylose is formed in plant cells by elimination of a water molecule. The linkage in amylose is an α-1-4-glycoside as in maltose. Hence, enzymic hydrolysis of amylose with amylase, yields maltose units.

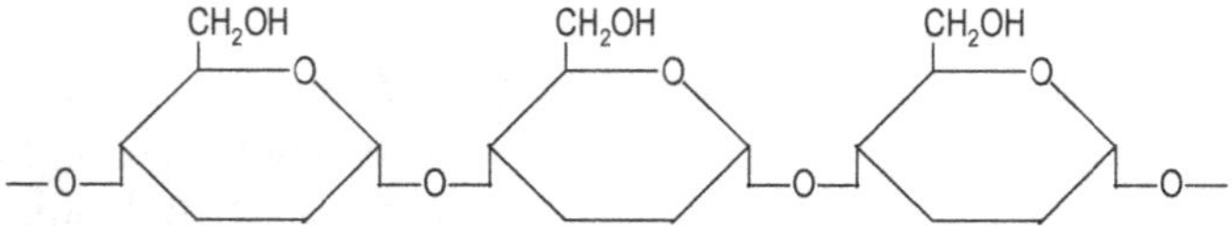

Amylose

Amylopectin possesses nearly 300–5500 glucose units per molecule. It has the same basic chain of α-1-4-glycoside linkage as in amylose but has, in addition, many side chains attached to the basic chain by α-1-6-glycoside linkage.

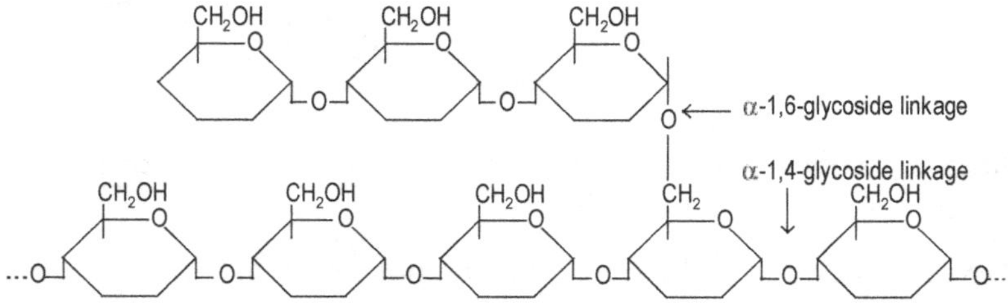

α-1,6-glycoside linkage

α-1,4-glycoside linkage

Amylopectin

Inulin It is a storage unit polysaccharide and is composed of large number of fructofuranose molecules condensed together in chains. In inulin, the fructose residues are joined together in straight chain by β-2,1 glycoside linkages.

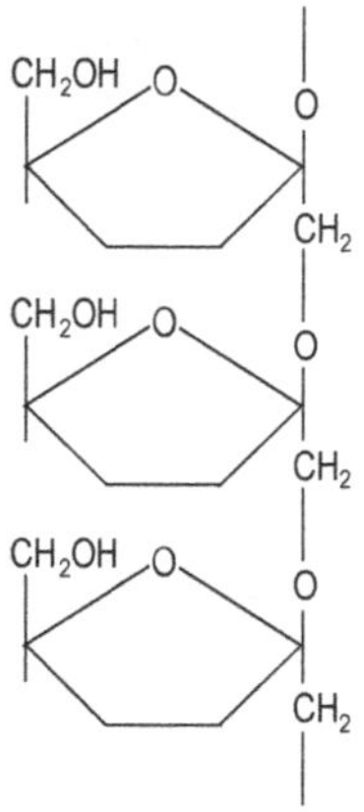

Inulin

Glycogen It is the main carbohydrate storage substance of animals and fungi. It is made up of molecules similar to amylopectin but with more numerous side chains.

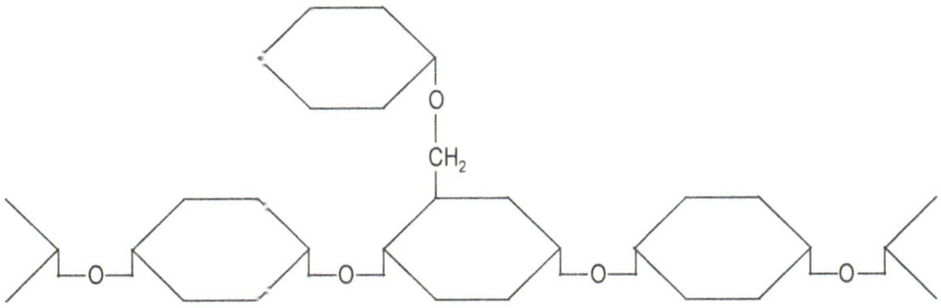

O
CH_2
O O O O

Glycogen

Cellulose Cellulose is a fibrous, tough, water-insoluble substance, found in the protective cell walls of plants. It is a linear, unbranched homopolysaccharide of 10,000 or more D-glucose units connected by β-1-4 glycosidic bonds. Due to β linkages, D-glucose chains in cellulose assume an extended conformation and undergo side-by-side aggregation by cross-links of hydrogen bonds into insoluble fibrils.

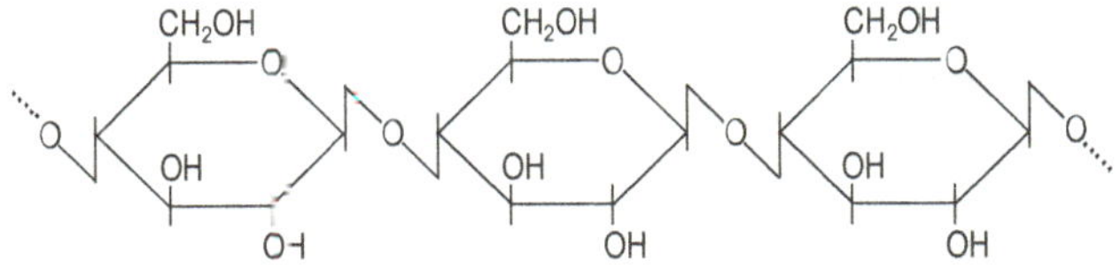

CH_2OH CH_2OH CH_2OH
O O O O O O O
OH OH OH
OH OH OH

Cellulose

Pectin Pectin is a polysaccharide of α-D-galacturonic acid; some of the free carboxyl groups are esterified with methyl alcohol and others are combined with calcium or magnesium ions. The structure of α-polygalacturonic acid is given below. The fruits of many plants contain pectin.

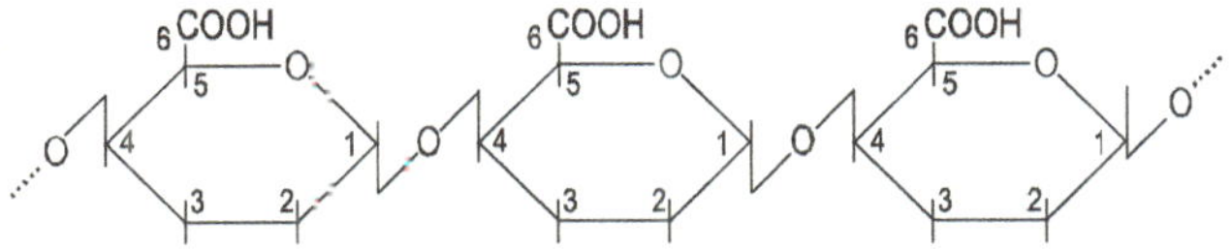

$_6$COOH $_6$COOH $_6$COOH
5 O 5 O 5 O
O 4 1 O 4 1 O 4 1 O
3 2 3 2 3 2

Pectin

Chitin Chitin is found in fungi but principally among the arthropods (crabs and insects). The armour of crabs and the exoskeleton of insects consists mostly of chitin. The chitin framework of lobster and crab shells is impregnated and hardened with calcium carbonate.

Chitin is closely related to cellulose. Here the alcoholic OH group on carbon atom 2 of β-D-glucose units is joined together by β-1,4 glycosidic linkages.

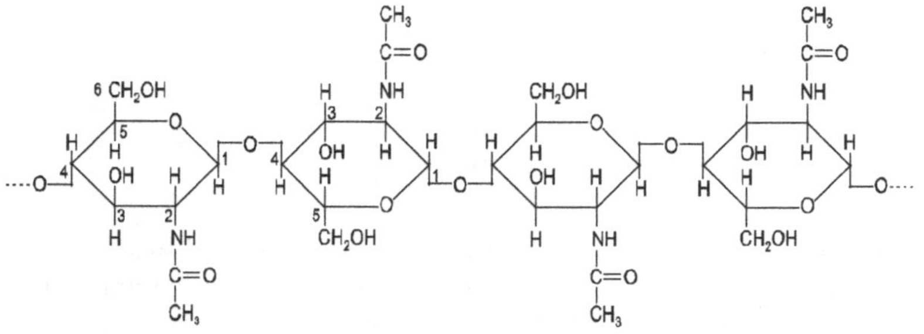

Chitin

On hydrolysis with mineral acids, chitin yields two final end products, namely glucosamine and acetic acid. Glucosamine is an important component of some glycoproteins (mucoproteins) such as mucin of saliva. Chitinase (from the gastric juice of snails or from bacteria), however, decomposes the chitin to N-acetylglucosamine.

Heteropolysaccharides These are polysaccharides which on hydrolysis, yield mixtures of monosaccharides. The most important heteropolysaccharides are the following:

Peptidoglycan Bacterial cell walls contain peptidoglycans. The rigid component of bacterial cell walls is a heteropolymer of alternating ($\beta1\rightarrow4$)-linked N-acetylglucosamine and N-acetylmuramic acid residues. The linear polymers lie side by side in the cell wall, cross-linked by short peptides, the exact structure of which depends on the bacterial species. The peptide cross-links link the polysaccharide chains into a strong sheath that envelops the entire cell and prevents cellular swelling and lysis due to osmotic entry of water.

Glycosaminoglycan The heteropolysaccharides called glycosaminoglycans are a family of linear polymers composed of repeating disaccharide units. One of the two monosaccharides is always either N-acetylglucosamine or N-acetylgalactosamine; the other is in most cases of uronic acid, usually D-glucuronic or L-iduronic acid. In some glycosaminoglycans, one or more of the hydroxyls of the amino sugar is esterifed with sulphate. Glycosaminoglycans are attached to extracellular proteins to form proteoglycans. The glycosaminoglycan hyaluronic acid (hyaluronate at physiological pH) contains alternating residues of D-glucuronic acid and N-acetylglucosamine. Hyaluronates form clear, highly viscous solutions that serve as lubricants in the synovial fluid of joints and give the vitreous humor of the vertebrate eye, its jelly-like consistency. Glycosaminoglycan includes hyaluronic acid, chondroitin and heparin.

Hyaluronic acid It is the most abundant member of polysaccharides and is found in higher animals as a component of various tissues such as the vitreous body of the eye, the umbilical cord and the synovial fluid of joints. The high viscosity of the synovial fluid and its role as biological lubricant is largely due to the presence of its hyaluronic acid content (about 0.03%). Frequently, it is prepared from umbilical cord.

Hyaluronic acid has the least complicate structure among mucopolysaccharides. It is a straight chain polymer of D-glucuronic acid and N-acetyl D-glucosamine (NAG) alternating in the chain.

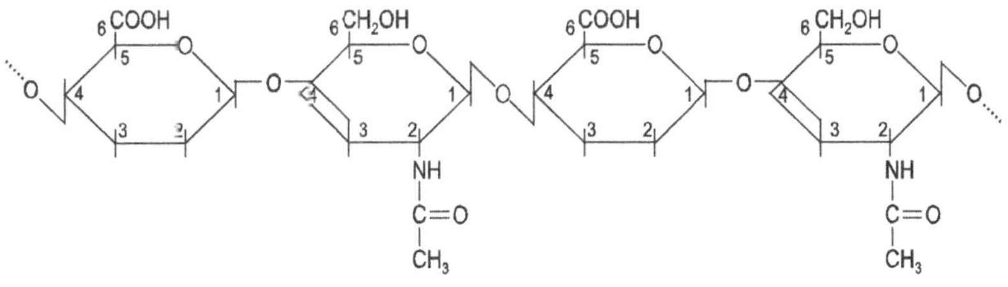

Hyaluronic acid

Hyaluronic acid, upon hydrolysis yields an equimolar mixture of D-glucuronic acid, D-glucosamine and acetic acid.

Chondroitin Chondroitin is found in cartilage and is a component of cell coats. It is the parent substance for two more widely distributed mucopolysaccharides, chodroitin sulphate A and chodroitin sulphate C.

Chodroitin is similar in structure to hyaluronic acid except that it contains galactosamine rather than glucosamine. It is thus, a polymer of β-D-glucuronido-1,3-N-acetyl-D-galactosamine joined by β-1→4 linkages.

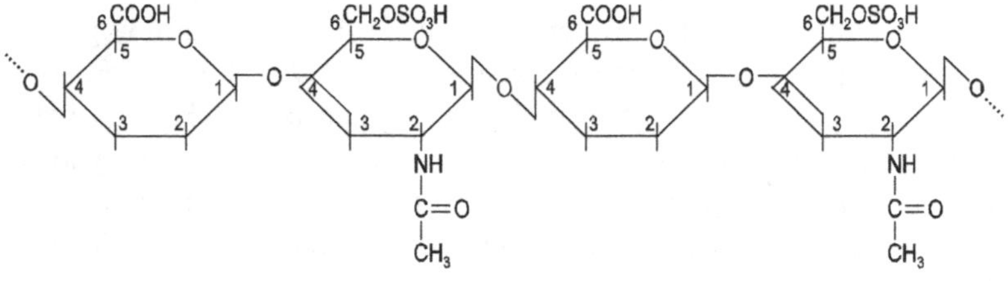

Chondroitin sulphate C

On hydrolysis, chondroitin produces equimolar mixture of D-glucuronic acid, D-galactosamine and acetic acid. Chodroitin sulphate are derivatives of chodroitin.

Heparin Heparin is present in liver, lungs, arterial walls and indeed, wherever mast cells are found, possibly for the purpose of neutralising biogenic amines (e.g. histamine).

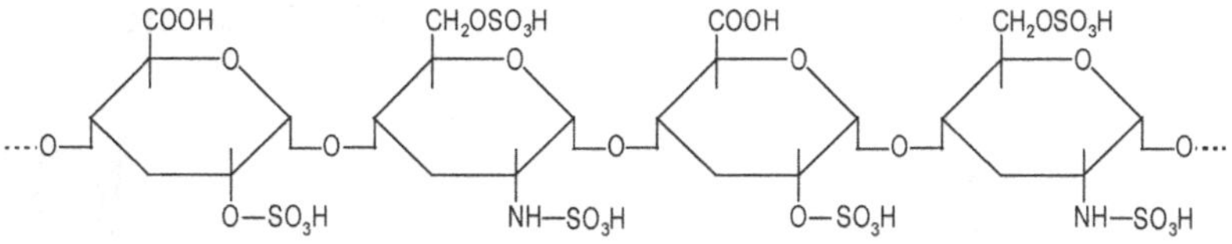

Heparin

Heparin is a heteropolysaccharide composed of D-glucuronic acid units, most of which (about 7 out of every 8) are esterified at C_2 and D-glucosamine-N-sulphate (sulphonylaminoglucose) units with an additional O-sulphate group at C_6. Both the linkages of the polymer are alternating α-1→4.

Heparin acts as an anticoagulant. It prevents coagulation of blood by inhibiting the prothrombin–thrombin conversions. This eliminates the effect of thrombin on fibrinogen.

Proteoglycan Proteoglycans are macromolecules of the cell surface or extracellular matrix in which one or more glycosaminoglycan chains are joined covalently to a membrane protein or a secreted protein. The glycosaminoglycan moiety commonly forms the greater fraction (by mass) of the proteoglycan molecule, dominates the structure and is often the main site of biological activity. Proteoglycans are major component of connective tissues such as cartilage, in which many noncovalent interactions with other proteoglycans, proteins and glycosaminoglycans provide strength and resilience.

Glycoprotein Glycoproteins are carbohydrate–protein conjugates in which the carbohydrate moieties are smaller and more structurally diverse than glycosaminoglycans of proteoglycans. The carbohydrate is attached at its anomeric carbon through a glycosidic link to the-OH of a serine or threonine residue (O-linked), or through an N-glycosyl link to the amide of nitrogen of an asparagine residue (N-linked) muramic. Some glycoproteins have a single oligosaccharide chain, but many have more than one. Muramic acid and neuraminic acid are components of cell wall and are present as glycoproteins. They are nine-carbon amino sugar derivatives. N-Acetyl derivatives of neuraminic acid are called sialic acids.

Many of the proteins secreted by eukaryotic cells are glycoproteins, including most of the proteins of blood. For example, immunoglobulins (antibodies) and certain hormones such as follicle stimulating hormone, lutenising hormone and thyroid stimulating hormone, are glycoproteins. Many milk proteins, including lactalbumin and some of the proteins secreted by the pancreas (e.g. ribonuclease) are glycosylated.

Glycolipids Glycolipids are membrane lipids in which the hydrophilic groups are oligosaccharides which, as in glycoproteins, act as specific sites for recognition by carbohydrate-binding protein.

3.5 CHEMICAL PROPERTIES OF MONOSACCHARIDES

i. Reactions in acid solutions When monosaccharides are treated with a strong mineral acid, the product furfural is obtained. This is the characteristic reaction of sugars and is known as dehydration. If the furfurals are treated with a variety of phenolic compounds, coloured products are produced.

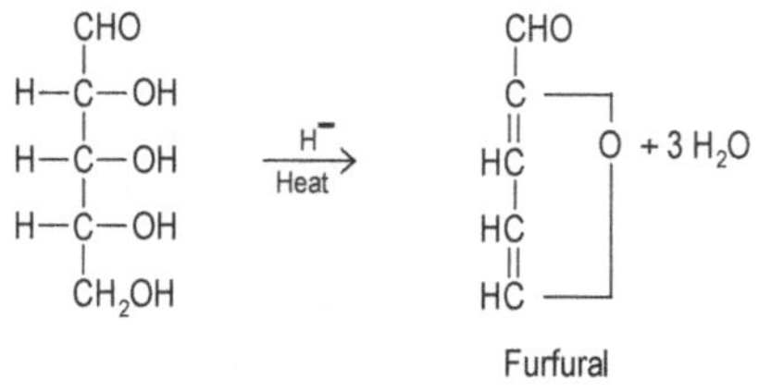

CHO | H—C—OH | H—C—OH | H—C—OH | CH$_2$OH $\xrightarrow[\text{Heat}]{H^-}$ Furfural + 3 H_2O

Furfural

Molisch test When sugar solution is treated with molisch reagent and concentrated H_2SO_4, a reddish violet zone appears between the acid and sugar layer.

Anthrone Test When a sugar is mixed with anthrone and concentrated sulphuric acid, blue or green colour is formed.

Seliwanoff's Test When ketoses are heated with HCl and resorcinol, a bright red colour is formed.

Tollen's Orcinol Test (Bial's test) This is a test for pentoses; when pentoses are treated with orcinol reagent, a blue colour is formed.

Tollen's Phloroglucinol Test This is a test for galactose; when galactose is treated with phloroglucinol and HCl, a red colour is formed.

ii. Reactions in alkaline solutions When the sugars are treated with dilute alkaline solution, they change to cyclic α and β forms. Sugars with strong alkali solution produce decomposition products, and yellow and brown pigments develop. Weak alkaline solutions of sugars undergo molecular change known as tautomerization whereby the 'H' atoms migrate from one carbon to another to form a mixture of enolic compounds.

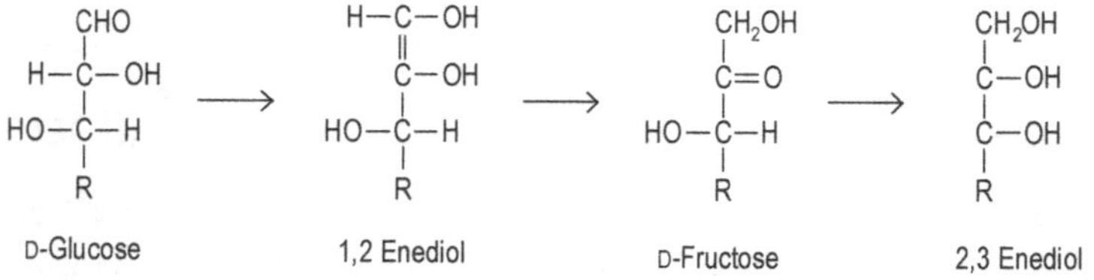

D-Glucose (CHO | H—C—OH | HO—C—H | R) → 1,2 Enediol (H—C—OH ‖ C—OH | HO—C—H | R) → D-Fructose (CH$_2$OH | C=O | HO—C—H | R) → 2,3 Enediol (CH$_2$OH | C—OH ‖ C—OH | R)

D-Glucose 1,2 Enediol D-Fructose 2,3 Enediol

iii. Reduction When monosaccharides are treated with hydrogen gas under pressure in the presence of a metal catalyst, or with an active metal, such as Ca, in water, the carbonyl group is reduced to an alcoholic hydroxyl group, yielding polyhydric alcohol.

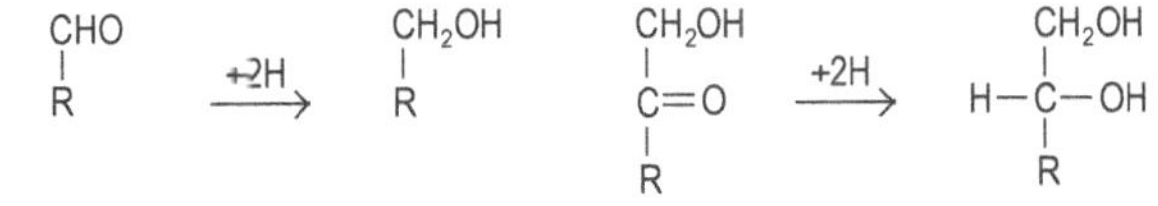

$$\underset{\text{Aldo sugar}}{R-CHO} \xrightarrow{+2H} \underset{\text{Alcohol}}{R-CH_2OH} \qquad \underset{\text{Ketosugar}}{CH_2OH-C(=O)-R} \xrightarrow{+2H} \underset{\text{Alcohol}}{CH_2OH-CH(OH)-R}$$

D-Glucose under these circumstances yields sorbitol.

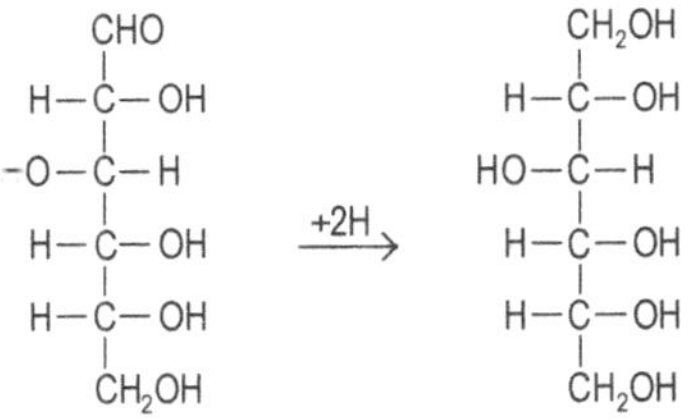

$$\underset{\text{D-Glucose}}{CHO-(H-C-OH)-(HO-C-H)-(H-C-OH)-(H-C-OH)-CH_2OH} \xrightarrow{+2H} \underset{\text{Sorbitol}}{CH_2OH-(H-C-OH)-(HO-C-H)-(H-C-OH)-(H-C-OH)-CH_2OH}$$

iv. Oxidation Monosaccharides are readily oxidised in acids in the presence of mild oxidising agents. Under this condition when only the aldehyde group is oxidised, an aldonic acid is formed. When primary alcohol is oxidised, uronic acid is formed. When both the groups are oxidised saccharic acids are formed.

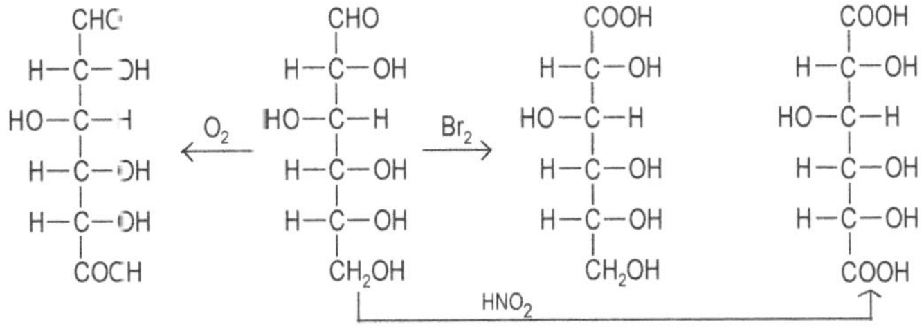

$$\underset{\text{Glucuronic acid}}{CHO-(H-C-OH)-(HO-C-H)-(H-C-OH)-(H-C-OH)-COOH} \xleftarrow{O_2} \underset{\text{Glucose}}{CHO-(H-C-OH)-(HO-C-H)-(H-C-OH)-(H-C-OH)-CH_2OH} \xrightarrow{Br_2} \underset{\text{Gluconic acid}}{COOH-(H-C-OH)-(HO-C-H)-(H-C-OH)-(H-C-OH)-CH_2OH}$$

$$\underset{\text{Glucose}}{CHO-(H-C-OH)-(HO-C-H)-(H-C-OH)-(H-C-OH)-CH_2OH} \xrightarrow{HNO_2} \underset{\text{Glucosaccharic acid}}{COOH-(H-C-OH)-(HO-C-H)-(H-C-OH)-(H-C-OH)-COOH}$$

v. Ester Formation Simple sugars or monosaccharides, on treatment with an acid anhydride form sugar acetate, benzoate, etc. During this reaction hydroxyl groups are esterified. The phosphoric acid esters of sugars are important as intermediate products formed during metabolism in the body. Glucose-1-phosphate, glucose-6-phosphate, fructose-6-phosphate, fructose,1,6-diphosphate are some examples.

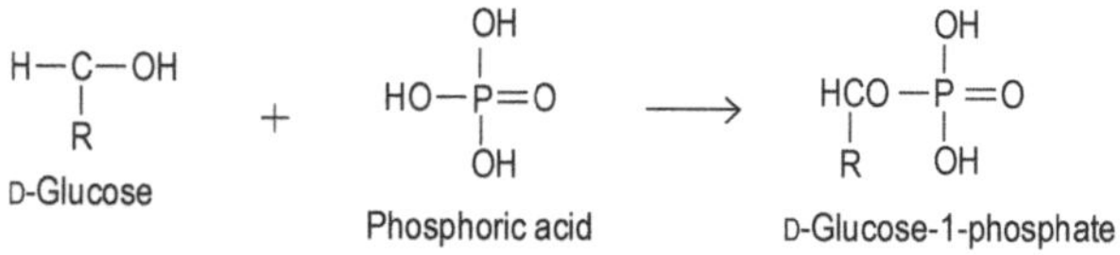

$$\underset{\text{D-Glucose}}{\text{H}-\underset{|}{\overset{}{\text{C}}}-\text{OH} \atop \text{R}} + \underset{\text{Phosphoric acid}}{\text{HO}-\text{P}(\text{OH})_2=\text{O}} \longrightarrow \underset{\text{D-Glucose-1-phosphate}}{\text{HCO}(\text{R})-\text{P}(\text{OH})_2=\text{O}}$$

vi. Osazone Formation When monosaccharides are treated with phenyl hydrazine, yellow crystalline compounds are formed which are called osazones. The osazone compounds of sugars are coloured and possess characteristic crystalline forms, melting points and precipitation times. During this reaction, one molecule of phenylhydrazine first reacts with one molecule of aldose or ketose sugar to form a hydrazone. The hydrazone thus formed is oxidised in the presence of excess phenylhydrazine to form glucosazone, and phenylhydrazine is reduced to form aniline and ammonia.

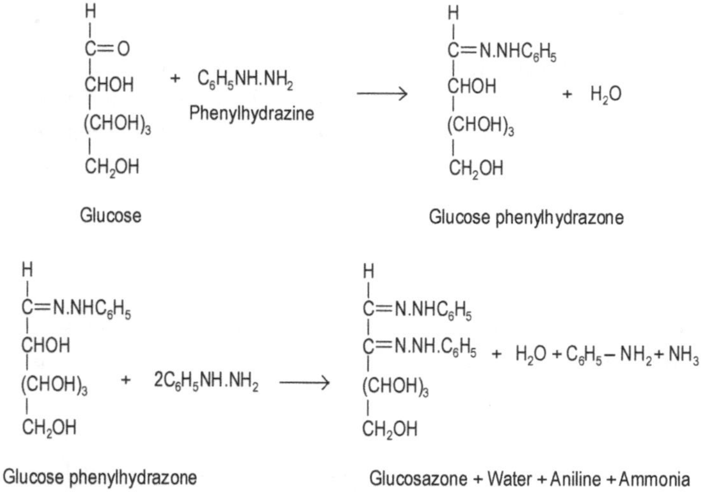

$$\underset{\text{Glucose}}{\text{H}-\text{C}=\text{O}-\text{CHOH}-(\text{CHOH})_3-\text{CH}_2\text{OH}} + \underset{\text{Phenylhydrazine}}{C_6H_5NH.NH_2} \longrightarrow \underset{\text{Glucose phenylhydrazone}}{\text{H}-\text{C}=\text{N.NHC}_6\text{H}_5-\text{CHOH}-(\text{CHOH})_3-\text{CH}_2\text{OH}} + H_2O$$

$$\underset{\text{Glucose phenylhydrazone}}{\text{H}-\text{C}=\text{N.NHC}_6\text{H}_5-\text{CHOH}-(\text{CHOH})_3-\text{CH}_2\text{OH}} + 2C_6H_5NH.NH_2 \longrightarrow \text{H}-\text{C}=\text{N.NHC}_6\text{H}_5-\text{C}=\text{N.NH.C}_6\text{H}_5-(\text{CHOH})_3-\text{CH}_2\text{OH} + H_2O + C_6H_5-NH_2 + NH_3$$

Glucosazone + Water + Aniline + Ammonia

Osazone formation

Fermentation Fermentation is a general term denoting the anaerobic degradation of glucose or other organic nutrients into various products for the purpose of obtaining energy in the form of ATP. Glucose, fructose, mannose, maltose and sucrose are readily fermented by baker's yeast, whereas galactose, lactose and pentose are not fermented. Alcohol and carbon dioxide are formed as products.

REVIEW QUESTIONS

1. Classify carbohydrates giving examples for each class.
2. What are homo- and heteropolysaccharides?
3. Explain the chemical properties of monosaccharides.
4. Write a brief note on:
   i. optical activity of carbohydrates
   ii. disaccharides
   iii. Haworth projection formula
5. Define or explain the following:
   i. Epimers
   ii. Stereoisomers
   iii. Pyranose and furanose
   iv. Glycosidic linkages
   v. Anthrone test
   vi. Molisch test
   vii. Seliwanoff's test
   viii. Tollen's orcinal test

4

CARBOHYDRATE METABOLISM

Carbohydrates are the principal source of energy for the cell. Carbohydrates are converted in the digestive tract, into absorbable glucose, galactose and fructose which enter the blood stream through villi. Subsequently, glucose may be used for the immediate release of energy or stored as glycogen mostly in the liver and muscles. Hepatic cells can also convert galactose and fructose into glucose. The process of glycogen formation from glucose is called *glycogenesis*. When the blood sugar level decreases, the liver glycogen is reconverted into glucose, and this process is called *glycogenolysis*. Synthesis of glucose from non-carbohydrate substances is called *gluconeogenesis*. The glucose that diffuses into the cells finds its way into the mitochondria and undergoes anaerobic and aerobic reactions. *Glycolysis* is a sequence of enzyme-catalysed anaerobic reactions that convert glucose into pyruvate. Pyruvate is decarboxylated into a two-carbon compound called acetyl, which is fed into the *Kreb's cycle* and undergoes aerobic reactions producing the energy-rich compound namely ATP.

4.1 GLYCOGENESIS

During glycogenesis, glucose is first phosphorylated in the presence of the enzyme **glucokinase** and **ATP**. This results in the formation of glucose-6-phosphate, which becomes glucose-1-phosphate by the transfer of the phosphate molecule to the first carbon atom of glucose. This reaction is catalysed by **phosphoglucomutase**. Glucose-1-phosphate is converted to glycogen in the presence of a key enzyme called **glycogen synthetase** or **phosphorylase**. This reaction involves a high-energy compound known as **uridine triphosphate (UTP)**.

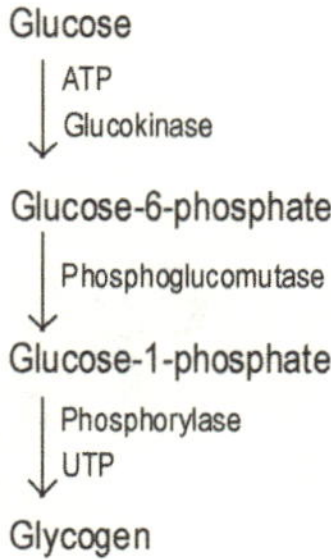

4.2 GLYCOGENOLYSIS

When the level of blood sugar decreases, the liver glycogen is reconverted into glucose and this process is called *glycogenolysis*. During this process, the reactions occurring in glycogenesis are reversed. Glycogen is converted into glucose-1-phosphate in the presence of **phosphorylase**. Glucose-1-phosphate, in the presence of the enzyme **phosphoglucomutase**, is converted into glucose-6-phosphate. This is converted into glucose by the enzyme **glucose-6-phosphatase**, which splits the phosphate away from glucose-6-phosphate.

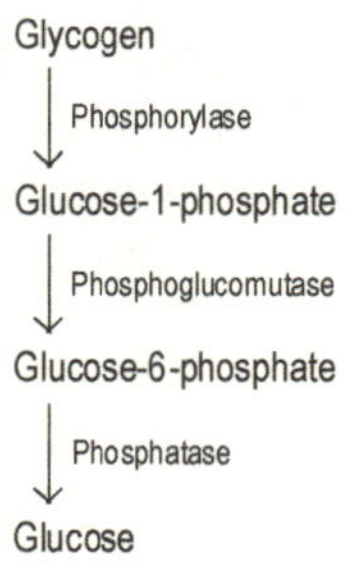

Glycogenolysis proceeds in the same manner in the muscles too, but the reaction stops with the formation of glucose-6-phosphate, because the enzyme glucose-6-phosphatase is absent in the muscles.

The glucose that diffuses into the cells finds its way into the mitochondria and undergoes anaerobic reactions. The anaerobic reactions occur in the absence of oxygen and these changes constitute *glycolysis*. During glycolysis, glucose is split into two molecules of pyruvic acid. In muscles, glycolysis results in lactic acid. In yeast and anaerobic bacteria, glycolysis results in the formation of ethyl alcohol and carbon dioxide. Aerobic reactions involve the Kreb's cycle and oxygen is utilised.

The splitting of glucose in organisms may take place in three different pathways. Based on the pattern of reactions undergone by the glucose molecules, there are three pathways:

i. Embden–Meyerhof Parners Pathway (EMP pathway)
ii. Hexose Monophosphate shunt (HMP shunt) or Hexose Monophosphate pathway (HMP pathway)
iii. Entner–Doudoroff pathway

While the first and third pathways constitute the preparatory step in aerobic and anaerobic respirations, the second one differs from them in being an independent respiratory cycle.

4.3 GLYCOLYSIS (EMBDEN–MEYERHOF PARNERS PATHWAY)

During glycolysis each molecule of glucose, after phosphorylation, is converted into two molecules of pyruvic acid (Figure 4.1).

The chemical reactions involved in Embden–Mayerhof Pathway are as follows:

1. The first step in glycolysis is the conversion of glucose to glucose-6-phosphate in the presence of **hexokinase**. ATP provides the phosphate as well as the energy for the synthetic reaction and is converted to ADP. Magnesium ions are also required as activators.

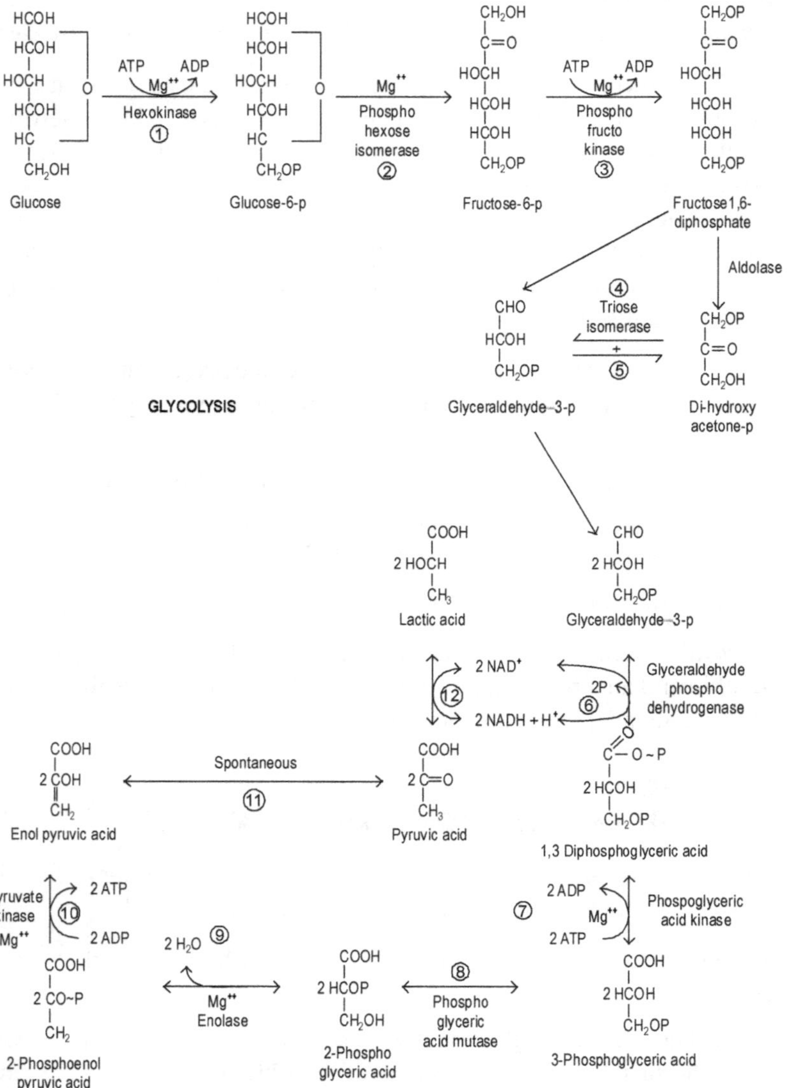

Fig. 4.1 EMP Pathway

2. The glucose-6-phosphate is converted to its isomer, fructose-6-phosphate by the action of **phosphohexose isomerase** in the presence of magnesium ions.
3. The fructose-6-phosphate is converted to fructose 1-6-diphosphate in the presence of **phosphofructokinase**. One more molecule of ATP is broken down to ADP and magnesium ions are required as activators.
4. The fructose-1-6-diphosphate is split up into two triose phosphate molecules—glyceraldehyde-3-phosphate and dihydroxyacetone phosphate—by the action of the enzyme **aldolase**.
5. The triose phosphates are readily interconvertible by the action of **phosphotriose isomerase**. Since subsequent steps utilise only glyceraldehyde-3-phosphate, it can be considered that two molecules of that substance are formed as a result of reactions 4 and 5.
6. Glyceraldehyde-3-phosphate undergoes dehydrogenation and phosphorylation to form 1-3-diphosphoglyceric acid by the action of the enzyme **glyceraldehyde-3-phosphodehydrogenase**.
7. **Phosphoglycerate kinase** in the presence of Mg^{++} transfers the energy-rich phosphate from the C-1 of 1-3-diphosphoglyceric acid to ADP to form ATP, leaving 3-phosphoglyceric acid.
8. The 3-phosphoglyceric acid is converted to 2-phosphoglyceric acid by the action of **phosphoglyceromutase**.
9. The enzyme **enolase** removes a molecule of water from 2-phosphoglyceric acid to form 2-phosphoenol pyruvic acid. This results in a redistribution of energy to make the enolic phosphate bond, an energy-rich one.
10. 2-phosphoenol pyruvic acid is converted to enol pyruvic acid in the presence of the enzyme **pyruvate kinase**. The energy rich phosphate of phosphoenol pyruvate is transferred to ADP to form ATP. Magnesium ions are required as activators.
11. Enol pyruvic acid being unstable is spontaneously converted to pyruvic acid (keto form). The reaction is non-enzymic.

12. If conditions are aerobic, pyruvic acid will be further converted to active acetate and gets oxidised in citric acid cycle or used for lipogenesis. If conditions are anaerobic, it is converted to lactic acid by the action of **lactic dehydrogenase** by taking up hydrogen from reduced NAD, which is formed in reaction 6. This will enable glycolysis to proceed under anaerobic conditions by providing a continuous supply of NAD (oxidised form) for the action of **glyceraldehyde-3-phosphate dehydrogenase** in reaction 6.

4.3.1 ENERGETICS OF EMP PATHWAY

During glycolysis (EMP pathway) ATP is produced in three different sites.

1. During the conversion of 3 phosphoglyceraldehyde into 1-3-diphosphoglyceric acid one mole of NADH is produced, which on oxidation by electron transport chain yields three moles of ATP.
2. During the conversion of 1-3-diphosphoglyceric acid into 3-phosphoglyceric acid one mole of ATP is produced.
3. The conversion of phosphoenol pyruvic acid into enol pyruvic acid yields one mole of ATP.

Thus 5 moles of ATP are produced during the breakdown of one mole of triose phosphate. Since, two moles of triose phosphate are formed from one mole of glucose, 10 moles of ATP are produced during the EMP pathway.

There are sites in glycolysis, where ATP is utilised for activation purposes. One mole of ATP is utilised during the activation of glucose to glucose-6-phosphate. The second site of ATP utilisation is in the formation of fructose-1-6-diphosphate from fructose-6-phosphate. One mole of ATP is utilised per mole of fructose-6-phosphate phosphorylated. Net gain of ATP during glycolysis is shown in Table 4.1.

Table 4.1 Net gain of ATP during glycolysis

Reaction	ATP used	ATP gained
(1) Formation of glucose-6-P	1	-
(3) Formation of fructose-1-6-diphosphate	1	-
(6) Formation of 1,3-diphosphoglyceric acid		2x3=6
(7) Formation of 3-phosphoglyceric acid		2x1=2
(10) Formation of enol pyruvic acid		2x1=2
	–2	**+10**

There is thus a net gain of 8 ATP molecules in glycolysis.

4.4 CITRIC ACID CYCLE (KREB'S CYCLE)

Citric acid cycle starts with the formation of citrate by the condensation of oxaloacetate with acetate, both of which can be formed from pyruvate. Thus, the conversion of pyruvate to acetate is an obligatory step in the utilisation of carbohydrate by this pathway. This is an oxidative step and occupies a key position in glucose metabolism. One of the carbons of pyruvate is removed as CO_2 in the conversion. The remaining two carbons are removed as CO_2 in the citric acid cycle. There is a simultaneous oxidation of hydrogen through coenzymes in both. All the enzymes involved in the aerobic pathway are located mainly in the mitochondria along with the respiratory chain. A few of them occur in cytoplasm also.

4.4.1 CONVERSION OF PYRUVATE TO ACETATE

This reaction is called **oxidative decarboxylation** and is brought about by the enzyme pyruvate **dehydrogenase**. During this reaction coenzyme A molecule is attached to pyruvic acid to form acetyl coenzyme A, otherwise known as active acetate.

$$\underset{\text{Coenzyme A}}{\text{HS.CoA}} + \underset{\text{Pyruvic acid}}{\begin{matrix}\text{COOH}\\ |\\ \text{C=O}\\ |\\ \text{CH}_3\end{matrix}} \xrightarrow[\text{NAD}^+\text{ lipoic acid, FAD, TPP, Mg}^{++}]{\text{Pyruvate dehydrogenase}} \underset{\text{Acetyl CoA}}{\text{CH}_3\text{CO}\sim\text{ScoA}} + \text{CO}_2$$

The reaction takes place in several steps as shown in Figure 4.2. Pyruvate dehydrogenase is an enzyme complex located in the mitochondrial matrix and consists of three enzymes and five coenzymes.

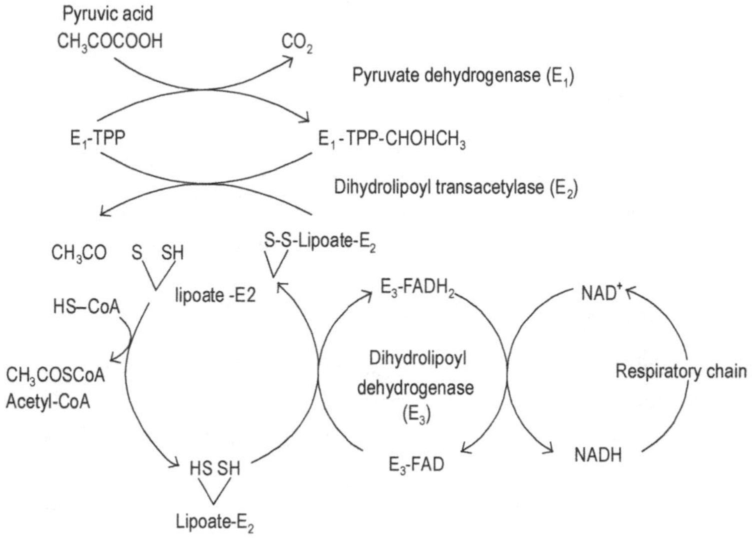

Fig. 4.2 Oxidation of pyruvate

The enzymes are

i. Pyruvate dehydrogenase

ii. Lipoate acyltransferase (dihydrolipoyl transacetylase) and

iii. Lipoamide dehydrogenase (dihydrolipoyl dehydrogenase)

The coenzymes are (i) TPP, (ii) lipoic acid, (iii) FAD, (iv) Coenzyme A and (v) NAD^+.

Closely associated with this complex are also two other enzymes—pyruvate dehydrogenase kinase and pyruvate dehydrogenase phosphatase.

Pyruvate dehydrogenase kinase will inactivate the enzyme, pyruvate dehydrogenase, by phosphorylating it. It requires ATP and Mg^{++} and is inactivated by high levels of ATP. Pyruvate dehydrogenase phosphatase removes the phosphate from the inactive enzyme and activates it. This also requires Mg^{++}. It is stimulated by Ca^{++}. These two enzymes help to regulate the pyruvate dehydrogenase activity.

The pyruvate dehydrogenase complex is capable of bringing about several stages in the conversion of pyruvate to acetyl-CoA and in the re-oxidation of reduced lipoic acid.

The reduced lipoic acid is re-oxidised to S-S form and the two hydrogen atoms are taken up by FAD to be later oxidised through the respiratory chain to a molecule of water.

The thioester bond in acetyl-CoA is energy rich and can readily make the energy available for synthetic reactions. Hence, it is called an active acetate molecule. In addition, the oxidation of the two hydrogen through the coenzymes and respiratory chains will yield 3 molecules of ATP for each pyruvate converted to acetate.

4.4.2 CONVERSION OF ACETATE TO OXALOACETATE

Active acetate can be formed not only from pyruvate but also by oxidation of fatty acids and from other sources. The active acetate combines with a molecule of oxaloacetate to form a molecule of citrate and this can be taken to the starting point of Kreb's cycle. The reactions involved in the cycle are outlined in Figure 4.3.

The various reactions involved in the citric acid cycle are as follows:

1. The cycle starts with the condensation of a molecule of acetyl-CoA with oxaloacetate to form citric acid. The condensation is brought about by **citrate synthetase**. Citrinyl-CoA is first formed and later hydrolysed to citric acid and coenzyme A.
2. A molecule of water is removed from citrate to form cis aconitic acid.
3. The molecule of water is added again, but the H^+ and OH^- are added at different sites to form a molecule of isocitric acid. Reaction 2 and 3 are brought about by the same enzyme **aconitase** which requires iron as activator.

4 &5. Isocitric acid now undergoes dehydrogenation brought about by **isocitrate dehydrogenase** to form oxalosuccinic acid.

Mammalian tissues contain two distinct enzymes—an enzyme requiring $NADP^+$ and Mn^{++} and another requiring NAD^+ and Mg^{++}. The NAD^+ dependent enzyme is located in the mitochondria and is the one concerned in the citric acid cycle. The dehydrogenation step produces oxalosuccinate. But this is never released from the enzyme. While still in the E-S complex, it is further decarboxylated and α keto glutaric acid is released. Isocitrate dehydrogenase is a regulatory enzyme in citric acid cycle. It is an allosteric enzyme inhibited by ATP and activated by ADP.

6. The α ketoglutarate now undergoes oxidative decarboxylation by the action of an enzyme complex **α ketoglutarate dehyrogenase**. It requires thiamine pyrophosphate (TPP), lipoic acid, NAD^+, FAD, coenzyme A and magnesium ions. The reaction is similar to the oxidative decarboxylation of pyruvate to form acetyl-CoA. The product in this case is succinyl-CoA. The bond linking CoA to succinic acid is energy rich.

7. **Succinate thiokinase** converts succinyl CoA to succinic acid. The energy released by hydrolysis of the CoA bond is utilised for the conversion of a molecule of guanosine diphosphate or inosine diphosphate (GDP or IDP) to the respective triphosphates, GTP or ITP. These can later interact with ADP to form ATP. This is the only step in citric acid cycle resulting in substrate phosphorylation.

8. Succinic acid loses two atoms of hydrogen to form fumaric acid. The enzyme concerned is **succinic acid dehydrogenase**. The hydrogen is transferred directly to FAD contained in the flavoprotein of the enzyme without the intervention of NAD^+.

9. Fumaric acid takes up a molecule of water by the action of **fumarase** to form malic acid.

10. Malic acid now undergoes dehydrogenation to form finally oxaloacetic acid which is one of the two components that formed citric acid to start the cycle. The other component acetate has been oxidised in the cycle.

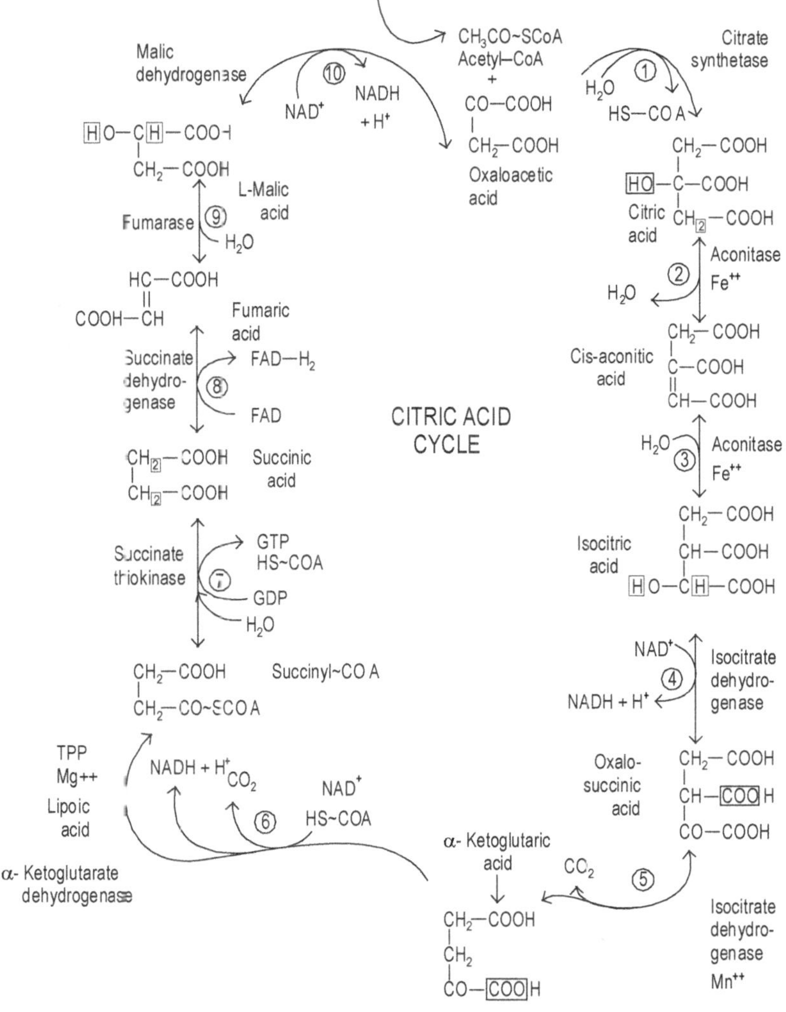

Fig. 4.3 Citric acid cycle

4.4.3 ELECTRON TRANSPORT CHAIN

The oxidation of reduced substrate is done by a number of electron acceptors constituting the electron transport system. The different components of the system are nicotinamide adenine dinucleotide (NAD), flavin adenine dinucleotide (FAD), and the cytochromes. When a reduced substrate loses hydrogen, it breaks up into proton and electron. NAD and FAD accept both these components of hydrogen molecule, while the cytochromes allow only the electrons to pass through and the protons are liberated in the substrate. Oxygen forms the terminal constituent of the electron transport system; it is the ultimate recipient of electrons and picks up the hydrogen proton from the substrate, and water is formed. Electron transport system plays a significant role in the energetics of cellular respiration. During this process, energy is dissipated, which when sufficient for the synthesis of ATP, brings about oxidative phosphorylation. Mitochondria contain most of the hydrogen and electron carriers. The most accepted sequence of electron carriers in the mitochondria is shown in Figure 4.4.

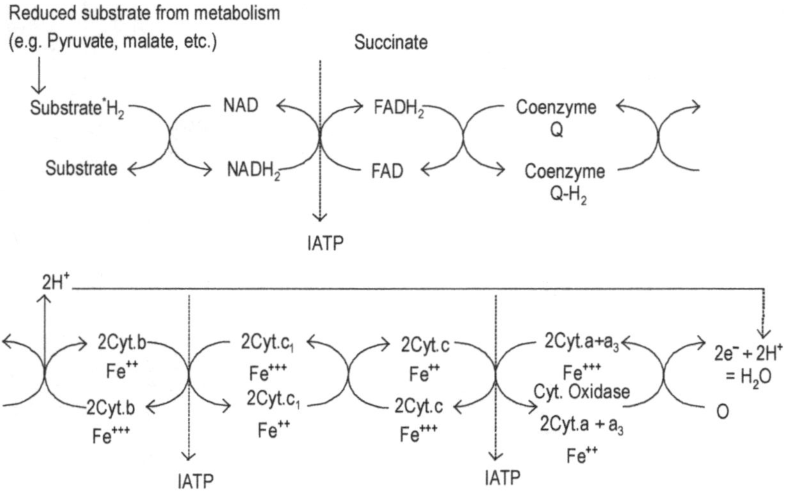

Fig. 4.4 Transport of hydrogen and electron through different carriers of electron transport chain

The electron is initially accepted by NAD and gets reduced to NADH. This reduced NADH is re-oxidised by FAD, which accept the electron from NADH and gets reduced to FADH. Flavoproteins may also be directly reduced by the substrate without involvement of pyridine system. For example, succinic dehydrogenase, which is a flavoprotein bearing FAD as its prosthetic group, is reduced directly by succinic acid without involving pyridine nucleotides.

In the normal course, reduced flavins are re-oxidised by ubiquinone (coenzyme Q), which is the next member of the electron transport chain. Ubiquinone is reduced to dihydroubiquinone. In case of aerobic dehydrogenases, which are also flavoproteins, the reduced flavins may be directly re-oxidised by molecular oxygen with the formation of hydrogen peroxide. Several flavoproteins contain firmly bound metallic ions such as Fe^{+++} in their prosthetic group. Such metallic ions are believed to be involved in the carriage of one electron to the next carrier. In some other cases, flavoproteins are known to carry electrons from pyridine nucleotides to cytochrome c directly. The examples are NADH-cytochrome-c-reductase and NADPH-cytochrome-c-reductase, identified in heart muscle, and yeast respectively.

Dihydroubiquinone may be oxidised by ferricytochrome b. At this level of oxidation, hydrogen atoms undergo ionisation to form two protons and two electrons. The protons cannot be carried further by the electron transport chain; however, the electrons can be carried easily by the cytochromes. The electron is accepted by ferricytochrome at its iron (Fe^{+++}), which is reduced to ferrocytochrome (Fe^{++}). One molecule of each cytochrome is capable of carrying only one electron at a time. Thus, for each molecule of hydrogen, which releases two electrons, two molecules of each cytochrome are required. At the level of cytochrome b, two protons per molecule of hydrogen are released in the mitochondrial system. Ferrocytochrome b is oxidised in the presence of ferricytochrome c_1. Ferricytochrome c_1 is reduced to ferrocytochrome c_1, which can be re-oxidised by ferricytochrome c. Ferrocytochrome c is oxidised by **cytochrome oxidase** ($a+a_3$). Reduced cytochrome oxidase, is autoxidisable. In the presence of oxygen, two protons (which are released at cytochrome b level), and two electrons (carried by cytochrome system), combine to form a water molecule.

4.4.4 OXIDATIVE PHOSPHORYLATION

As a result of biological oxidations, the energy initially existing in the chemical bonds of various metabolites is released as free energy, a substantial portion of which is trapped under suitable conditions in the form of high-energy phosphate bond in ATP. This process is known as oxidative phosphorylation. In some of the biological oxidations some other high-energy phosphate bond compound is formed in the beginning, which is later capable of transferring its bond-energy together with the phosphate group to ADP forming ATP. In some other oxidations, the energy is not trapped at all and it is lost as heat energy. A major portion of biochemically useful energy, contained in the ATP molecules, is mainly derived from electron transport chain oxidation.

In the mitochondria, the energy liberated is immediately trapped in the presence of adenosine diphosphate (ADP) and inorganic phosphate (Pi) resulting in the formation of adenosine triphosphate (ATP). Formation of ATP is thus an endergonic reaction, the energy for which is supplied by the electron transport chain oxidation. Formation of one mole of ATP from ADP requires approximately 8 Kcals energy under physiological conditions. Obviously, ATP formation is not possible at the sites where energy release is lesser than this value. Under physiological conditions, ATP is formed at three sites in the entire respiratory chain and these sites are

i. between NAD and flavins,

ii. between cytochrome b and cytochrome c_1 and

iii. between cytochrome c and cytochrome a_3.

Substrates involving the entire respiratory chain for oxidation can form three moles of ATP per mole of hydrogen removed. Most of the biological oxidations involve the entire respiratory chain. Succinate requires FAD as initial oxidising agent; hence, during oxidation of succinate by the respiratory chain, the first site of phosphorylation is bypassed, and hence, only two moles of ATP are produced per mole of succinate oxidised.

Energetics of Kreb's cycle In Kreb's cycle, energy is generated at six different sites (Table 4.2).

1. During oxidative decarboxylation of pyruvic acid, one mole of NADH is produced which on oxidation by electron transport chain yields 3 moles of ATP.
2. During the conversion of isocitric acid to oxalosuccinic acid 1 mole of NADH or NADPH is produced which on oxidation produces 3 moles of ATP.
3. During the conversion of α-ketoglutaric acid into succinyl-CoA 3 moles of ATP are produced.

4. Conversion of succinyl-CoA to succinic acid yields 1 mole of ATP due to substrate level of phosphorylation.
5. During the oxidation of succinic acid one mole of $FAD.H_2$ is produced. This, on oxidation by electron transport chain, leads to formation of 2 moles of ATP.
6. Conversion of malic acid to oxaloacetic acid produces 1 mole of NADH which on oxidation produces 3 moles of ATP.

Table 4.2 Formation of ATP during the oxidation of pyruvic acid by Kreb's cycle

Site of synthesis of ATP	Coenzyme oxidised	No. of moles of ATP produced
Formation of acetyl-CoA	$NADH^+ + H^+$	3
Formation of oxalosuccinic acid	$NADH^+ + H^+$	3
Formation of succinyl CoA	$NADH^+ + H^+$	3
Formation of succinic acid	Substrate phosphorylation	1
Formation of fumaric acid	$FAD.H_2$	2
Formation of oxaloacetic acid	$NADH^+ + H^+$	3
	Total	**15**

When one mole of pyruvic acid is oxidised to CO_2 and H_2O by Kreb's cycle 15 moles of ATP are produced. Since, two moles of pyruvic acid may be obtained per mole of glucose entering into glycolytic reactions, 30 moles of ATP are synthesised.

Under aerobic conditions, complete oxidation of 1 mole of glucose through glycolysis and Kreb's cycle may produce 38 moles of ATP.

4.5 ALTERNATE AEROBIC PATHWAY: PENTOSE METABOLISM

The alternate aerobic pathway, also known as hexose monophosphate shunt, Warburg–Dickens–Lipmann pathway or pentose phosphate pathway is a cyclic mechanism analogous to tricarboxylic acid cycle in which O_2 is utilised in the early part of the reactions. Energetically also, this pathway is considered more efficient because, it yields more than 30 molecules of ATP for each molecule of glucose oxidised. It has been postulated that this could be a major pathway for utilisation of glucose in the cornea, lens of the eye, lactating mammary gland, adipose tissue, liver and foetal heart.

The sequence of reactions begins with glucose–6–phosphate, which follows an aerobic pathway, instead of the usual glycolytic pathway (Figure 4.5). The reactions of the pathway can be divided into two broad phases. (i) Conversion of hexose to pentose (ii) Conversion of pentose to hexose.

4.5.1 CONVERSION OF HEXOSE TO PENTOSE

i. Glucose-6-phosphate is oxidised to 6-phosphogluconic acid by **glucose-6-phosphodehydrogenase** and coenzyme NADP.

$$\text{Glucose-6-P} + \text{NADP} \xrightarrow[\text{dehydrogenase}]{\text{Glucose-6-phospho}} \text{6-Phosphogluconic acid} + \text{NADPH} + \text{H}^+$$

ii. 6-Phosphogluconic acid is then oxidised by **6-phosphogluconate dehydrogenase** and NADP and subsequently decarboxylated to form ribulose-5-phosphate.

$$\text{6-Phosphogluconic acid} + \text{NADP} \xrightarrow[\text{dehydrogenase}]{\text{Phosphogluconate}} \text{Ribulose-5-phosphate} + \text{NADPH} + \text{H}^+$$

Ribulose-5-phosphate is converted to ribose-5-phosphate through an enediol, catalysed by **phosphoriboisomerase**.

$$\text{Ribulose-5-phosphate} \underset{\text{Isomerase}}{\overset{\text{Phosphoribo}}{\rightleftharpoons}} \text{Ribulose-5-phosphate}$$

iii. Ribulose-5-Phosphate is converted to xylulose-5-phosphate by **epimerase**.

$$\text{Ribulose-5-phosphate} \xrightarrow{\text{Epimerase}} \text{Xylulose-5-phosphate}$$

4.5.2 CONVERSION OF PENTOSE TO HEXOSE

iv. Xylulose-5-phosphate undergoes transketolation to form glyceraldehyde-3- phosphate. During this reaction a ketol group (active glycolaldehyde) is removed from xylulose-5-phosphate by enzyme **transketolase**. TPP and Mg^{++} ion act as cofactors.

$$\text{Xylulose-5-phosphate} \xrightarrow[\text{TPP, Mg}^{++}]{\text{Transketolase}} \text{Glyceraldehyde-3-phosphate} + \text{active glycolaldehyde}$$

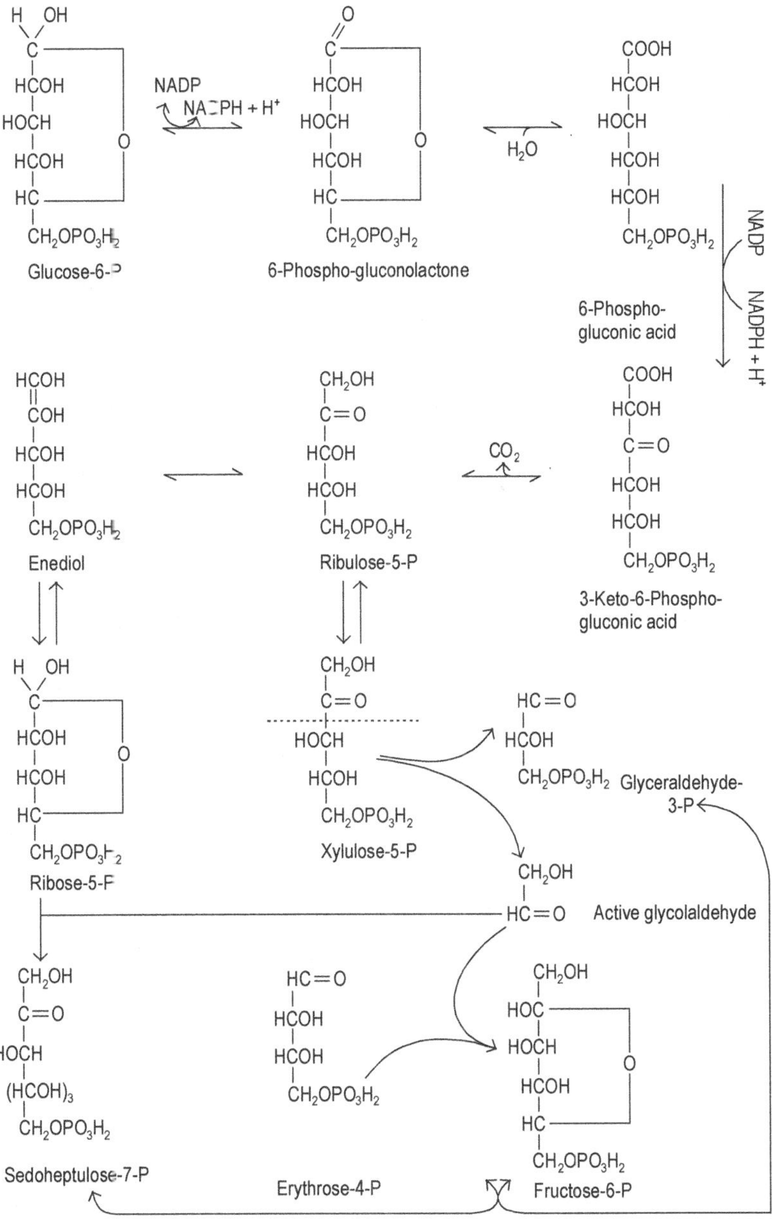

Fig. 4.5 Hexose monophosphate shunt

v. The ketol fragment removed from xylulose-5-phosphate condenses with ribose-5-phosphate to form a seven carbon sugar sedoheptulose, by the action of a **transketolase**.

$$\text{Ketol} + \text{Ribose-5-phosphate} \xrightarrow{\text{Transketolase}} \text{Sedoheptulose}$$

vi. Finally, a transaldolase transfers the dihydroxy-acetone moiety of sedoheptulose to glyceraldehyde-3-phosphate to form the hexose, fructose-6-phosphate leaving a tetrose residue, erythrose-4-phosphate. Fructose-6-phosphate can be readily converted to glucose-6-phosphate.

$$\text{Glyceraldehyde-3-phosphate} + \text{Sedoheptulose} \longrightarrow \text{Fructose-6-phosphate} + \text{erythrose-4-phosphate}$$

vii. Addition of a ketol fragment to erythrose-4-phosphate by transketolase reaction, will give fructose-6-phosphate again.

$$\text{Ketol} + \text{Erythrose-4-phosphate} \longrightarrow \text{Fructose-6-phosphate}$$

$$\text{Fructose-6-phosphate} \longrightarrow \text{Glucose-6-phosphate}$$

The cycle repeats starting with glucose-6-phosphate, which is regenerated at the end of each cycle.

4.5.3 METABOLIC SIGNIFICANCE OF HEXOSE MONOPHOSPHATE SHUNT

i. The major function of hexose monophosphate shunt pathway is to furnish and maintain a continuous supply of reduced NAD and NADP, which are required for various metabolic reactions.

ii. Reduced NADP is required for the enzyme system concerned with the synthesis of fatty acids.

iii. It forms an important pathway by which pentoses are synthesised and metabolised in the body. Pentoses are essential for synthesis of nucleotides and nucleic acids.

4.5.4 ENERGETICS OF HMP SHUNT

There are only two energy-generating sites in this pathway. During the conversion of glucose-6-phosphate to 6-phosphogluconic acid, 6 moles of NADPH are generated. Further, during conversion of 6-phosphogluconic acid into 3-keto-6-phosphogluconic acid, 6 moles of

NADPH are generated. Since, six moles of glucose enter in the cycle and five out of these are regenerated, it can be assumed that 1 mole of glucose has been completely oxidised resulting in the formation of 12 moles of NADPH. If the reduced NADP is completely oxidised by electron transport chain, this would result in the formation of 36 moles of ATP (equivalent to 288 Kcal energy). One mole of ATP (8 Kcals) is used in the initial hexokinase reaction. Thus, there occurs a net gain of 35 moles of ATP. Practically the reduced NADP obtained in this pathway is mostly used for reductive biosynthetic processes going on in the body.

4.6 CORI'S CYCLE (LACTIC ACID CYCLE)

Glycogen is synthesised in all the tissues of the body but the major sites are liver and muscle. Liver glycogen may get converted into muscle glycogen or vice versa. The interrelationship is shown by Cori's cycle (Lactic acid cycle) (Figure 4.6). Liver glycogen is broken down into glucose (glycogenolysis) which diffuses into blood stream and is carried to the muscle where it is converted to muscle glycogen (glycogenesis). Muscle glycogen breaks down into lactic acid (by glycogenolysis and glycolysis), which diffuses out of muscles into the blood stream. It is carried to the liver tissue where it can synthesise liver glycogen. It is evident that liver glycogen may be converted into muscle glycogen by intermediate formation of glucose, whereas muscle glycogen may be converted into liver glycogen, by intermediate formation of lactic acid.

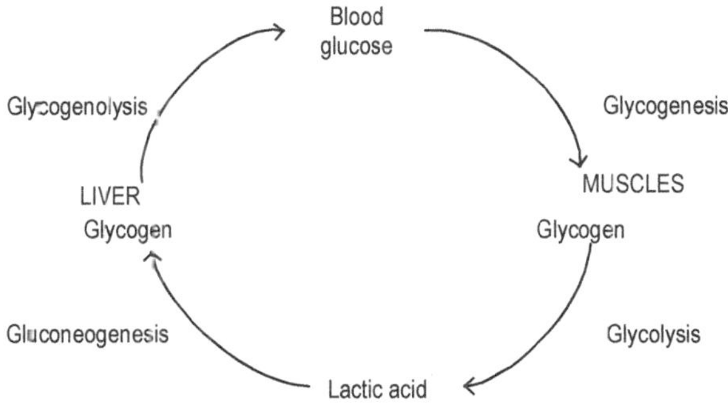

Fig. 4.6 Cori's cycle

This difference is due to the fact that muscle tissues lack the enzyme **glucose-6-phosphatase** and hence formation of free glucose is not possible in the muscles. Liver cells contain glucose-6-phosphatase, which converts glucose-6-phosphate into glucose. Glucose is freely permeable through cellular membrane of liver as well as muscle tissues.

4.7 GLUCONEOGENESIS

The process of synthesis of glucose from non-carbohydrate sources is known as **gluconeogenesis.** This process mostly occurs in the liver and kidney tissues at a basal rate but becomes very active when diet is not able to meet the carbohydrate requirement of the body at the desired rate. This process is particularly required for most tissues which are exclusively dependent on glucose for their energy supply, e.g. central nervous system, RBC and adrenal medulla. The daily requirement of CNS and RBC for glucose is about 140g and 30g respectively.

The most common gluconeogenic substances are lactic acid and glycerol. Besides these, other substances such as propionic acid, certain amino acids such as glutamic acid, glycine, serine, aspartic acid, arginine and ornithine, certain α-keto acids such as pyruvic acid, α- ketoglutaric acid and oxaloacetic acid are also converted into glucose. These substances at some stage of their metabolism are linked with glycolytic or citric acid cycle reactions and by reversal of these reactions, these can be ultimately converted into glucose or glycogen. In the normal course, when the demand of sugar is fulfilled, the rate of gluconeogenesis is decreased. Glucocorticoids, glucagons and catecholamines increase gluconeogenesis, whereas insulin suppresses it.

The process of gluconeogenesis plays various important roles in the body.

i. It helps in the regulation of the blood sugar level at times when dietary carbohydrates are not able to meet body carbohydrate requirement fully. By regulating blood sugar level, it also regulates the minimum required level of glycogen in the liver and muscle tissues. Also, it protects delicate organs like the brain against the harmful effects that might occur due to hypoglycemia.

ii. Gluconeogenesis brings about proper disposal of lactic acid produced by the muscles during and after exercise and glycerol produced in the adipose tissue due to turnover of the fats and prevents their wastage.

The linking site of various gluconeogenic substances in glycolysis and citric acid cycle are reversible, hence at the time of emergency, biosynthesis of fats and proteins from carbohydrates or vice versa is possible through these sites. Thus, a dynamic equilibrium is established among carbohydrates, fats and proteins.

REVIEW QUESTIONS

1. Discuss the process of glycolysis and its energetics.
2. Describe Kreb's cycle and its energetics.
3. Describe hexose monophosphate shunt. Add a note on its significance.
4. Define oxidative phosphorylation. Explain how ATP is produced in biological systems.
5. Write short notes on:
   i. glycogenesis
   ii. glycogenolysis
   iii. gluconeogenesis
   iv. Cori's cycle
   v. electron transport chain

5

LIPIDS

Lipids or fats are greasy materials occurring widely in nature. They are generally insoluble in water but soluble in fat solvents. They include (a) naturally occurring fats, e.g. butter and oils (b) substances which are chemically related to fats but differ in certain common properties, e.g. lecithin, a waxy substance soluble in fat solvents and also mixing well with water to form a colloidal solution (c) substances which are related to fats because of certain common properties like solubility and biological origin, but differ from fat in appearance and chemical structure, e.g. cholesterol, which is in liquid state at room temperature.

5.1 OCCURRENCE

Fats are widely distributed in plant and animal tissues and can be extracted by using fat solvents like alcohol and ether. In plants, they are particularly present in nuts and seeds. In animals, nervous system is rich in lipids. The fat depots such as subcutaneous tissues, mesenteric tissues and fatty tissues around the kidneys and yellow bone marrow contain large amounts of fat. Animal sources rich in fat are milk, egg, meat, liver and fish oils.

5.2 BIOLOGICAL SIGNIFICANCE OF FATS

i. Fat is an essential dietary constituent; it performs the important function of supplying fuel to the body. It yields more heat and energy than carbohydrates.

ii. Certain normal breakdown products of fatty acids in the body such as acetic acid and bile acids form important building blocks of biologically active and complex materials like cholesterol, sex hormones and steroids.

iii. Another function of dietary lipid is to supply the essential dietary constituent known as the essential fatty acids (EFA).

iv. Deposits of fat underneath the skin provide an insulating effect to the body. They protect the body from excessive heat or cold.

v. Vitamins A, D, E and K are fat-soluble vitamins, which are dietary essentials.

5.3 CLASSIFICATION OF LIPIDS

Bloor classified lipids into (i) simple lipids, (ii) compound lipids, (iii) derived lipids and (iv) substances associated with lipids.

5.3.1 SIMPLE LIPIDS

These include fats, waxes and their components.

FATS

Fats represent the most abundant and widespread class of lipids occurring in nature. In plants they are particularly abundant in nuts and seeds, where they represent reserve food materials. Important vegetable fats are coconut oil, gingely oil, groundnut oil, olive oil, cotton seed oil and linseed oil. In animals, they are present in the subcutaneous tissue, nervous tissue, liver, mesentery, etc. Important animal fats are butter, fish oils, egg and milk.

Fat may be liquid or solid at room temperature depending upon its melting point. If the melting point is low, it exists as liquid at ordinary temperature. Fats which are liquid at room temperature are called oils.

Coconut oil is liquid at the temperature prevailing in tropical countries, whereas it is solid in cooler countries.

Chemical composition of fats Animal and vegetable fats are complex mixtures of glycerides, i.e. they are esters of glycerol and fatty acids. They are usually triglycerides and are called natural fats. Triglycerides or natural fats are composed of three molecules of fatty acid, esterified to glycerol. A triglyceride is formed by the condensation of one molecule of glycerol with three molecules of fatty acid. The fatty acids may be of the same type, when the resulting glyceride is called a simple glyceride.

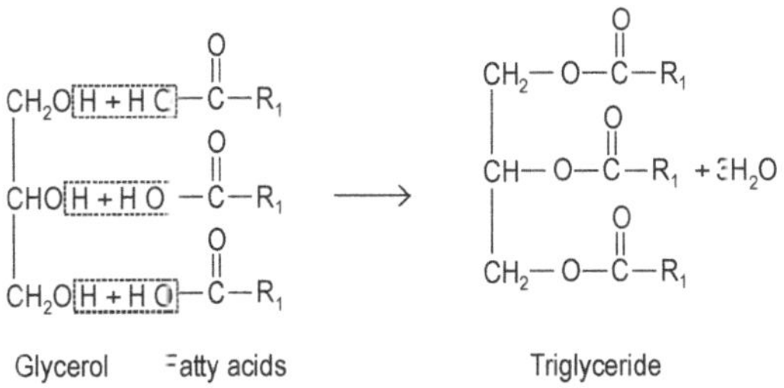

$$\begin{array}{l} CH_2O\boxed{H + HO}-\overset{\overset{O}{\|}}{C}-R_1 \\ | \\ CHO\boxed{H + HO}-\overset{\overset{O}{\|}}{C}-R_1 \\ | \\ CH_2O\boxed{H + HO}-\overset{\overset{O}{\|}}{C}-R_1 \end{array} \longrightarrow \begin{array}{l} CH_2-O-\overset{\overset{O}{\|}}{C}-R_1 \\ | \\ CH-O-\overset{\overset{O}{\|}}{C}-R_1 \\ | \\ CH_2-O-\overset{\overset{O}{\|}}{C}-R_1 \end{array} + 3H_2O$$

Glycerol Fatty acids Triglyceride

If the three fatty acids in a given glyceride are different, then the resulting glyceride is called a mixed glyceride. The common fatty acids present in natural fats are:

palmitic acid – $CH_3(CH_2)_{14}COOH$

stearic acid – $CH_3(CH_2)_{16}COOH$

oleic acid – $CH_3(CH_2)_7CH{=}CH(CH_2)_7COOH$

Physical properties of fats

i. They are greasy to touch.

ii. They are insoluble in water but are readily soluble in fat solvents like alcohol, ether, chloroform, etc.

iii. Pure glycerides are tasteless, odourless, colourless and neutral in reaction. But after exposure to air for sometime they become acidic and develop a yellow colour due to partial hydrolysis and oxidation of unsaturated fatty acids.

iv. The hardness or consistency of the fat depends upon the relative amount of the saturated and unsaturated fatty acids present in the fat. At room temperature, fats containing saturated fatty acids are in solid state and fats containing unsaturated fatty acids are in liquid state and these are oils.

v. Fats have definite melting points. Presence of saturated fatty acids in fats increases the melting point, but presence of unsaturated fatty acids lower the melting point.

vi. Fats have lesser specific gravity than water, and therefore float on water.

vii. *Spreading of fat* When a liquid fat is placed on water, it spreads uniformly over the surface of water and if the quantity is sufficiently small, it will form a layer of one molecule thick. The effect of this property is to lower the surface tension and help the transport of fat.

Two groups in the structure of fat exert their role in the mechanism of spreading. These are:

a. A hydrophilic carboxyl group which dissolves in water, referred to as C_1.

b. A hydrophobic hydrocarbon chain, which forms a layer on the surface referred to as C_2.

The two groups are arranged above and below the surface of water.

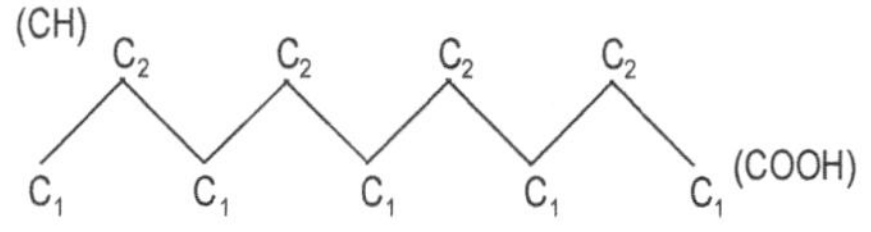

(CH) C_2 C_2 C_2 C_2

C_1 C_1 C_1 C_1 C_1 (COOH)

The hydrophobic hydrocarbon chain C_2 forms a layer on the surface and is anchored to its water-soluble carboxyl group C_1. The layer so formed, has the thickness of a mole, that is, it forms a monomolecular layer.

viii. Emulsification Though fats are insoluble in water, they can be broken down into minute droplets and dispersed in water. This process is known as emulsification. Naturally occurring emulsions are milk and yolk of egg. Emulsification is an essential requisite for digestion of fats in the intestine. Emulsification is brought about either by chemical or mechanical action.

Chemical Properties of fats

i. ***Hydrolysis*** Fats are readily hydrolysed to glycerol and fatty acids by acids, alkalies, super-heated steam and enzymes (lipase). Hydrolysis is of two types: (a) Enzyme hydrolysis and (b) Alkali hydrolysis.

Enzyme hydrolysis is governed by lipase. This is important in the digestion of fats. Alkali hydrolysis refers to hydrolysis of fats by boiling with alkalies. This is important in industries. The process of alkali hydrolysis is called saponification, which means 'soap-making'.

$$\begin{array}{l} CH_2-O-\overset{O}{\overset{\|}{C}}-R_1 \\ | \\ CH-O-\overset{O}{\overset{\|}{C}}-R_2 \\ | \\ CH_2-O-\overset{O}{\overset{\|}{C}}-R_3 \end{array} \xrightarrow{3\,KOH} \begin{array}{l} CH_2OH \\ | \\ CHOH \\ | \\ CH_2OH \end{array} + \begin{array}{l} R_1-\overset{O}{\overset{\|}{C}}-OK \\ R_2-\overset{O}{\overset{\|}{C}}-OK \\ R_3-\overset{O}{\overset{\|}{C}}-OK \end{array}$$

Triglyceride Glycerol Soap

Acid number It is the number of milligrams of KOH required to neutralise the free fatty acids present in 1 gm of fat. The acid number, thus, indicates the quantity of free fatty acid present in a fat. Obviously, a fat which has been both processed and stored properly has a very low acid number.

Saponification number This is a constant which is of special value in the identification and characterisation of neutral fat. It is defined as the number of milligrams of KOH required to saponify 1 gm of fat.

ii. ***Auto-oxidation*** Unsaturated fatty acids in fats react with ozone and oxygen to undergo a reaction called auto-oxidation and forms ozonides, peroxides, aldehydes and ketones.

Rancidity Auto-oxidation occurring in natural edible fats is called rancidification. The fat, which has become rancid has a disagreeable odour and taste and is unfit for consumption. Rancidification occurs more frequently in summer.

Rancidity can be considerably prevented by the addition of traces of organic compounds like phenols, gallic acid and vitamin C and E. Such substances are called antioxidants and these are present in many vegetable oils.

iii. ***Addition reaction*** Fats containing unsaturated fatty acids, readily add on elements such as halogens and hydrogen at their double bonds.

$$-CH{=}CH{=}CH{=} \xrightarrow{+3I} CHI-CHI-CHI$$

Iodine number or Iodine value The number of grams of iodine taken up by 100 gm of given fat is called its iodine value or iodine number. Iodine number is a measure of the degree of unsaturation of the fatty acid.

Polenske number It is the number of millilitres of 0.1 N KOH required to neutralise the insoluble fatty acids (i.e. those which are not volatile with steam distillation) obtained from 5 gm of fat.

Reichert–Meissl number It is the number of millilitres of 0.1 N KOH required to neutralise the soluble, volatile fatty acids derived from 5 gm of fat. The Reichert–Meissl number thus measures the quantity of short chain fatty acids (up to C-10 inclusive) in the fat molecule. The Reichert–Meissl numbers of coconut and palm oils range between 5 and 8. Butter fat is exceptional in having a high Reichert–Meissl number, ranging from 17 to 35. This high value makes possible the detection of any foreign fats which are, sometimes, adulterated in the manufacture of butter.

Acetyl number It is the number of milligrams of KOH required to neutralise the acetic acid obtained by saponification of 1 gm of fat after it has been acetylated. The treatment of fat or fatty acid mixture with acetic anhydride results in acetylation of all alcoholic OH groups. The acetyl number is thus a measure of the number of OH groups in the fat. For example, castor oil has a high acetyl number (146) since it has a high concentration of ricinoleic acid, a hydroxy acid.

Hydrogenation Unsaturated fats can combine with hydrogen under the influence of a suitable catalyst such as finely divided nickel, platinum and copper at a high temperature and become more saturated. For example, oleic acid, which is an unsaturated fatty acid, can be saturated by the addition of 2H atoms to give stearic acid.

$$CH_3(CH_2)_{14}CH{=}CH\ COOH \xrightarrow{+2H} CH_3(CH_2)_{16}COOH$$

Oleic acid — Stearic acid

WAXES

These are esters of fatty acids with alcohols other than glycerol. They are insoluble in water, but soluble in fat solvents. They are not easily hydrolysed or digested by lipase. The importance of waxes in plants and animals is that they serve as protective agents. The waxy coating on the surface of plants prevents excessive loss of moisture. In birds, the presence of wax on the surface of feathers helps to keep them soft and pliable. The following are some examples of waxes

i. ***Bees wax*** This is secreted by honey bee to form honey comb. The chief ingredient is myricyl palmitate.

$$CH_3(CH_2)_{14}.COO-CH_2(CH_2)_{28}.CH_3$$

Myricyl palmitate

ii. ***Sperm Whale wax*** Waxes serve as the chief storage form or fuel in planktons. Since marine organisms consume planktons in large quantities, waxes act as marine food and storage lipids in them. Sperm whale wax and bees wax are composed mainly of palmitic acid esterified with either hexacosanol ($CH_3(CH_2)_{24}.CH_2OH$) or triacontanol ($CH_3(CH_2)_{28}.CH_2OH$).

$$CH_3(CH_2)_{14}COO\boxed{H + HO}CH_2(CH_2)_{24}CH_3 \xrightarrow{-H_2O} CH_3(CH_2)_{14}.COO-CH_2(CH_2)_{24}.CH_3$$

Palmitic acid — Hexacosanol — Hexacosanyl palmitate

The sperm whale wax is rich in cetyl palmitate.

$$CH_3(CH_2)_{14}.COO-CH_2(CH_2)_{14}.CH_3$$

Cetyl palmitate

iii. ***Chinese wax*** This is the secretion of an insect.

iv. ***Carnauba wax*** This is found on the leaves of carnauba palm of Brazil. This is the hardest known wax which consists mainly of fatty acids esterified with tetracosanol.

$$CH_3(CH_2)_{32}CH_2OH$$

Tetracosanol

v. ***Lanoline*** This is also known as 'wool wax'. It is obtained from wool and is used in making ointments.

5.3.2 COMPOUND LIPIDS

These are lipids which contain certain chemical groups other than alcohol and fatty acid. Compound lipids include phospholipids, glycolipids, sulpholipids and lipoproteins.

PHOSPHOLIPIDS

These are also known as phosphotides. They are important compounds present in all vegetable and animal cells. They are abundantly present in heart, brain, kidneys, egg yolk and soya bean. They are all composed of fatty acid, glycerol, phosphoric acid and nitrogenous base. Phospholipids play an important role in the prevention of fatty liver and in the process of blood coagulation. Thromboplastin, a factor in blood clotting, is a phospholipid. The three important phospholipids are lecithin, choline and cephalin.

i. ***Lecithin*** A typical lecithin consists of one glycerol molecule esterified to 2 molecules of fatty acids. One of these fatty acids may be unsaturated and the other saturated. It contains phosphoric acid molecule and a nitrogenous base, choline. Lecithin is important in the metabolism of fat and in the prevention of fatty livers.

ii. ***Choline*** Choline is a component of lecithin and sphingomyelin and hence an important constituent of phospholipid. Choline is an important factor in the prevention of fatty liver. Absence of choline in the diet leads to accumulation of fat in the liver of rats, leading to fatty liver. Acetylcholine, the acetylated derivative of choline plays an important role in the transmission of nerve impulses. Choline is present in egg yolk, liver, heart, kidney and milk.

iii. ***Cephalin*** Cephalin occurs along with lecithin in all animal and plant cells. Sphingomyelins are found in brain and nerve tissues. They contain fatty acid, phosphoric acid, choline and a complex amino alcohol sphingosine or sphingol.

GLYCOLIPIDS

Glycolipids contain a sugar in addition to fatty acid and sphingosine. Glycolipids are further subdivided into cerebrosides and gangliosides. Cerebrosides are found in brain. Gangliosides are glycolipids containing glucose, galactose and a substance called neuraminic acid in addition to fatty acid and sphingosine. Gangliosides are important components of specific receptor sites on the surface of the cell membranes. For example, they are found in the specific sites on nerve endings to which neurotransmitter molecules are bound, during the chemical transmission of an impulse from one nerve cell to the next.

SULPHOLIPIDS

These are lipids containing sulphur. They are abundantly present in the white matter of the brain.

LIPOPROTEINS

Lipoproteins are formed by combination of proteins with lipids which include the phospholipids lecithin and cephalin, fatty acid, cholesterol, glycerides and fat soluble vitamins. Lipoproteins are widely distributed in the body tissues.

5.3.3 DERIVED LIPIDS

These include fatty acids and glycerol.

FATTY ACIDS

Fatty acid is an important constituent of fat. It is a monocarboxylic acid with a hydrocarbon chain. The length of the chain may vary from 2 to 34 carbon atoms. It is obtained by the hydrolysis of fats by acids, alkalies or enzymes. Almost all the fatty acids found in nature have straight chain and even number of carbon atoms. The chain may be saturated or unsaturated. The commonest fatty acids that occur in nature are palmitic, stearic and oleic acids. Among these fatty acids, oleic acid is unsaturated.

Fatty acids may be divided into 3 classes: (i) saturated fatty acids, (ii) unsaturated fatty acids and (iii) cyclic fatty acids.

i. ***Saturated fatty acids*** These are fatty acids which do not contain double bonds. They do not exhibit addition reactions and their iodine values are nil. Generally, presence of saturated fatty acid makes the fat solid, but there are exceptions to this rule. For example, butter, which is a solid fat, has a higher iodine value (35 to 50) than coconut oil, which is a liquid fat (iodine value 6–10).

   The general formula of saturated fatty acid is $CH_3(CH_2)_nCOOH$.

   Some of the saturated fatty acids are listed below

   Acetic acid – CH_3COOH
   Butyric acid – $CH_3(CH_2)_2COOH$
   Caproic acid – $CH_3(CH_2)_4COOH$
   Caprylic acid – $CH_3(CH_2)_6COOH$
   Capric acid – $CH_3(CH_2)_8COOH$
   Lauric acid – $CH_3(CH_2)_{10}COOH$
   Myristic acid – $CH_3(CH_2)_{12}COOH$
   Palmitic acid – $CH_3(CH_2)_{14}COOH$
   Stearic acid – $CH_3(CH_2)_{16}COOH$
   Arachidic acid – $CH_3(CH_2)_{18}COOH$
   Lignoceric acid – $CH_3(CH_2)_{22}COOH$

ii. ***Unsaturated fatty acids*** These are fatty acids which contain double bonds. Iodine values vary according to the degree of unsaturation. They are generally liquid at room temperature. Some of the unsaturated fatty acids are listed below:

Oleic acid–$CH_3(CH_2)_7CH=CH(CH_2)_7COOH$

Linoleic acid–$CH_3(CH_2)_4CH=CH\ CH_2\ CH=CH(CH_2)_7COOH$

Arachidonic acid–$CH_3(CH_2)_4CH=(CHCH_2CH=)_3CH(CH_2)_3COOH$

Linolenic acid–$CH_3CH_2CH=CHCH_2CH=CHCH_2CH=CH(CH_2)_7COOH$

iii. ***Cyclic acids*** These are fatty acids having a cyclic ring structure containing five carbon atoms. They are unsaturated. e.g. hydnocarpic acid and chaulmoogric acid.

```
     CH                                   CH
   //   \                               //   \
 HC      CH.(CH2)10COOH               HC      CH.(CH2)12COOH
  |       |                            |       |
 H2C ---- CH2                         H2C ---- CH2
```

Hydnocarpic acid   Chaulmoogric acid

Essential fatty acids Certain fatty acids, which have specific nutritional importance, are called essential fatty acids (EFA). They are long chain polyunsaturated fatty acids that cannot be synthesised by the body. They are required in small quantity, but are essential constituents of diet. A few common essential fatty acids are (1) linoleic acid, (2) linolenic acid, (3) arachidonic acid. Lack of EFA in adequate amounts in the diet of rats and some animals, produced deficiency symptoms like skin lesions, cessation of growth and abnormalities in pregnancy and lactation. These symptoms were reversed to normal condition by the administration of the EFA. Several workers have reported that essential fatty acids are nutritionally essential for humans also. Deficiency in infants and children produce skin lesions like eczema.

Vegetable oils are rich in EFA and animal fats are poor sources of EFA. Experiments on EFA in the diet of human have indicated that a diet containing animal fat tends to increase the serum cholesterol whereas a diet containing vegetable fats and EFA in adequate quantity, tend to lower the serum cholesterol. This indicates that unsaturated fatty acids and EFA tend to lower the serum cholesterol. This property of EFA is important in the management of hypercholesterolemia, which is associated with atherosclerosis or thickening of the arteries. This is considered to be the immediate cause of coronary disease of the heart.

EFA are also important for the growth of cells, reproductive function, to prolong clotting time and to increase the fibrinolytic activity.

GLYCEROL

Glycerol is a component of fat, and commonly called glycerine; it is the alcoholic component of triglycerides. It is closely allied to carbohydrates in its chemical structure and properties. But it is closely associated with fats in its occurrence and functions.

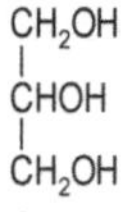

$$\begin{array}{c} CH_2OH \\ | \\ CHOH \\ | \\ CH_2OH \\ \text{Glycerol} \end{array}$$

Properties of glycerol

i. It is a colourless heavy liquid with a specific gravity of 1.26.

ii. It is sweetish in taste and is commonly known as glycerine.

iii. It is soluble in water and alcohol but insoluble in lipid solvents like ether, chloroform and benzene.

iv. When glycerol is oxidised with H_2O_2 in a slightly alkaline solution, a mixture of trioses glyceraldehyde and dihydroxyacetone are formed.

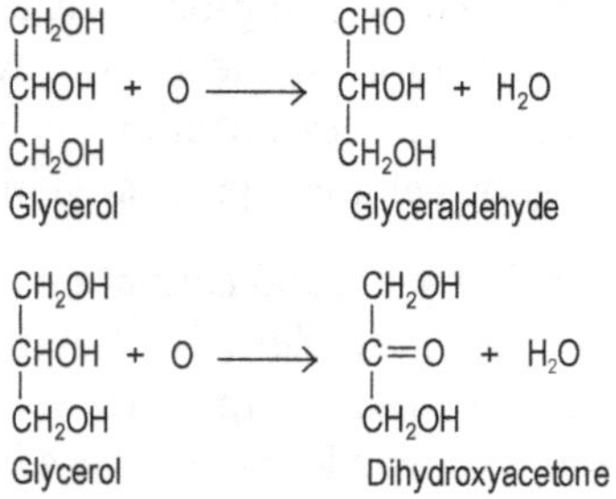

$$\begin{array}{c} CH_2OH \\ | \\ CHOH \\ | \\ CH_2OH \\ \text{Glycerol} \end{array} + O \longrightarrow \begin{array}{c} CHO \\ | \\ CHOH \\ | \\ CH_2OH \\ \text{Glyceraldehyde} \end{array} + H_2O$$

$$\begin{array}{c} CH_2OH \\ | \\ CHOH \\ | \\ CH_2OH \\ \text{Glycerol} \end{array} + O \longrightarrow \begin{array}{c} CH_2OH \\ | \\ C{=}O \\ | \\ CH_2OH \\ \text{Dihydroxyacetone} \end{array} + H_2O$$

v. *Acrolein formation* When glycerol is heated either alone or in the presence of a dehydrating agent like $KHSO_4$, a substance called acrolein, which is an unsaturated aldehyde, is obtained. Acrolein has an irritating acid odour.

$$\begin{array}{l} CH_2OH \\ CHOH \\ CH_2OH \\ \text{Glycerol} \end{array} \xrightarrow{KHSO_4} \begin{array}{l} CHO \\ | \\ CH \\ \| \\ CH_2 \\ \text{Acrolein} \end{array} + 2H_2O$$

vi. Glycerol is used as a solvent in the manufacture of cosmetics and in the preparation of medicines.

vii. Glycerol, when treated with nitric acid, forms nitroglycerine, which is an explosive. In medicine nitroglycerine is used as a vasodilator in the treatment of coronary heart disease.

5.3.4 SUBSTANCES ASSOCIATED WITH LIPIDS

STEROIDS

Steroids represent a large group of compounds, which exist in nature and have a common and characteristic structure based on the cyclopentanoper hydrophenanthrene ring.

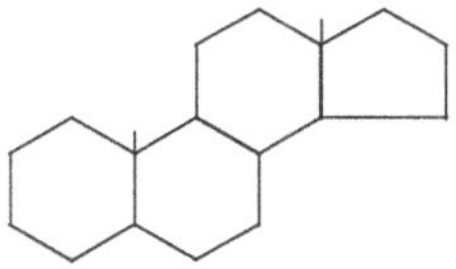

The various substances of physiological interest, having the same steroid structure are classified as follows:

a. Sterols which include Cholesterol, 7-Dehydrocholesterol, ergosterol, and calciferol.

b. Bile acids

c. Sex hormones

   Female – estradiol, progesterone
   Male – Androgen, testosterone, androsterone

d. Adrenocortical hormone

e. Saponin

f. Cardiac glucosides

g. Toad poison

a. Sterols Sterols are solid alcohol. They are widely present in animal and plant tissues. They include cholesterol, ergosterol, calciferol etc.

i. *Cholesterol* Cholesterol is widely present in animal and plant tissues. It occurs in large quantities in brain and nerve tissues. It is found in adrenal gland, glandular tissues, plasma and egg yolk. The concentration of cholesterol in human plasma is 150 to 250 mg in 100 ml of blood. Cholesterol consists of phenantherene and cyclopentane ring.

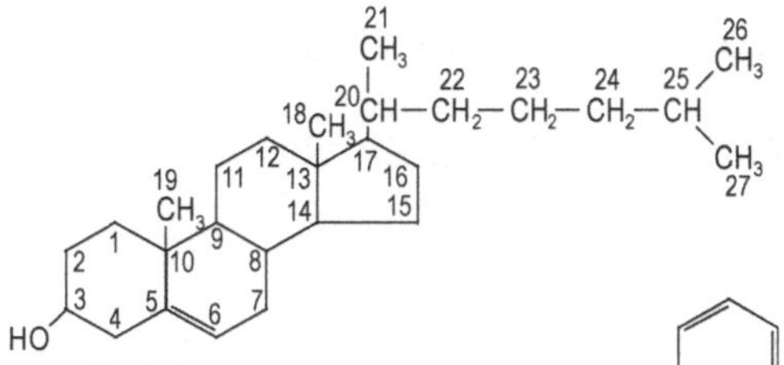

Cholesterol

Cyclopentane ring

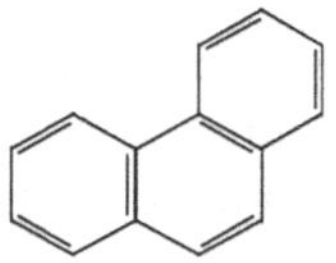

Phenantherene

ii. *7-Dehydrocholesterol* is present in the skin. It is converted to vitamin D when irradiated with ultraviolet light.

iii. *Ergosterol* is present in ergot and yeast. It is an important compound, because it is the precursor of vitamin D to which it can be converted by irradiation with ultraviolet light.

iv. *Calciferol* is vitamin D_2 obtained by irradiating the plant sterol ergosterol.

b. Bile acids These are synthesised from cholesterol. They are important factors in the digestion and absorption of lipids in the intestines. Human bile consists of cholic acid, deoxycholic acid, chenodeoxycholic acid and lithocholic acid.

c. Sex hormones Male hormones include testosterone and androsterone. Female hormones include estrogens and progesterone.

Testosterone and androsterone do not have the side chain in position 17. They possess either a keto or hydroxyl group in position 3 and 17.

Estrogen and progesterone also do not have the side chain in position 17. They possess hydroxyl groups in position 3. Estrone has a ketone group in position 17. Estriol has hydroxyl groups in position 16 and 17. Progesterone has ketone groups at C3 and C17.

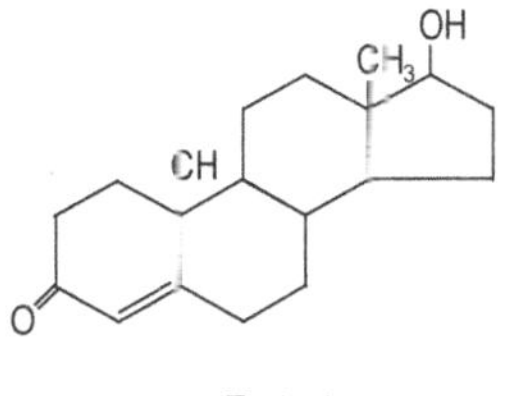

Testosterone

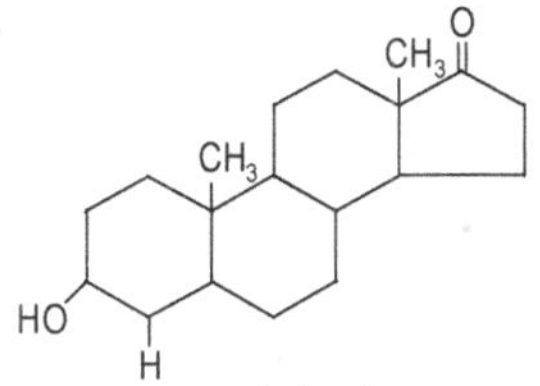

Androsterone

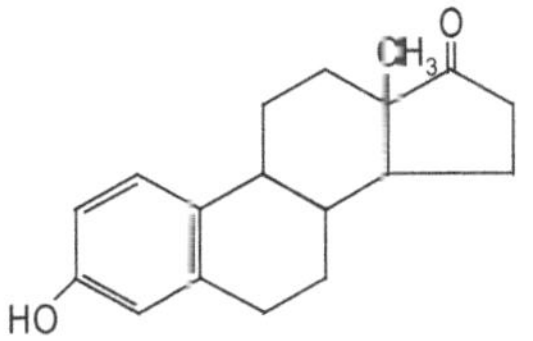

Estrone

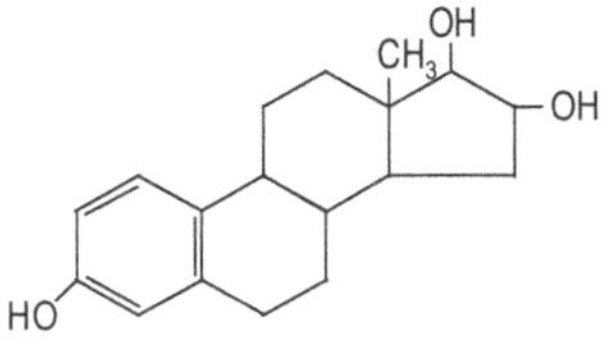

Estriol

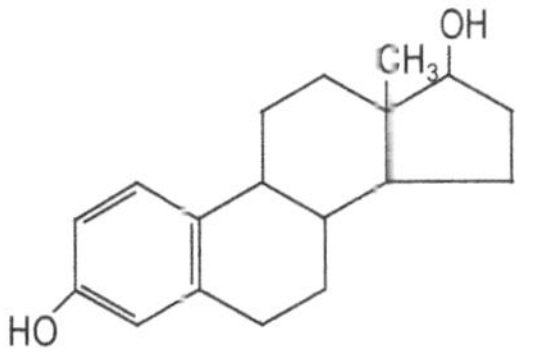

Estradiol

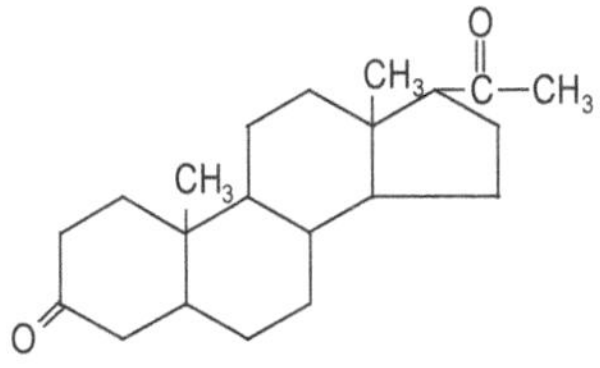

Progesterone

d. Adrenocortical hormones Corticosterone contains ketone groups in positions C3 and C17. It has an OH group at C11.

e. Saponins Saponins are hemolytic poisons widely distributed in plants. They are steroids.

f. Cardiac glucosides These are present in digitalis and squill. They have cyclopentanoper hydrophenanthrene ring structure.

g. Toad poisons These have structure of steroids. They are cardiac toxins secreted by the parotid glands of toads.

REVIEW QUESTIONS

1. Classify lipids.
2. Enumerate the physical and chemical properties of fats.
3. Give an account of fatty acids.
4. Write short notes on:
   i. biological significance of fats
   ii. saturated fatty acids
   iii. unsaturated fatty acids
   iv. steroids
   v. compound lipids
5. Define or explain
   i. emulsification
   ii. acid number
   iii. saponification number
   iv. iodine number
   v. Polenske number
   vi. Reichert–Meissl number
   vii. acetyl number

6

LIPID METABOLISM

Lipid substances form an essential component of protoplasm entering into the composition of cell membranes and mitochondrial membranes and stored in tissues to be used during starvation. Lipid plays an important role in metabolism as the fuel for the production of ATP. The first step in lipid metabolism is hydrolysis of lipid into fatty acids and glycerol through the action of lipolytic enzyme.

$$\begin{array}{l} CH_2-O-\overset{O}{\overset{\|}{C}}-R \\ | \\ CH-O-\overset{O}{\overset{\|}{C}}-R \\ | \\ CH_2-O-\overset{O}{\overset{\|}{C}}-R \end{array} \xrightarrow[+3H_2O]{\text{Lipase}} \begin{array}{l} CH_2OH \\ | \\ CHOH \\ | \\ CH_2OH \end{array} + 3\,R-\overset{O}{\overset{\|}{C}}-OH$$

Fat Glycerol Fatty acids

Both fatty acids and glycerol are independently oxidised to release energy.

6.1 OXIDATION OF GLYCEROL

The glycerol on enzymatic reaction with ATP is converted to glycerophosphate. This is followed by oxidation or dehydrogenation in which hydrogen is removed by glycerol phosphate dehydrogenase and a triose phosphate namely 3 phosphoglyceraldehyde is formed. This three carbon compound enters glycolysis and Kreb's cycle and gets completely oxidised to CO_2 and H_2O.

$$\begin{array}{ccccc} CH_2OH & & CH_2OH & & CHO \\ | & \xrightarrow[\text{Glycerokinase}]{\text{ATP} \quad \text{ADP}} & | & \xrightarrow[\text{Glycerol } PO_4 \text{ dehydrogenase}]{NAD^+ \quad NADH^+} & | \\ CHOH & & CHOH & & CHOH \\ | & & | & & | \\ CH_2OH & & CH_2OP & & CH_2OP \end{array}$$

Glycerol Glycerophosphate 3 Phosphoglyceraldehyde

$$\downarrow$$

$$CO_2 + H_2O \xleftarrow{15\ ATP} \text{Kreb's cycle} \longleftarrow \text{Pyruvic acid}$$

6.2 OXIDATION OF FATTY ACIDS

Several theories have been put forward to explain the mechanism of fatty acid oxidation in the tissues because of the formation of a variety of compounds during oxidation.

6.2.1 β-OXIDATION THEORY

Knoop, based on the results of feeding phenyl-labelled fatty acids to dogs and subsequent analysis of the urine samples, concluded that fatty acids are oxidised by β-oxidation. β-oxidation is restricted to fatty acids with more than 18 carbon atoms. Oxidation of fatty acids takes place at the carbon atom in the β position of the carboxyl group, resulting in the splitting of the two terminal carbon atoms, leaving a fatty acid chain shorter than the original acid by 2 carbon atoms.

$$R\ CH_2CH_2\overset{\beta}{CH_2}\,|\,\overset{\alpha}{CH_2}\ COOH \longrightarrow R\ CH_2CH_2COOH + CH_3\ COOH$$

Oxidation of phenyl-substituted fatty acids with odd number of carbon atoms in the chain gives rise to benzoic acid and even-numbered acids gives rise to phenylacetic acid. Such acids after being coupled with glycine are eliminated as hippuric acid and phenyl aceturic acid respectively.

$$C_6H_5CO\boxed{OH + H}HNCH_2COOH \longrightarrow C_6H_5CONHCH_2COOH + H_2O$$

Benzoic acid Glycine Hippuric acid

$$C_6H_5CH_2CO\boxed{OH + H}HNCH_2COOH \longrightarrow C_6H_5CH_2CONHCH_2COOH + H_2O$$

Phenylacetic acid Glycine Phenyl aceturic acid

6.2.2 ω-OXIDATION THEORY

According to Verkade, fatty acids are also degraded by omega oxidation where terminal carbon atom is oxidised to carboxyl group and hence, a long chain dicarboxylic acid is first formed. Thereafter, β-oxidation starts from both the ends leaving ultimately a dicarboxylic acid. The theory is based on the fact that dicarboxylic acids are also formed during oxidation of fatty acids.

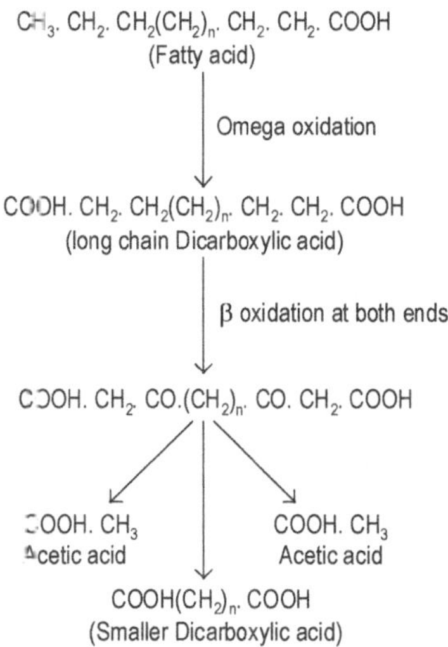

6.2.3 MULTIPLE-ALTERNATE OXIDATION THEORY

This theory was proposed by Hurtley. According to this theory, fatty acids are oxidised at every alternate β-carbon atom and cleavage occurs at every alternate site of oxidation, thus resulting in fragments of four carbon atoms namely acetoacetic acid.

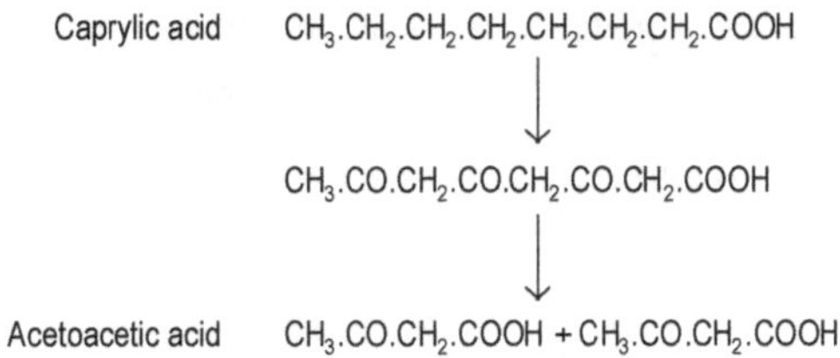

6.2.4 β-OXIDATION–CONDENSATION THEORY

According to Mackey, the fatty acids are subjected to β-oxidation and acetyl units are split off. The acetyl units then reunite to form acetoacetic acid molecules.

$$\underset{\text{Valeric acid}}{2CH_3.CH_2.CH_2.CH_2.COOH} \longrightarrow 2CH_3.COOH + 2CH_3.CH_2.COOH$$

$$\downarrow \text{condensation}$$

$$\underset{\text{Acetoacetic acid}}{CH_3.CO.CH_2.COOH}$$

6.2.5 α-OXIDATION THEORY

Brain microsomes have been shown to catalyse α-oxidation of the long chain fatty acids resulting in the formation of α-hydroxy fatty acids. These hydroxy fatty acids may then further be oxidised to form α-keto fatty acids which can be oxidatively decarboxylated to give CO_2 and fatty acids having one carbon less than the parent fatty acids.

$$\underset{\text{Fatty acid}}{R.CH_2.CH_2.COOH} \longrightarrow \underset{\alpha\text{-hydroxy fatty acid}}{R.CH_2.CHOH.COOH} \longrightarrow \underset{\alpha\text{-keto fatty acid}}{R.CH_2.CO.COOH}$$

$$\downarrow \searrow CO_2$$

$$\underset{\text{Fatty acid (one carbon less)}}{R.CH_2.COOH}$$

6.3 β-OXIDATION OF FATTY ACIDS

β-oxidation was first explained by Knoop; but later it was modified by Hymen, Green and Kennedy. β-oxidation of fatty acids is done in five steps (Figure 6.1). (i) Activation (ii) Desaturation (iii) Hydration (iv) Oxidation (v) Thiolysis.

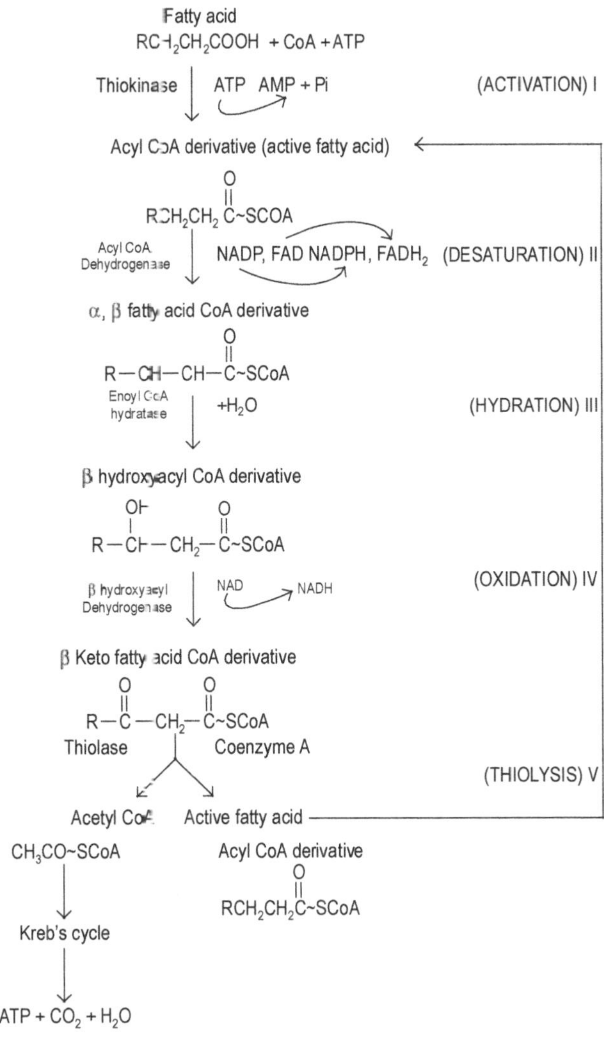

Fig. 6.1 β-oxidation

i. **Activation** Fatty acids are activated in the presence of **thiokinase**. Activated fatty acid is known as acyl CoA derivative. This requires the supply of ATP, CoA and Mg^{++} ions. This reaction is also known as ATP-CoA priming.

$$\text{Fatty acid} + \text{CoA} + \text{ATP} \xrightarrow[\text{Mg}^{++}]{\text{Thiokinase}} \text{Acetyl CoA derivative}$$

ii. **Desaturation** Once the fatty acid has been activated it can be dehydrogenated by **acyl CoA dehydrogenase** into α, β-fatty acid CoA derivative.

$$\text{Acyl CoA} \xrightarrow{\text{Acyl CoA dehydrogenase}} \alpha, \beta\text{-fatty acid CoA}$$

iii. **Hydration** α, β-fatty acid CoA undergoes hydration and combines with a molecule of water under the influence of **hydratase (enoyl CoA hydratase)** to form β-hydroxy acyl CoA derivative.

$$\alpha, \beta\text{-fatty acid CoA} + H_2O \xrightarrow{\text{Hydratase}} \beta\text{-hydroxy acyl CoA}$$

iv. **Oxidation** β-hydroxy acyl CoA derivative undergoes oxidation to form keto fatty acid in the presence of **β-hydroxy acyl dehydrogenase** and NAD.

$$\beta\text{-hydroxy acyl CoA} + \text{NAD} \xrightarrow{\beta\text{ hydroxy acyl dehydrogenase}} \beta\text{-keto fatty acid CoA}$$

v. **Thiolysis** The final step is the process of β-oxidation in which β-keto fatty acid CoA is split into a molecule of acetyl CoA and active fatty acid and the enzyme involved in this reaction is **thiolase**.

$$\beta\text{-keto fatty acid CoA} \xrightarrow{\text{Thiolase}} \text{acetyl CoA} + \text{active fatty acid}$$

The active fatty acid again may be recycled and once again, acetyl CoA and active fatty acid can be formed. Acetyl CoA enters Kreb's cycle, combines with oxaloacetic acid to form citric acid and gets completely oxidised to CO_2 and H_2O.

6.3.1 KETOGENESIS

When the metabolism of carbohydrates is impaired, two molecules of acetyl CoA condense to form acetoacetyl CoA, which in turn is hydrolysed by **deacylase** in the liver to yield acetoacetic acid which may be reduced to β-hydroxy butyric acid in the presence of

β-hydroxy butyric dehydrogenase and reduced NAD or decarboxylated to form acetone.

$$2CH_3-\overset{O}{\overset{\|}{C}}\sim SCoA \longrightarrow CH_3-\overset{O}{\overset{\|}{C}}-CH_2-\overset{O}{\overset{\|}{C}}\sim SCoA \longrightarrow CH_3-\overset{O}{\overset{\|}{C}}-CH_2-COOH$$

Acetyl CoA — Acetoacetyl CoA — Acetoacetic acid

$$CH_3-\overset{OH}{\overset{|}{C}H}-CH_2-COOH \qquad CH_3-\overset{O}{\overset{\|}{C}}-CH_3+CO_2$$

β-hydroxy butyric acid — Acetone

All these substances are collectively known as *ketone bodies*. The synthesis of ketone bodies is known as *ketogenesis*. They are disposed off by oxidation, which is called *ketolysis*. This takes place in the extra hepatic tissues especially in the muscle. If the ketone bodies are accumulated in the blood, it is known as *keto anaemia*. If they are accumulated in urine, such condition is called *ketourea*. Excretion of large amount of acetoacetic acid and β-hydroxy butyric acid in the urine may lead to a disease called *acidosis*.

6.3.2 ENERGETICS OF FATTY ACID OXIDATION

If an even-numbered fatty acid is taken into consideration, e.g. palmitic acid (C_{16}), it is split completely into 8 acetyl units in seven rounds. On completion of each cycle, one mole of FAD. H_2 and one mole of NADH are produced which are equivalent to 5 moles of ATP. In seven cycles therefore, 35 moles of ATP are synthesised. Each acetyl unit is oxidised by Kreb's cycle, generating 12 moles of ATP. Thus, the total number of ATP generated during oxidation of 8 moles of acetyl coenzyme A units would be 96. In total, 131 moles of ATP are generated per mole of palmitic acid oxidised. One ATP is used in the preliminary thiokinase reaction. Hence, there occurs a net gain of 130 moles of ATP.

6.4 BIOSYNTHESIS OF LIPIDS

In the biosynthesis of lipids, the reverse of β-oxidation occurs. During this process, the acetyl CoA is converted to fatty acids such as palmitic, oleic and stearic acids. These fatty acids will combine with glycerol to form neutral fats or triglycerides, which are stored up in the various fat deposits of the body.

6.4.1 SYNTHESIS OF FATTY ACIDS

Synthesis of fatty acids can occur in mitochondria as well as in cytoplasm. Synthesis in microsomes has also been reported. However, most of the synthesis occurs in the cytoplasm.

Mitochondrial System The synthesis is mostly restricted to lengthening of an existing fatty acid by a reversal of β-oxidation. Long-chain fatty acids especially stearic and palmitic acids are synthesised. The coenzymes used in the reductive steps are both NADH and NADPH. All enzymes of β-oxidation can act in the reverse direction except the acyl-CoA dehydrogenase. This is catalysed by α, **β-unsaturated acyl-CoA reductase (enoyl CoA reductase)** and requires NADPH as hydrogen donor. Pyridoxal phosphate is required in the initial step of condensation of acetyl-CoA with acyl CoA.

Extra-mitochondrial system This can synthesise fatty acids de-novo starting from acetyl-CoA (Figure 6.2). Acetyl CoA serves as precursor for the de-novo synthesis of fatty acids in the extra mitochondrial system (cytoplasm). Acetyl CoA condenses with oxaloacetate to form citrate in the mitochondria. Acetyl CoA and oxaloacetate are released from mitochondria in the cytoplasm by the action of an enzyme **ATP-citrate lyase**. Subsequent steps are brought about by another multi-enzyme complex. In yeast, mammals and birds, the synthase system called the "fatty acid synthase complex" is a multi-enzyme complex. It is a dimer with two identical subunits—monomer I and monomer II. Each monomer contains six enzymes and an acyl carrier protein (ACP) molecule. The ACP has an–SH group in the 4-phosphopantothene moiety of the enzyme **ketoacyl synthetase** or the **condensing enzyme**. The pantothenyl–SH of monomer I is in close proximity to the cysteinyl–SH group of monomer II and vice versa.

Acetyl CoA is carboxylated to malonyl-CoA by the action of an enzyme **acetyl CoA carboxylase**. Malonyl-CoA reacts with acetyl CoA and gets decarboxylated in the presence of condensing enzyme to form acetoacetyl CoA. Acetoacetyl CoA is reduced to β-hydroxy butyryl-CoA in the presence of an enzyme **ketoacyl-CoA reductase**. β-hydroxy butyryl-CoA, by the removal of water molecule, is converted to α, β-unsaturated butyryl CoA in the presence of **hydratase**; this

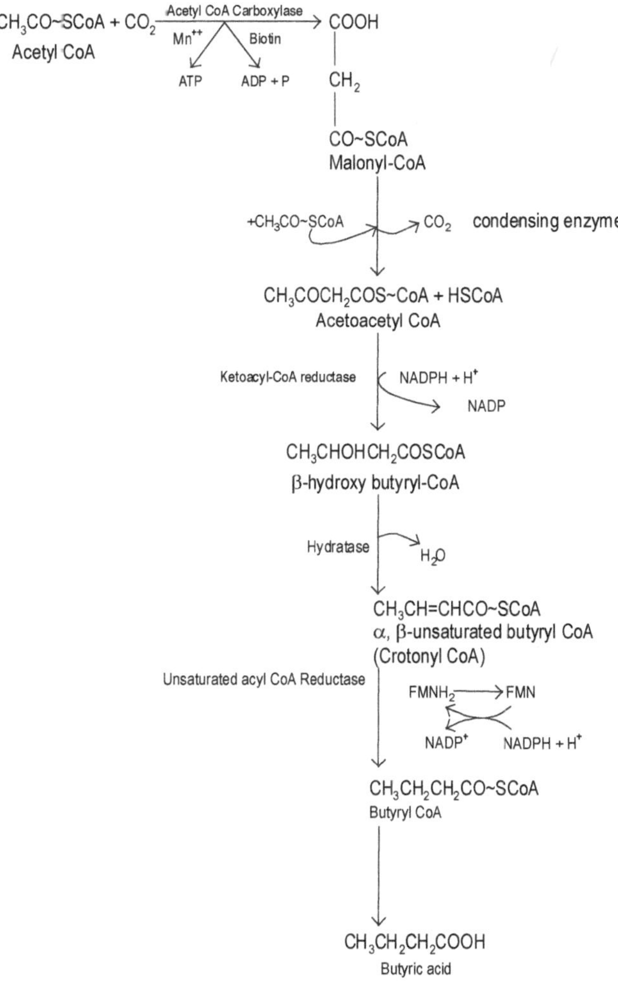

Fig. 6.2 Biosynthesis of fatty acid

again, by the action of an enzyme **unsaturated acyl CoA reductase**, is converted to butyryl CoA. All these reactions occur on the pantothenyl-SH of monomer II. The butyrate is transferred to cysteinyl-SH of monomer I. A fresh molecule of malonate is taken up on to the pantothenyl-SH monomer II and transferred to cysteinyl-SH of monomer I. The set of reactions are repeated till a 16-carbon palmitate is formed on the pantothenyl-SH monomer II which is then released into the cytosol by the action of **thioesterase (deacylase)**. Table 6.1 lists the enzymes of the fatty acid synthase complex.

$$\text{Citrate} + \text{ATP} + \text{HS}{\sim}\text{CoA} \xrightarrow{\text{ATP citrate lyase}} \text{Acetyl-CoA} + \text{Oxaloacetate} + \text{ADP} + \text{P}$$

Table 6.1 Fatty acid synthase complex

Monomer I	Monomer II
Ketoacyl synthase *	Phospho ACP
Cysteinyl-SH	Pantothenyl–SH
Transacylase	Ketoacyl reductase
Enoyl reductase **	Hydratase
Hydratase	Enoyl reductase
Ketoacyl reductase	Transacylase
ACP–phospho pantothenyl–SH	Ketoacyl synthase
Thioesterase	Cysteinyl–SH

* a condensing enzyme

** an unsaturated acyl-CoA reductase

6.4.2 SYNTHESIS OF TRIGLYCERIDES

Fatty acids combine with glycerol to form triglycerides. The glycerol can be derived from dihydroxyacetone phosphate obtained during glycolysis. Figure 6.3 depicts the mode of synthesis of triglycerides. Both glycerol and fatty acids are activated by ATP. In liver, kidneys, heart and intestinal mucosa, glycerol is activated and phosphorylated by ATP, in the presence of the enzyme **glycerokinase**. In muscles and adipose tissues, where glycerokinase is absent, α glycerophosphate is formed from dihydroxyacetone phosphate by reduction with NADH in the presence of **α-glycerophosphate dehydrogenase**. Fatty acids are activated to form acyl-CoA by ATP and CoA, in the presence of the enzyme **thiokinase**.

Two molecules of acyl-CoA combine with α-glycerophosphate to form α, β, diglyceride phosphate (phosphotidic acid) in the presence of **glycerophosphate acyltransferase**. The phosphatidic acid is dephosphorylated by phosphatase to form α-β diglyceride. Another molecule of acyl-CoA is esterified with the diglyceride, to form a triglyceride.

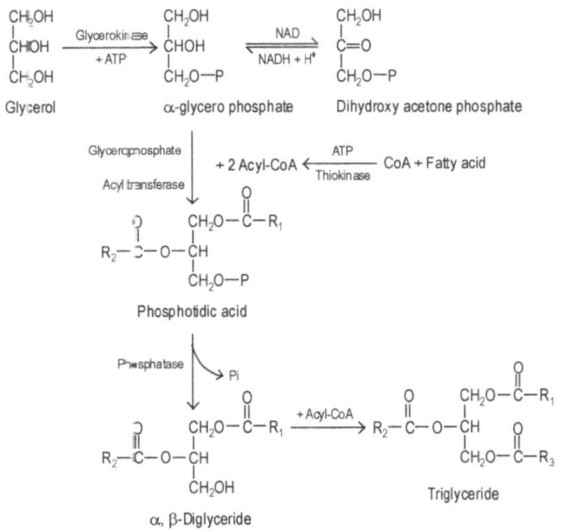

CH_2OH–CHOH–CH_2OH (Glycerol) —Glycerokinase, + ATP→ CH_2OH–CHOH–CH_2O–P (α-glycero phosphate) ⇌ (NAD / NADH + H^+) CH_2OH–C=O–CH_2O–P (Dihydroxy acetone phosphate)

Glycerophosphate Acyl transferase; + 2 Acyl-CoA ← (ATP, Thiokinase) CoA + Fatty acid

CH_2O–C(=O)–R_1 / R_2–C(=O)–O–CH / CH_2O–P (Phosphotidic acid)

Phosphatase → Pi

CH_2O–C(=O)–R_1 / R_2–C(=O)–O–CH / CH_2OH (α, β-Diglyceride) —+ Acyl-CoA→ CH_2O–C(=O)–R_1 / R_2–C(=O)–O–CH / CH_2O–C(=O)–R_3 (Triglyceride)

Fig. 6.3 Synthesis of triglyceride

6.5 METABOLISM OF CHOLESTEROL

Cholesterol is of major significance because of its relationship to many physiologically active steroids, sex hormones, adrenal cortex hormones, bile salts, etc which are present in our body. It is an insoluble substance and along with other substances, tends to precipitate in and along the lining of the blood vessels, thereby restricting the flow.

Ingested cholesterol is absorbed along with other lipids. It is normally present in blood to the extent of 150 to 250 mg per 100 ml, being equally distributed between the cells and the plasma. In the cells, cholesterol occurs in free form. While in the plasma about 75% is found in the form of cholesterol esters. Figure 6.4 depicts the biosynthesis of cholesterol.

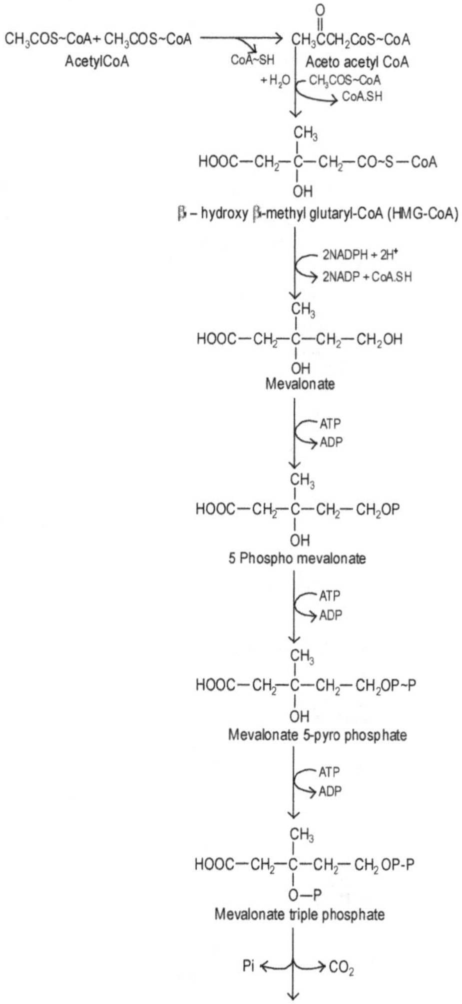

$CH_3COS{\sim}CoA + CH_3COS{\sim}CoA \longrightarrow CH_3\overset{O}{\overset{\|}{C}}CH_2CoS{\sim}CoA$

AcetylCoA (− CoA~SH) → Aceto acetyl CoA

$+H_2O$, $CH_3COS{\sim}CoA$ → CoA.SH

$HOOC-CH_2-\underset{OH}{\overset{CH_3}{C}}-CH_2-CO{\sim}S-CoA$

β – hydroxy β-methyl glutaryl-CoA (HMG-CoA)

$2NADPH + 2H^+$ → $2NADP + CoA.SH$

$HOOC-CH_2-\underset{OH}{\overset{CH_3}{C}}-CH_2-CH_2OH$

Mevalonate

ATP → ADP

$HOOC-CH_2-\underset{OH}{\overset{CH_3}{C}}-CH_2-CH_2OP$

5 Phospho mevalonate

ATP → ADP

$HOOC-CH_2-\underset{OH}{\overset{CH_3}{C}}-CH_2-CH_2OP{\sim}P$

Mevalonate 5-pyro phosphate

ATP → ADP

$HOOC-CH_2-\underset{O-P}{\overset{CH_3}{C}}-CH_2-CH_2OP\text{-}P$

Mevalonate triple phosphate

Pi ← → CO_2

↓

-Contd-

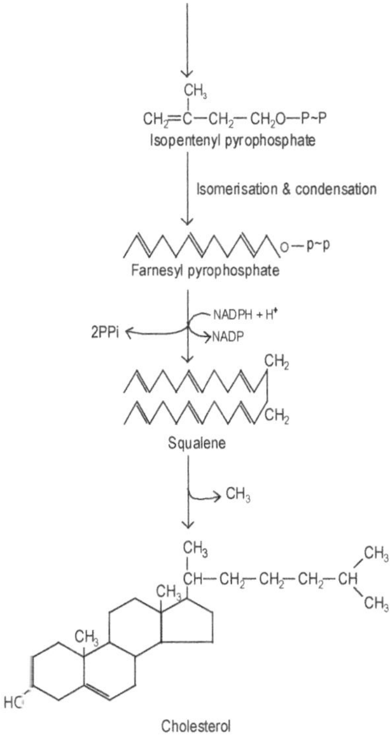

Fig. 6.4 Biosynthesis of cholesterol

Cholesterol is synthesised in the body from two-carbon units in the form of acetyl CoA formed either from fatty acid or from the metabolism of carbohydrate through pyruvate. Two molecules of acetyl CoA condense to form acetoacetyl CoA which react with a third molecule of acetyl CoA to form β-hydroxy β-methyl glutaryl CoA which in turn, gives rise to the intermediate compound called mevalonic acid. Mevalonic acid is phosphorylated three times in succession by ATP, forming first a monophosphomevalonic acid, next a diphosphomevalonic acid and finally triple phosphorylated mevalonic acid, a transient intermediate which simultaneously loses a molecule of phosphate and

CO_2 to form isopentenyl pyrophosphate which can also exist in an isomeric form 3,3-dimethyl ellyl pyrophosphate. These compounds are said to be the forerunners of many important biological compounds including carotenoid pigments and cholesterol. One molecule of 3, 3 methyl ellyl pyrophosphate now reacts with one of isopentenyl pyrophosphate to yield geranyl pyrophosphate which with another molecule of isopentenyl pyrophosphate forms farnesyl pyrophosphate with the removal of inorganic pyrophosphate at each stage. The two molecules of farnesyl pyrophosphate finally condense to form the hydrocarbon squalene which, by ring closure and loss of methyl groups, is readily converted into cholesterol by enzymes present in the liver.

6.5.1 BILE ACIDS

Bile acids are important end products of the metabolism of cholesterol in liver. The four bile acids isolated from the human bile are cholic acid, deoxycholic acid, chenodeoxycholic acid and lithocholic acid.

Cholic acid is present in largest amounts in the bile itself, forming a maximum of 60% of the total. It is a saturated sterol having three OH groups on the nucleus in positions 3, 7 and 12.

Deoxycholic acid lacks the OH group in position 7 and constitutes 25% of the total.

Chenodeoxycholic acid lacks the OH group in position 12 and constitutes about 15 to 20% of the total.

Lithocholic acid has only one OH group in position 3.

Route of Formation of Bile acids Approximately 80% of the cholesterol metabolised is converted by liver tissues to bile acids, which later get conjugated in the liver with glycine and taurine by enzymes of the liver. Prior to conjugation, the bile acids are activated to acyl CoA.

$$\text{Cholic acid} \xrightarrow[\text{ATP, Mg}^{++}]{\text{CoA}} \text{Cholyl-CoA} \xrightarrow[\text{Glycine}]{\text{Taurine}} \begin{cases} \rightarrow \text{Taurocholic acid} \\ \rightarrow \text{Glycocholic acid} \end{cases}$$

Intestinal bacteria act upon the unabsorbed bile acids and the products are excreted in the faeces. Thus, it may be seen that bile represents the main route for the excretion of cholesterol, which is held in solution by the emulsifying action of bile salts. Under abnormal conditions, this can lead to the formation of cholesterol stones (calculi) in the biliary passages like gall bladder and common bile duct and cause obstruction to the circulation of bile pigments resulting in obstructive jaundice.

REVIEW QUESTIONS

1. Explain the theories of oxidation of fatty acids.
2. Discuss β-oxidation of fatty acids and its energetics.
3. Give an account of fatty acid biosynthesis and the role of acyl carrier proteins.
4. Explain the biosynthesis of cholesterol.
5. Discuss briefly the biosynthesis of triglyceride.
6. Define/explain the following:
    - i. β-oxidation of fatty acids
    - ii. α-oxidation of fatty acids
    - iii. ω-oxidation of fatty acids
    - iv. ketone bodies
    - v. bile acids

7

PROTEINS

The term protein is derived from the Greek word, which means 'of prime importance'. Proteins are found in all living cells. They form essential constituents of protoplasm, cell membrane and nuclear material. They may be present as simple soluble protein or complex protein combined with lipids and nucleic acids. Proteins form about 15% of the total body weight. The proteins in different tissues such as connective tissue, bone, muscles, brain, blood and other biological fluids differ in composition and properties. The protein content of plants is relatively lesser than that of animals. In plants, seeds and cereals contain the highest protein content while the leaves, stem and tubers contain much less protein.

7.1 BIOLOGICAL SIGNIFICANCE

The biological significance and functions of proteins are varied but vital. They regulate and integrate the numerous physiological and metabolic processes in the body through hormones, enzymes and nucleoproteins. The plasma proteins help in the maintenance of acid–base balance and osmotic pressure of the body fluids. Transport proteins in blood plasma bind and carry specific molecules or ions from one organ to another. Hemoglobin of red blood cells binds oxygen as the blood passes through the lungs and carries it to the tissues, where oxygen is released to carry out the energy-yielding oxidation of nutrients. Actin, myosin and tubulin are contractile proteins that help the cells and organisms to change shape or to move about. Many proteins like collagen, keratin, etc. serve as supporting filaments to give biological structures strength or protection. The immunoglobulins or antibodies of vertebrates are specialised proteins that defend organisms against invasion by other species or protect them from injury. Fibrinogen and thrombin are blood-clotting proteins that prevent loss of blood when the vascular system is injured.

7.2 GENERAL PROPERTIES OF PROTEINS

- Proteins are complex substances of high molecular weight consisting of large number of α-amino acids linked by peptide linkages.
- They contain C, H, O, N and sometimes P and S. Elements such as Fe, Cu, I and Zn are occasionally present.
- The molecular weight of the proteins varies from 6000 to many millions. Because of the high molecular weight most proteins are not diffusible through membranes like cellophane.
- Proteins are generally soluble in water, salt solution, dilute acids and alkalies. Owing to their large size, they form colloids and exhibit colloidal properties.
- Proteins possess free ionic or electrically charged groups, so that they can migrate in an electrical field. Owing to their charges, they combine with ionic reagents giving rise to insoluble compounds.
- Proteins are precipitated by salts of heavy metals like silver, mercury and lead in an alkaline medium. Here, the metals combine with the carboxyl groups to form metalloproteins. One important application of this property is the use of proteins as an antidote in metallic poisoning.
- Certain alkaloidal reagents precipitate proteins.
- Proteins are amphoteric substances; they contain both NH_2 and COOH groups. Hence they can react with acids and bases.

- Some proteins coagulate on heating. They are called heat coagulable proteins. e.g. albumin and globulin.
- Colour reaction: All proteins give colour reaction when treated with certain reagents.
- Proteins are hydrolysed to their constituent amino acids by boiling with acids and alkali, and by the action of appropriate proteolytic enzymes.

7.3 AMINO ACIDS

Amino acids are the simplest units of protein molecule and they form the building blocks of protein structure. The general formula of an amino acid can be written as:

$$\begin{array}{c} \text{H} \\ | \\ \text{NH}_2\text{—C—COOH} \\ | \\ \text{R} \end{array}$$

All amino acids are α-amino acids because the NH_2 group is attached to the α-carbon atom which is next to the COOH group. Examination of the structure of an amino acid except glycine, reveals that the α-carbon atom has four different groups attached to it, thus making it asymmetric. Because of the presence of asymmetric carbon atoms, amino acids exist in optically active forms. Therefore, it is possible to have dextrorotatory and levorotatory forms in such amino acid. The distinction between dextro and levo rotation in amino acids is represented by symbols + (for dextro) and – (for levo). In view of the presence of asymmetric carbon atoms, the configuration of an amino acid may be represented to belong to 'D' or 'L' series.

$$\begin{array}{cc} \text{R} & \text{R} \\ | & | \\ \text{HCNH}_2 & \text{NH}_2\text{CH} \\ | & | \\ \text{COOH} & \text{COOH} \\ \text{D-amino acid} & \text{L-amino acid} \end{array}$$

It has been found that L-amino acids are more common than D-forms and most of the naturally occurring amino acids are L-amino acids. Therefore L-amino acids are called natural amino acids.

7.3.1 CLASSIFICATION OF AMINO ACIDS

Amino acids are classified into different groups:

i. ***Monoamino monocarboxylic acids*** These amino acids have one amino group and one carboxylic group in the structure. They are neutral amino acids.

e.g.

a. Glycine (α-amino acetic acid)

CH_2NH_2COOH

b. Alanine (α-amino propionic acid)

$CH_3{-}CH(NH_2){-}COOH$

c. Serine (α-amino β-hydroxy propionic acid)

$CH_2OH{-}CH(NH_2){-}COOH$

d. Threonine (α-amino β-hydroxy *n*-butyric acid)

$CH_3{-}CHOH{-}CH(NH_2){-}COOH$

e. Valine (α-amino isovaleric acid)

$(CH_3)_2CH{-}CH(NH_2){-}COOH$

f. Leucine (α-amino isocaproic acid)

$(CH_3)_2CH{-}CH_2{-}CH(NH_2){-}COOH$

g. Isoleucine (α-amino β-methyl-*n*-valeric acid)

$CH_3{-}CH_2{-}CH(CH_3){-}CH(NH_2){-}COOH$

ii. **Monoamino dicarboxylic acids** These amino acids contain acidic amino acids.

e.g.

a. Aspartic acid (α-amino succinic acid)

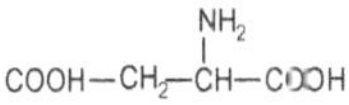

$$COOH-CH_2-CH(NH_2)-COOH$$

b. Glutamic acid (α-amino glutaric acid)

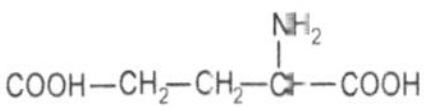

$$COOH-CH_2-CH_2-CH(NH_2)-COOH$$

iii. ***Diamino monocarboxylic acids*** These amino acids are α-amino acids containing two amino groups and one carboxylic group. They are also called basic amino acids.

a. Arginine (α-amino-δ-guanidino-*n*-valeric acid)

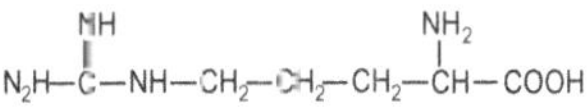

$$N_2H-C(=NH)-NH-CH_2-CH_2-CH_2-CH(NH_2)-COOH$$

b. Lysine (α-ε-diamino-*n*-caproic acid)

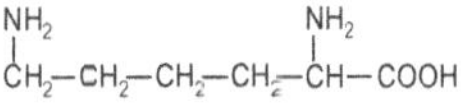

$$CH_2(NH_2)-CH_2-CH_2-CH_2-CH(NH_2)-COOH$$

iv. **Sulphur-containing amino acids** These amino acids are α-amino acids containing aliphatic side chain with sulphur atoms.

a. Cysteine (α-amino-β-thiopropionic acid)

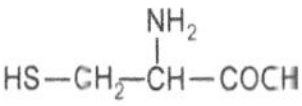

$$HS-CH_2-CH(NH_2)-COOH$$

b. Cystine (α-amino di cysteine)

Cystine consists of two cysteine residues linked by a disulphide linkage.

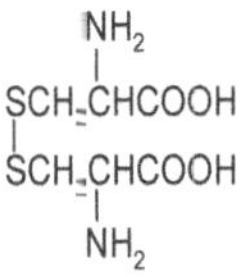

$$\begin{array}{l} SCH_2CH(NH_2)COOH \\ | \\ SCH_2CH(NH_2)COOH \end{array}$$

c. Methionine (α-amino-γ-methyl thio-*n*-butyric acid)

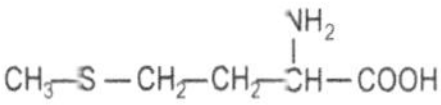

$$CH_3-S-CH_2-CH_2-CH(NH_2)-COOH$$

v. ***Aromatic amino acids*** The aromatic amino acids contain aromatic rings.

a. Phenylalanine (α-amino-β-phenyl-propionic acid)

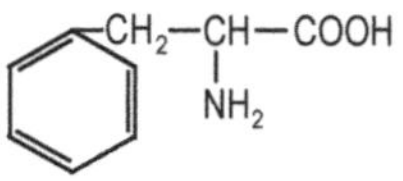

$C_6H_5-CH_2-CH(NH_2)-COOH$

b. Tyrosine (α-amino-β-para-hydroxy-phenyl-propionic acid)

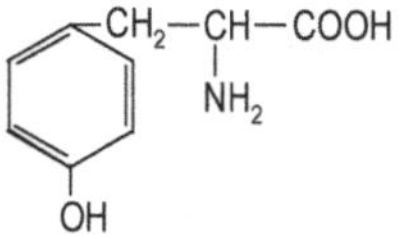

$HO-C_6H_4-CH_2-CH(NH_2)-COOH$

c. Tryptophan (α-amino-β-indole-propionic acid)

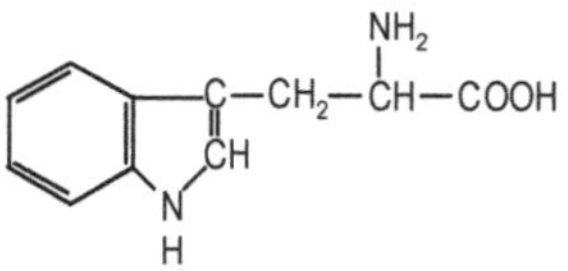

NH_2

$C-CH_2-CH-COOH$

CH

N

H

vi. ***Heterocyclic amino acids*** These amino acids contain either imidazole or indole ring.

a. Proline (Pyrrolidine-2-carboxylic acid)

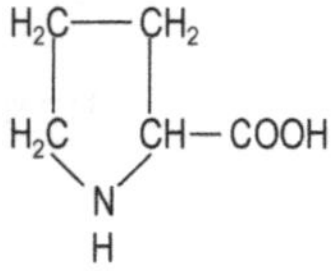

H_2C-CH_2

$H_2C \quad CH-COOH$

N

H

b. Hydroxyproline (4-Hydroxyproline)

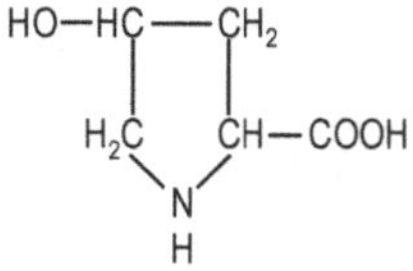

$HO-HC-CH_2$

$H_2C \quad CH-COOH$

N

H

c. Histidine (α-amino-β-imidazole-propionic acid)

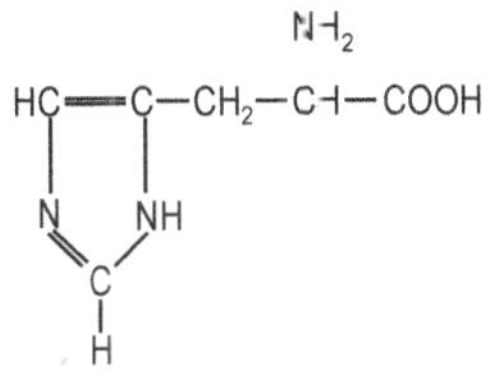

$$HC{=}C{-}CH_2{-}CH(NH_2){-}COOH$$
N NH
C
H

vii. ***Amino acids formed as metabolic intermediaries***

a. Diiodotyrosine

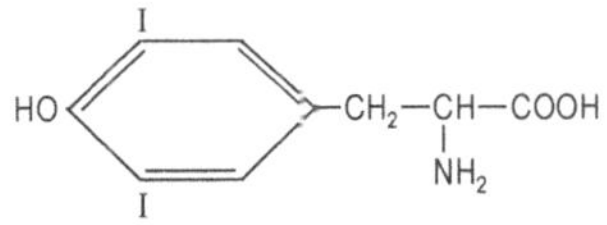

I
HO —CH₂—CH—COOH
NH₂
I

b. Thyroxine

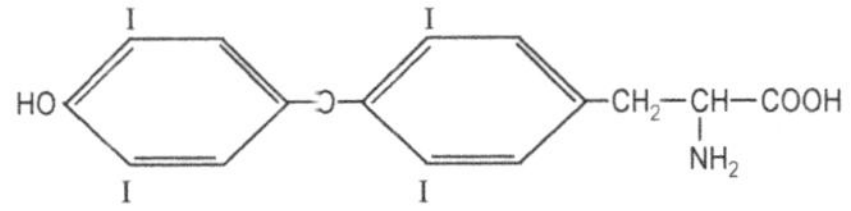

I I
HO —O— —CH_2—CH—COOH
NH_2
I I

c. Ornithine

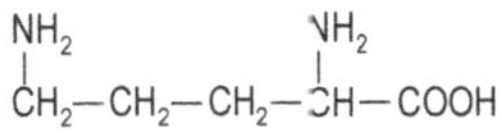

$$\underset{\displaystyle CH_2}{\overset{\displaystyle NH_2}{|}}{-}CH_2{-}CH_2{-}\underset{\displaystyle CH}{\overset{\displaystyle NH_2}{|}}{-}COOH$$

d. Citrulline

$$O{=}\underset{\displaystyle C}{\overset{\displaystyle NH_2}{|}}{-}NH{-}CH_2{-}CH_2{-}CH_2{-}\underset{\displaystyle CH}{\overset{\displaystyle NH_2}{|}}{-}COOH$$

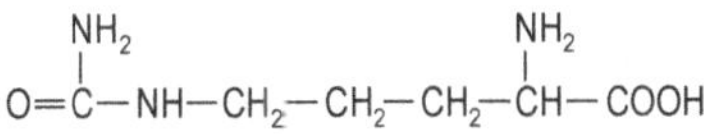

Essential and non-essential amino acids In the protein systems of the various life forms there are about twenty amino acids. Plants and bacteria can generally make all of them but mammals such as rats and humans do not have this capacity; amino acids that cannot be synthesised by humans and other vertebrates, and obtained from the diet are called **essential amino acids**, whereas, amino acids that can be synthesised by humans and other vertebrates and are thus not required in diet are called **non-essential amino acids**.

Table 7.1 Essential and non-essential amino acids (for man)

Essential amino acids	Non-essential amino acids
Arginine	Glycine
Histidine	Alanine
Isoleucine	Serine
Leucine	Glutamate, glutamine
Lysine	Aspartate, aspargine
Methionine	Tyrosine
Phenylalanine	Cysteine, Cystine
Threonine	Proline
Tryptophan	Hydroxyproline
Valine	

Non-protein associated amino acids Several other amino acids are found in the body freely or in combined states (i.e. not associated with peptides or proteins). These non-protein associated amino acids perform specialised functions. Several of the amino acids found in proteins also serve functions distinct from the formation of peptides and proteins, e.g. tyrosine in the formation of thyroid hormones or glutamate acting as a neurotransmitter.

7.3.2 PHYSICAL PROPERTIES OF AMINO ACIDS

i. Amino acids are crystalline substances.
ii. They are generally soluble in water and insoluble in organic solvents.
iii. They have high melting point (200° C to 300° C).
iv. Amino acids may be tasteless, sweet or bitter.
v. All amino acids except glycine are optically active.
vi. Amino acids are amphoteric substances, as they contain both acidic (COOH) and basic (NH_2) groups. They can react with both alkalies and acids to form salts.

In an acid solution amino acids carry positive charges and hence they move towards the cathode in an electric field. In an alkaline solution, the amino acids carry negative charges and therefore move towards the anode. But during certain reactions, amino acids are found to be electrically neutral. This occurs as a result of dissociation of H^+ ion which passes from the COOH to the NH_2 group. The amino acid is said to exist as **zwitter ion** under such conditions.

$$H-\underset{H}{\overset{NH_2}{\underset{|}{\overset{|}{C}}}}-COOH$$

Glycine

$$H-\underset{H}{\overset{NH_3^+}{\underset{|}{\overset{|}{C}}}}-COO^-$$

Glycine as zwitter ion

In this stage, the amino acid will migrate neither to the cathode nor the anode, in an electrical field. The pH at which the amino acid has no tendency to move either to the positive or negative electrode is called its **isoelectric pH**. At this pH, the amino acid molecule bears a net charge of zero. The isoelectric point is symbolised by pI.

7.3.3 CHEMICAL PROPERTIES OF AMINO ACIDS

1. Reaction of amino acids due to amino groups

i. ***Reaction with formaldehyde*** This reaction is called **Sorenson's formal titration**. Amino acids dissolve in water to form a neutral solution. The carboxyl groups react with the basic amino groups to form zwitter ions. Because of this, the carboxyl groups of α-amino acids in aqueous solution cannot be accurately titrated with alkalies. Sorenson noted that, if amino acid solutions are treated with an excess of neutralised formaldehyde solution, the mixture becomes acidic and can be titrated sharply with standard alkali, using phenolphthalein as indicator. The amino group combines with formaldehyde to form dimethylol amino acid which is an amino acid–formaldehyde complex. In this way, the existence of all the amino groups are eliminated, resulting in an acidic solution, due to the presence of the carboxyl groups. This solution can be titrated with standard alkali, using phenolphthalein as indicator. The amount of alkali required indicates the amount of carboxyl groups present.

$$R-\underset{H}{\overset{NH_3^+}{C}}-COO^- + OH^- + 2CH_2O \rightleftharpoons R-\underset{H}{\overset{HOH_2C-N-CH_2OH}{C}}-COO^- + H_2O$$

Zwitter ion — Dimethylol amino acid

ii. ***Reaction of glycine with benzoic acid*** Amino acids act as bases towards acids, and form salts. This may be shown by the reaction of glycine with benzoic acid to form benzoyl glycine. The amino groups condense with benzoic acid to form hippuric acid. This phenomenon occurs in man as a process for detoxification of benzoic acid.

$$\underset{\text{Benzoic acid}}{C_6H_5COOH} + \underset{\text{Glycine}}{HNHCH_2COOH} \longrightarrow \underset{\text{Benzoyl glycine}}{C_6H_5CONHCH_2COOH} + H_2O$$

iii. ***Reaction with nitrous acid (Van Slyke reaction)*** Nitrous acid reacts with the amino group of amino acids to form the corresponding hydroxy acids and liberates nitrogen gas.

$$R-\underset{H}{\overset{NH_2}{C}}-COOH + HONO \longrightarrow R-\underset{H}{\overset{OH}{C}}-COOH + N_2 + H_2O$$

Amino acid + Nitrous acid — Hydroxy acid

iv. ***Reaction with carbon dioxide (Siegfried's carbamino reaction)*** The free amino groups of amino acids can condense with CO_2 to form carbamino compounds. This reaction is called Siegfried's carbamino reaction.

$$\underset{\text{Amino acid}}{R-NH_2} + CO_2 \longrightarrow \underset{\text{Carbamino compound}}{R-\overset{H}{N}-COO^-} + H^+$$

2. Reactions due to carboxyl groups

i. ***Formation of esters*** Amino acids react with alcohols in the presence of HCl to form ester. HCl combines with zwitter ion to form amino acid hydrochloride. This will react with alcohol to form ester.

$$CH_3\langle^{NH_3^+}_{COO^-} + HCl \longrightarrow CH_3\langle^{NH_3Cl}_{COOH} + \underset{\text{Alcohol}}{ROH} \longrightarrow CH_3\langle^{NH_3Cl}_{COOR} + H_2O$$

Zwitter ion — Aminoacid hydrochloride — Ester hydrochloride

ii. ***Formation of amine*** When amino acids are treated preferably in the presence of barium hydroxide, CO_2 is lost and the corresponding amine is formed. Decarboxylation of histidine to histamine, which takes place in the large intestine by the bacterial enzymatic action is an example of this reaction.

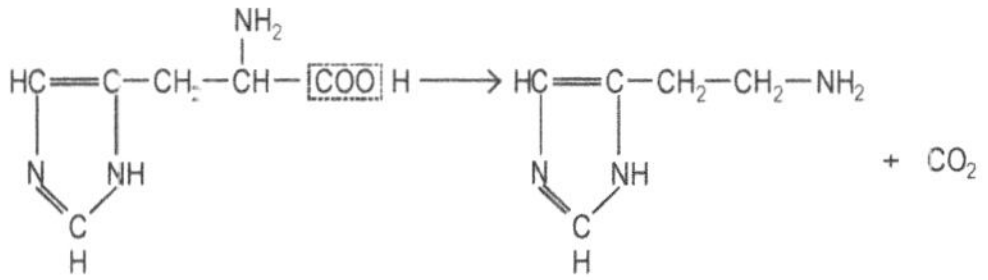

$$\underset{\text{Histidine}}{\text{HC}=\text{C}-\text{CH}_2-\text{CH}(\text{NH}_2)-\text{COOH}\ (\text{imidazole ring: N, NH, CH})} \longrightarrow \underset{\text{Histamine}}{\text{HC}=\text{C}-\text{CH}_2-\text{CH}_2-\text{NH}_2\ (\text{imidazole ring: N, NH, CH})} + CO_2$$

iii. ***Amide formation*** Amino acids form amides when treated with anhydrous or alcoholic ammonia. The amides of aspartic acid and glutamic acids, aspargine and glutamine, are important in the transport of ammonia in the body.

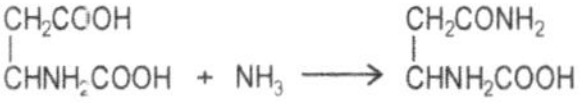

$$\underset{\text{Aspartic acid}}{\begin{array}{l} CH_2COOH \\ | \\ CHNH_2COOH \end{array}} + NH_3 \longrightarrow \underset{\text{Aspargine}}{\begin{array}{l} CH_2CONH_2 \\ | \\ CHNH_2COOH \end{array}}$$

3. Colour reactions Proteins on being treated with certain chemical reagents, exhibit characteristic colour reactions, which help in qualitative identification of proteins.

i. ***Biuret reaction*** All proteins except dipeptides give biuret reaction when treated with strong alkali, such as caustic soda and dilute $CuSO_4$ solutions, giving pinkish violet colour. Biuret ($NH_2CONHCONH_2$) also gives pinkish violet colour with NaOH and $CuSO_4$ solution. Owing to this similarity in colour reaction, it is assumed that there is a similar linkage (=CH–CO–NH–CH) in the protein molecules too.

ii. ***Millon's reaction*** When a solution of protein or a solid protein is treated with Millon's reagent (a mixture of several nitrates of mercury in nitric acid solution) a brick red colour is obtained after heating. This is due to the presence of the hydroxyphenyl group (C_6H_4OH) in the protein molecules.

iii. ***Xanthoproteic reaction*** When strong HNO_3 is added either to solid proteins or to a solution of protein, a yellow colour is obtained which changes to orange on the addition of alkali. The reaction is primarily due to the presence of the benzene ring or phenyl group in the protein molecules. Tyrosine, tryptophan and phenylalanine contain this phenyl group.

iv. ***Glyoxylic acid reaction*** (Hopkins–Cole reaction): When a protein solution is treated with glyoxylic acid, a violet colour is produced on addition of H_2SO_4. This reaction is due to the presence of indole group which is present in tryptophan.

v. ***Ninhydrin reaction*** When ninhydrin is added to protein solution, a deep blue or violet-pink or red colour is produced. This reaction is given by proteins and α-amino acids.

vi. ***Diazo reaction*** When a protein solution is treated with diazo-benzene sulphonic acid in the presence of mild alkali, a red colour is obtained. This colour is due to the presence of histidine or tyrosine.

7.4 CHEMICAL BONDS INVOLVED IN PROTEIN STRUCTURE

Proteins in general are macromolecules, consisting of about twenty different amino acids. The union of these amino acids to each other forming a chain and also among various amino acid residues of different chains involves various types of chemical bonds.

7.4.1 PRIMARY BOND (-CO–NH-)

The primary bond found in all proteins is the **peptide bond**. The C atom of–COOH group of one amino acid is linked with the N atom of $-NH_2$ group of the adjacent amino acid.

```
   H  H  O  H
   |  |  ‖  |
 —C—N—C—C—
   |        |
   R        R
```

Peptide bond

7.4.2 SECONDARY BOND

Proteins are held in their natural configuration by secondary bonds like **disulphide bond** (-S–S-), **hydrogen bond** (**>CO...HN<**), **nonpolar** or **hydrophobic bond**, **electrostatic** or **ionic bond**, **van der Waals forces,** etc.

i. ***Disulphide bond (-S–S-)*** The thio (-SH) group of two cysteine molecules may be oxidised to yield a molecule of cystine, an amino acid with a disulphide bridge.

$$O + \begin{array}{c} NH_2 \\ | \\ HSCH_2CHCOOH \\ \\ HSCH_2CHCOOH \\ | \\ NH_2 \end{array} \underset{\text{reduction}}{\overset{\text{oxidation}}{\rightleftharpoons}} \begin{array}{c} NH_2 \\ | \\ SCH_2CHCOOH \\ | \\ SCH_2CHCOOH \\ | \\ NH_2 \end{array} + H_2O$$

Cysteine Cystine

Oxytocin, a peptide hormone, consists of internal disulphide bond between two cysteine units separated from each other in the peptide chain by 4 other amino acid units.

```
┌──S────────────S──┐
Cys— Tyr — Ile — Gln — Asn — Cys — Pro — Leu —
```

ii. ***Hydrogen bond (>CO....HN<)*** The formation of a hydrogen bond is due to the tendency of hydrogen atom to share electrons with two neighbouring atoms, especially, O and N.

iii. ***Hydrophobic bonds*** Many amino acids (like alanine, valine, leucine, isoleucine, methionine, tryptophan, phenylalanine and tyrosine) have the side chains or R groups, which are essentially hydrophobic; such R groups can unite among themselves with elimination of water to form linkages between various segments of a chain or between different chains. The hydrophobic bonds play an important role in protein interactions. e.g. the formation of enzyme–substrate complexes and antigen–antibody interactions.

iv. ***Ionic or electrostatic bond*** Ions possessing similar charge repel each other; whereas, the ions having dissimilar charge attract each other. These ionic bonds are responsible for maintaining the folded structure (tertiary structure) of the globular proteins.

v. ***Van der Waals forces*** Van der Waals bond is formed between two atoms approaching each other at about 0.3–0.4 nm distance. A single van der Waals bond is weak in itself but a cluster of them in a restricted area can produce significant cohesion between molecules. e.g. silk fibroin.

7.5 CONFIGURATION AND CONFORMATION OF PROTEINS

Configuration denotes the arrangement in space of an organic molecule that is conferred by the presence of either (1) double bonds, around which there is no freedom of rotation, or (2) chiral centres around which substituent groups are arranged in a specific sequence. For example in L and D isomers of alanine, the substituent groups have two different configurations around a chiral centre. The identifying characteristic of configurational isomers is that they cannot be interconnected without breaking one or more covalent bonds.

The term conformation, on the other hand, refers to the spatial arrangement of substituent groups of organic molecules that are free to assume different positions in space without breaking any bonds, because of the freedom of rotation about their carbon–carbon single bonds. In the simple hydrocarbon ethane, for example, there is complete freedom of rotation around the C–C single bond. Many different conformations of the ethane molecule are therefore possible, depending upon the degree of rotation of one of the carbon atoms in relation to the other, but they are all freely interconvertible through rotation.

Since the covalent backbone of polypeptide chains contain only a single bond, a polypeptide assumes an infinite number of conformation in space. The covalent peptide bonds and the amino acid sequence of polypeptide chains are referred to as their primary structure.

The basic primary structure of a protein is relatively simple and consists of one or more linear chains of a number of amino acid units. This linear, unfolded structure or the polypeptide chain often assumes a helical shape to produce the secondary structure. This, in turn, may fold in certain specific patterns to produce the twisted three-dimensional or the tertiary structure of the protein molecule. Finally, certain other proteins are made up of subunits of similar or dissimilar types of the polypeptide chains. These subunits interact with each other in a specific manner to give rise to the quaternary structure of the protein. This, in fact, defines the degree of polymerisation of a protein unit.

7.5.1 PRIMARY STRUCTURE

The primary structure of a protein refers to the sequence of amino acids, the constituent units of the polypeptide chain. The main mode of linkage of the amino acids in proteins is the peptide bond, which links the α-carboxyl group of one amino acid residue to the α-amino group of the other.

Peptide bond

$$NH_2-\underset{H}{\overset{R_1}{C}}-CO-NH-\underset{H}{\overset{R_2}{C}}-CO-NH-\underset{H}{\overset{R_3}{C}}-CO-NH-\underset{H}{\overset{R_4}{C}}-CO-NH-\underset{H}{\overset{R_5}{C}}-COOH$$

7.5.2 SECONDARY STRUCTURE

If the peptide bonds were the only type of linkage present in proteins, these molecules would have behaved as irregularly coiled peptide chains of considerable length. But the globular proteins, however, do show some regular characteristic properties, indicating the presence of a regular coiled structure in these molecules. This involves the folding of the chain, which is mainly due to the presence of hydrogen bonds. Thus, folding and hydrogen bonding between neighbouring amino acids results in the formation of a rigid and tubular structure called a helix. This constitutes the secondary structure of proteins. Two types of helix conformations have been recognised—α type and β type. The α-helix has a series of amino acids woven into spiral chain which is held together by hydrogen bonds between each >CO group and the–NH group of the third amino acid residue along the chain. The α-helix is so named

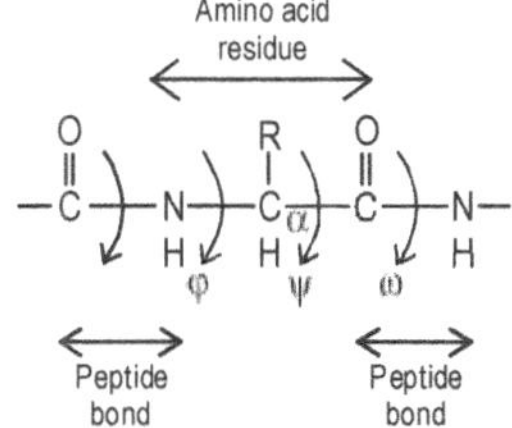

Fig. 7.1 Section of a polypeptide chain showing two peptide bonds (-CONH-) and one amino acid residue (-NH–C_α; (R) (H)–CO-). The functional group R varies, depending on the type of amino acid.

because of the mobility of α-carbon atoms. Besides the hydrogen bonds, it also shows the presence of many disulphide bonds. Another type of protein conformation called β-helix is found in silk, muscle and contractile fibres. Here two or more peptide chains are linked together laterally by hydrogen bonds.

The formation of the α helix and β strands is the key process in the folding of main chain of the protein molecule, and therefore, the formation of its secondary structure. The main chain can rotate on either side of each rigid peptide unit. The degree of rotation at the bond between nitrogen and a carbon atom of the main chain is called phi (ϕ). The degree of rotation between a carbon and carbonyl carbon atoms is called psi (ψ).

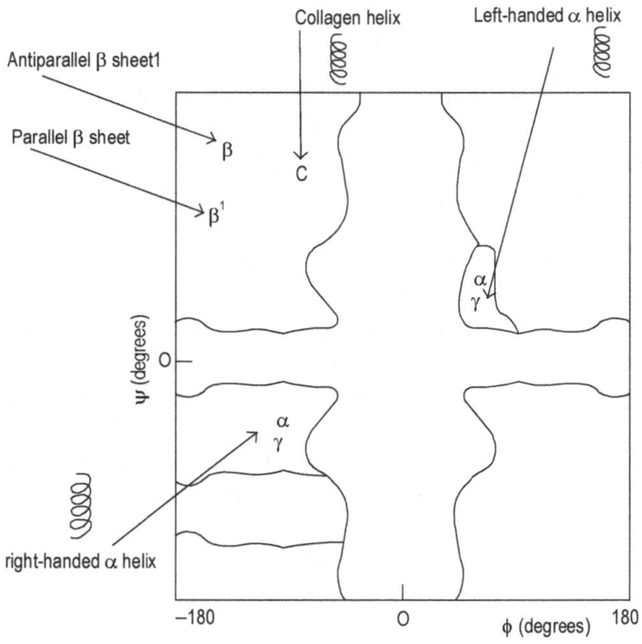

Fig. 7.2 Ramachandran plot

G.N.Ramachandran recognised that a residue in a polypeptide chain can have only a specific range of values for ϕ and ψ. The allowed ranges of ϕ and ψ can be predicted and visualised in steric contour diagrams called Ramachandran Plots. Ramachandran plot for any amino acid except glycine and proline shows three separate allowed ranges: (a) ϕ and ψ values that produce antiparallel β sheet, the parallel β sheet and the collagen helix. (b) ϕ and ψ values that produce a right handed a helix. (c) ϕ and ψ values that produce a left-handed helix (Figure 7.2). Thus the Ramachandran plots display allowed conformations of the main chain of a protein molecule.

7.5.3 TERTIARY STRUCTURE

The tertiary structure, involves the folding of the helices of globular proteins. e.g. myoglobin.

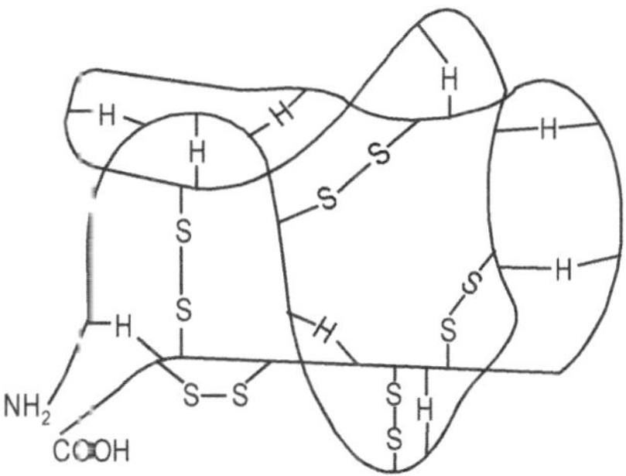

Fig. 7.3 A diagrammatic representation of tertiary structure of protein molecule stabilised by hydrogen bonds, disulphide bridges and other non-covalent forces

7.5.4 QUATERNARY STRUCTURE

Some globular proteins consist of two or more interacting peptide chains. These chains may be identical or different in their primary structure. This specific association of a number of subunits into complex large-sized molecules is referred to as the quaternary structure. The same forces (disulphide, hydrogen, hydrophobic and ionic bonds) involved in the formation of tertiary structure of proteins are also involved here to link the various polypeptide chains. Tobacco mosaic virus (TMV), with 158 amino acid residues, is an instance of the protein-protein interactions.

7.6 CLASSIFICATION OF PROTEINS

Proteins are classified on the basis of their structure and physical properties.

7.6.1 CLASSIFICATION OF PROTEINS BASED ON THEIR STRUCTURE

Based on the higher levels of structure, proteins are classified into two groups—(1) fibrous proteins, having polypeptide chains arranged in long strands or sheets, and (2) globular proteins, having polypeptide chains folded into a spherical or globular shape.

Fibrous proteins Fibrous proteins share properties that give strength and/or flexibility to the structures in which they occur. e.g. α-keratin, collagen and silk fibroin.

i. ***α-keratin*** The α-keratins have evolved for strength. e.g. hair, wool, nails, claws, quills, horns, hooves and much of the outer layer of skin. The α-keratin helix is right-handed α-helix. Two strands of α-keratin, oriented in parallel (with their amino terminals at the same end) are wrapped around each other to form a supertwisted coiled coil. α-keratin is rich in the hydrophobic residues—Ala, Val, Leu, Ile, Met and Phe. An individual polypeptide in the α-keratin coiled coil has a relatively simple tertiary structure, dominated by an α-helical secondary structure with its helical axis twisted in a left-handed super helix. The intertwining of two α-helical polypeptide is an example of quaternary structure. Coiled coils of this type are common structural elements in filamentous proteins and in the muscle protein myosin. The quaternary structure of α-keratin can be quite complex. Many coiled coils can be assembled into large supramolecular complexes such as the arrangement of α-keratin to form the intermediate filament of hair. In α-keratins, the cross-links stabilising quatenary structure are disulphide bonds. In the hardest and toughest α-keratins, such as those of rhinoceros horn, up to 18% of the residues are cysteines involved in disulphide bonds.

ii. ***Collagen*** Like the α-keratin, collagen has evolved to provide strength. It is found in connective tissue such as tendons, cartilage, the organic matrix of bone and the cornea of the eye. The collagen helix is unique secondary structure quite distinct from the α-helix. It is left-handed and has three amino acid residues per turn.

iii. ***Silk fibroin*** Fibroin, the protein of silk, is produced by insects and spiders. Its polypeptide chains are predominantly in the β-conformation. Fibroin is rich in alanine and glycine residues, permitting a close packing of β-sheets and an interlocking arrangement of R groups. The overall structure is stabilised by extensive hydrogen bonding between all peptide linkages in the polypeptides of each β-sheet and by the optimisation of van der Walls interactions between sheets. Silk does not stretch because the β-conformation is already highly extended. However, the structure is flexible because the sheets are held together by numerous weak interactions rather than by covalent bonds such as the disulphide bonds in α-keratin.

Globular Proteins In a globular protein, different segments of a polypeptide chain (or multiple polypeptide chains) fold back on each other. This folding generates a compact form relative to polypeptides in a fully extended conformation. The folding also provides the structural diversity necessary for proteins to carry out a wide array of biological functions. Globular proteins include enzymes, transport proteins like hemoglobin, motor proteins, regulatory proteins, immunoglobulins and proteins with many other functions.

Hemoglobin Hemoglobin is formed of two distinct components, the heme and globin (Figure 7.4). The heme is the iron-containing component or the prosthetic group of iron while the globin is the conjugated protein. Each molecule of hemoglobin contains four heme groups, each of molecular weight 872. An oxygen molecule may unite reversibly by combining with one of the four iron atoms that is attached by valency bonding to four pyrrole groups that make up the heme molecule. A remaining sixth valency bond of iron apparently attaches with the globin.

X-ray analysis has revealed that the hemoglobin molecule is roughly spherical, with a diameter of about 5.5 nm. Human hemoglobin protein consists of four polypeptide chains of two types, two α chains and two β chains.The polypeptide portion is collectively called as globin. The α chain has valine at the N-terminal and arginine at the C-terminal whereas in the β chain, valine is situated at the N-terminal and histidine at the C-terminal. Each chain has a heme prosthetic group in a crevice on its surface. The heme groups are involved in the binding of oxygen. The α chain has 141 residues and the β chain, which is more acidic, has

146 residues. The protein, thus, has a total 574 amino acid residues. Each of the four chains has a characteristic tertiary structure, in which the chain is folded. Like myoglobin, the α and β chains are held together as a pair by ionic and hydrogen bonds. The two pairs are then joined to each other by additional ionic bonds, hydrogen bonds and the hydrophobic forces. Thus, the four polypeptide chains fit together almost tetrahedrally to produce the characteristic quaternary structure.

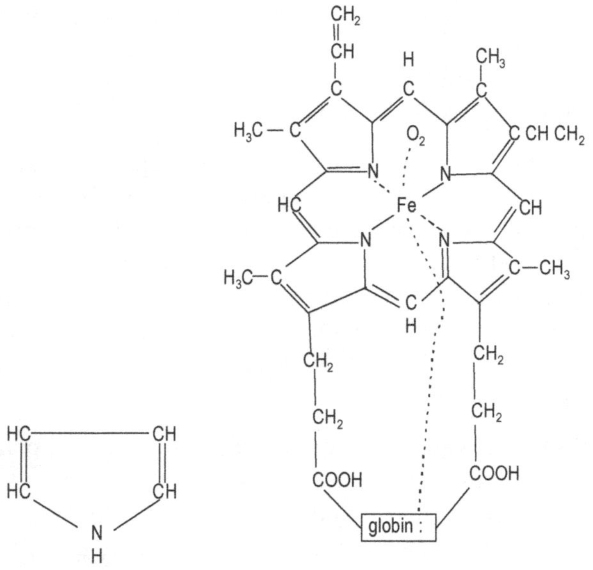

Fig. 7.4 Chemical structure of hemoglobin (a) pyrrole ring (b) hemoglobin molecule

7.6.2 CLASSIFICATION OF PROTEINS BASED ON THEIR PHYSICAL PROPERTIES

Proteins are classified on the basis of their physical properties like solubility and composition. There are three main groups—(1) simple proteins, (2) conjugated proteins and (3) derived proteins.

Simple Proteins

The simple proteins include the following:

i. ***Albumin and Globulin*** Albumin is soluble in water, whereas, globulin is insoluble in water. Both albumin and globulin are soluble in dilute neutral solutions of salts and alkalies.

ii. ***Prolamines*** Glutelin and gliadin are the proteins of wheat. Both contain large quantity of amino acids. They are insoluble in water and alcohol and soluble in dilute acids and alkalies.

iii. ***Scleroproteins or Albuminoids*** They are similar to albumins and globulins. They form most of the supporting structures of animals e.g. collagen in cartilage and white fibrous connective tissues, ossein in bones, dentine in teeth, and keratin in hair, nails, hoofs and horns.

iv. ***Histones*** They occur in the globin of hemoglobin, nucleoproteins and in spermatozoa of fish.

v. ***Protamines*** These are the simplest of proteins and contain about eight amino acids. They are found in association with nucleic acids in the sperm cells of certain fish.

Conjugated Proteins

These are proteins composed of simple proteins combined with some non-protein substances known as prosthetic groups.

i. ***Nucleoproteins*** These are compounds of protein with nucleic acid. Nucleoproteins are important compounds found in the protoplasm and nuclei. They are also constituents of chromatin.

ii. ***Phosphoprotein*** These are proteins containing phosphoric acid in organic combination i.e. phosphoric acid is linked to the hydroxyl group of certain amino acids like serine in the protein. e.g. casein in milk, vitellin of egg.

iii. ***Glycoproteins and Mucoproteins*** Simple proteins combined with carbohydrates are called glycoproteins and mucoproteins. The distinction between glycoproteins and mucoproteins is based on the amount of carbohydrates. Mucoprotein contains more than 4% hexosamine. Glycoproteins contain less than 4% hexosamine. They are found in mucin, mucoids and certain proteins in the plasma, and are also found in gonadotropic hormone.

iv. ***Chromoproteins*** These proteins contain certain heterocyclic compounds like porphyrins as the prosthetic group. The porphyrins combine with metal giving rise to coloured proteins or chromoproteins. e.g. hemoglobin, melanoprotein, flavoprotein, cytochrome and chlorophyll.

v. ***Lipoproteins*** These are proteins conjugated with lipids such as neutral fat, phospholipids and cholesterol.

vi. ***Metalloproteins*** These proteins contain metals as their prosthetic group. e.g. ferritin contains iron, ceruloplasm contains copper, and anhydrase, an enzyme protein, contains zinc.

Derived proteins

These are proteins derived from the simple proteins or conjugated proteins by the action of acids, alkalies or enzymes. They are products resulting from partial to complete hydrolysis of the proteins. They include two types of derivatives.

i. ***Primary derivatives*** Primary derivatives like metaproteins, are denaturation products of proteins resulting from the action of heat, acids and alkalies.

ii. ***Secondary derivatives*** Secondary derivatives are obtained at a later stage of hydrolysis. e.g. proteoses, peptones, peptides and diketopiperazines.

   a. *Proteoses* These are not coagulated by heat but precipitated by saturated salt solutions. Proteoses are further subdivided into primary proteoses, which are precipitated by half saturation with ammonium sulphate and secondary proteoses, which are precipitated by full saturation with ammonium sulphate.

   b. *Peptones* These are products of further hydrolytic decomposition. They are soluble in saturated salt solutions.

   c. *Peptides* These are products derived at a still later stage of hydrolysis. They are made up of two or more amino acids.

   d. *Diketopiperazines* These are cyclic anhydrides of two amino acids.

7.7 REACTIONS CHARACTERISTIC OF PROTEINS

7.7.1 HYDROLYSIS OF PROTEINS

Proteins are complex molecules composed of a number of amino acid units, joined to one another by peptide linkages. Hydrolysis of protein involves breaking up of the complex protein molecule at the peptide linkages. The mechanism of hydrolysis involves addition of a molecule of water across the peptide bond resulting in the formation of free amino acids.

$$\begin{array}{c}\text{Proteolytic enzyme}\\ \downarrow\\ NH_2-\underset{H}{\overset{R_1}{C}}-CO-NH-\underset{H}{\overset{R_2}{C}}-CO-NH-\underset{H}{\overset{R_3}{C}}-CO-NH-\underset{H}{\overset{R_4}{C}}-CO-NH-\underset{H}{\overset{R_5}{C}}-COOH\\ \text{Proteolytic enzyme}\\ \downarrow\\ NH_2-\underset{H}{\overset{R_1}{C}}-CO-NH-\underset{H}{\overset{R_2}{C}}-COOH + NH_2-\underset{H}{\overset{R_3}{C}}-CO-NH-\underset{H}{\overset{R_4}{C}}-CO-NH-\underset{H}{\overset{R_5}{C}}-COOH\\ \downarrow\\ NH_2-\underset{H}{\overset{R_1}{C}}-COOH + H_2N-\underset{H}{\overset{R_2}{C}}-COOH\\ \text{Amino acid 1} + \text{Amino acid 2}\end{array}$$

The sequence of events and the various products obtained during hydrolysis are

Protein ⟶ Metaproteins ⟶ Proteoses ⟶ Peptones ⟶ Peptides ⟶ Amino acids

7.7.2 DENATURATION

Denaturation is the disorganisation of the native (naturally occurring) protein molecule, by which the specific configuration or regular arrangement of the protein molecule is altered to an irregular diffuse arrangement. This leads to changes in the physical and chemical properties of the affected protein. Denaturation may involve a mere unfolding of the peptide chain or splitting of the protein into smaller units.

Denaturation may be brought about by physical agents like heat, UV light, X-rays, high pressure and violent shaking of the protein solution, or by chemical agents like acids, alkalies, enzymes, organic solvents, strong urea solution and high concentration of salt or heavy metals.

7.7.3 RENATURATION

Some proteins, when denatured, cannot be brought back to their original state. In that case, denaturation is described as 'irreversible type'. On the other hand, denaturation in other proteins is of 'reversible' type. For example, if trypsin is exposed to a temperature of 80–90°C, it denatures and when this solution is cooled at 37°C, the solubility and the activity of this protein-enzyme is regained. The process of regaining normal protein properties by a denatured protein is called renaturation or refolding. During renaturation, certain antibodies may cause a re-rolling of the protein bundles so that most of the original bonds are recovered. The recovery of the renatured protein is, however, never complete.

7.7.4 PRECIPITATION

Precipitation is brought about by neutralising the charges of the metaprotein and bringing it to the isoelectric point. This may be brought about by the action of heat, acid, alkali or enzymes. This is also called flocculation. In other words, precipitation is clumping together of the denatured metaprotein by neutralisation of the charges.

REVIEW QUESTIONS

1. Explain the biological significance of proteins.
2. Classify amino acids with examples.
3. Enumerate physico-chemical properties of amino acids.
4. Explain primary, secondary, tertiary and quarternary structure of proteins with a note on the type of forces which stabilises their structures.
5. Discuss the chemistry of hemoglobin.
6. Give an account of classification of proteins.
7. Write short notes on:
   i. Essential and non-essential amino acids
   ii. Non-protein associated amino acids
   iii. Zwitter ion
   iv. Sorenson's formal titration
   v. Peptide bond
   vi. Van der Waals forces
   vii. Ramachandran plot
   viii. Keratin
   ix. Silk fibroin
   x. Hydrolysis of proteins
   xi. Denaturation of proteins

PROTEIN METABOLISM

Proteins are hydrolysed by proteases into amino acids. Amino acids contain a nitrogen moiety, which is not completely oxidised in the body. Generally, the amino group is transferred to a keto acid to synthesise some non-essential amino acids or is removed or converted to urea.

The catabolic pathway of amino acids involves the following reactions:

1. Removal of amino group
2. Metabolism of ammonia
3. Metabolism of carbon skeleton

8.1 REMOVAL OF AMINO GROUP FROM AMINO ACID

This reaction is accomplished by (a) oxidative deamination (b) transamination (c) transdeamination.

8.1.1 OXIDATIVE DEAMINATION

The process of oxidative deamination takes place in two stages.

The first stage is oxidation or in other words dehydrogenation. This is followed by the second stage, hydrolysis, during which ammonia and keto acid are liberated. The next reaction may be described as transformation of an amino acid, RCH NH_2 COOH to the corresponding keto acid, RCO COOH with the liberation of NH_3.

This reaction, which is catalysed by the specific **amino acid oxidase** with FAD as coenzyme may be represented as:

$$\underset{\text{Amino acid}}{RCH\,NH_2\,COOH} + FAD \xrightarrow[-2H]{} \underset{\text{Imino acid}}{RC=NH\,COOH} + FADH_2$$

The imino acid which is formed as an intermediate, is unstable and spontaneously undergoes hydrolysis to form the keto acid, liberating ammonia.

$$\underset{\text{Imino acid}}{RC=NH\,COOH} + H_2O \longrightarrow \underset{\text{Keto acid}}{RC=O.COOH} + NH_3$$

The reduced FAD ($FADH_2$) is reoxidised spontaneously by O_2, and H_2O_2 is formed.

$$FADH_2 + O_2 \longrightarrow H_2O_2 + FAD$$

H_2O_2 is decomposed by the catalase present in cells.

The overall reaction of oxidative deamination by **amino acid oxidases** may be written as:

$$\underset{\text{Amino acid}}{R-CH(NH_2)-COOH} \xrightarrow{-2H} \underset{\text{Imino acid}}{R-C(=NH)-COOH} \xrightarrow{+H_2O} \underset{\text{Keto acid}}{R-C(=O)-COOH} + NH_3$$

$$\underset{\text{Alanine}}{CH_3-CHNH_2-COOH} \xrightarrow{-2H} \underset{\text{Imino acid}}{CH_3-C(=NH)-COOH} \xrightarrow{+H_2O} \underset{\text{Pyruvic acid}}{CH_3-C(=O)-COOH} + NH_3$$

$$\underset{\text{Aspartic acid}}{COOH-CHNH_2-CH_2-COOH} \xrightarrow{-2H} \underset{\text{Imino acid}}{COOH-C(=NH)-CH_2-COOH} \xrightarrow{+H_2O} \underset{\text{Oxaloacetic acid}}{COOH-C(=O)-CH_2-COOH} + NH_3$$

Oxidative deamination of glutamic acid is catalysed by **glutamate dehydrogenase**, which has NAD as coenzyme. The reduced NAD is oxidised via the oxidative chain by the nicotinamide-linked dehydrogenase.

$$\begin{array}{lcl c l}
\text{COOH} & & \text{COOH} & & \text{COOH} \\
| & & | & & | \\
\text{CHNH}_2 & & \text{C=NH} & & \text{C=O} + \text{NH}_3 \\
| & \xrightarrow{-2H} & | & \xrightarrow{+H_2O} & | \\
\text{CH}_2 & & \text{CH}_2 & & \text{CH}_2 \\
| & & | & & | \\
\text{CH}_2 & & \text{CH}_2 & & \text{CH}_2 \\
| & & | & & | \\
\text{COOH} & & \text{COOH} & & \text{COOH} \\
\text{Glutamic acid} & & \text{Imino acid} & & \text{Keto glutaric acid}
\end{array}$$

By oxidative deamination, amino acids are split to form α ketoacids and ammonia. Ketoacids are metabolites of carbohydrate metabolism and are oxidised in Kreb's cycle to yield energy. NH_3, which is highly toxic, may be converted to urea or detoxified to glutamine.

8.1.2 TRANSAMINATION

The process of transfer of an amino group from an amino acid to an original α-keto acid, resulting in the formation of a new amino acid and keto acid is known as transamination. In other words, it is deamination of an amino acid coupled with amination of a keto acid. It is an important method by which many amino acids undergo interconversion in metabolic reactions.

Transamination is catalysed by **transaminases**, with pyridoxal phosphate functioning as coenzyme. There are two active transaminases in tissues, catalysing interconversions. These are (i) **glutamate–oxaloacetate transaminase (GOT)** and (ii) **glutamate–pyruvate transaminase (GPT).**

GOT catalyses the transfer of NH_3 from glutamic acid to oxaloacetic acid resulting in the formation of α keto glutaric acid and aspartic acid.

$$\begin{array}{lclclcl}
\text{COOH} & & \text{COOH} & & \text{COOH} & & \text{COOH} \\
| & & | & & | & & | \\
\text{CH}_2 & & \text{CH}_2 & & \text{CH}_2 & & \text{CH}_2 \\
| & + & | & \longrightarrow & | & + & | \\
\text{CH}_2 & & \text{C=O} & & \text{CH}_2 & & \text{CHNH}_2 \\
| & & | & & | & & | \\
\text{CHNH}_2 & & \text{COOH} & & \text{C=O} & & \text{COOH} \\
| & & & & | & & \\
\text{COOH} & & & & \text{COOH} & & \\
\text{Glutamic acid} & & \text{Oxaloacetic acid} & & \alpha\ \text{keto glutaric acid} & & \text{Aspartic acid}
\end{array}$$

GPT catalyses the transfer of NH_3 from glutamic acid to pyruvic acid resulting in the formation α-keto glutaric acid and alanine.

$$
\begin{array}{ccccccc}
\begin{array}{c} COOH \\ | \\ CH_2 \\ | \\ CH_2 \\ | \\ CHNH_2 \\ | \\ COOH \end{array} & + & \begin{array}{c} CH_3 \\ | \\ C{=}O \\ | \\ COOH \end{array} & \longrightarrow & \begin{array}{c} COOH \\ | \\ CH_2 \\ | \\ CH_2 \\ | \\ C{=}O \\ | \\ COOH \end{array} & + & \begin{array}{c} CH_3 \\ | \\ CHNH_2 \\ | \\ COOH \end{array} \\
\text{Glutamic acid} & & \text{Pyruvic acid} & & \alpha\text{ keto glutaric acid} & & \text{Alanine}
\end{array}
$$

8.1.3 TRANSDEAMINATION

This is a combination of transamination and deamination. Transdeamination has been suggested as a possible method by which some amino acids lose their amino group. The amino group is first removed from glutamic acid by transamination with pyruvic acid to form alanine, which next gets deaminated by a specific amino acid oxidase as in oxidative deamination. As an example, transdeamination of glutamic acid is given here:

i. Transamination

$$
\begin{array}{ccccccc}
\begin{array}{c} COOH \\ | \\ CH_2 \\ | \\ CH_2 \\ | \\ CHNH_2 \\ | \\ COOH \end{array} & + & \begin{array}{c} CH_3 \\ | \\ C{=}O \\ | \\ COOH \end{array} & \longrightarrow & \begin{array}{c} COOH \\ | \\ CH_2 \\ | \\ CH_2 \\ | \\ C{=}O \\ | \\ COOH \end{array} & + & \begin{array}{c} CH_3 \\ | \\ CHNH_2 \\ | \\ COOH \end{array} \\
\text{Glutamic acid} & & \text{Pyruvic acid} & & \alpha\text{ keto glutaric acid} & & \text{Alanine}
\end{array}
$$

ii. Deamination

$$
\begin{array}{ccccc}
\begin{array}{c} CH_3 \\ | \\ CHNH_2 \\ | \\ COOH \end{array} & \xrightarrow{-2H} & \begin{array}{c} CH_3 \\ | \\ C{=}NH \\ | \\ COOH \end{array} & \xrightarrow{+H_2O} & CH_3-\overset{\overset{\displaystyle O}{\|}}{C}-COOH + NH_3 \\
\text{Alanine} & & \text{Imino acid} & & \text{Pyruvic acid}
\end{array}
$$

8.1.4 DECARBOXYLATION

It is a process by which –COOH group leaves a primary amine as CO_2. Pyridoxal phosphate is required as a cofactor.

Histidine $\xrightarrow{\text{Decarboxylase}}$ Histamine + CO_2

(Histidine: CH=C–CH₂–CH(NH₂)–COOH with imidazole ring N, NH, CH; Histamine: CH=C–CH₂–CH₂–NH₂ + CO₂)

8.1.5 TRANSMETHYLATION

Transmethylation is a process where methyl groups are transferred from one compound to another. In this way, the body can synthesise the essential compounds. Methionine is an efficient supplier of methyl groups.

$$CH_3\text{–}S\text{–}CH_2\text{–}CH_2\text{–}HCNH_2\text{–}COOH \longrightarrow SH\text{–}CH_2\text{–}CH_2\text{–}HCNH_2\text{–}COOH + CH_3$$

Methionine → Homocysteine

These methyl groups may be used in the formation of choline, creatine and other important compounds needed by the body.

$$NH_3.CH_2.CH_2.OH \xrightarrow[\text{Transmethylase}]{CH_3} CH_3.NH.CH_2.CH_2.OH$$

Ethanolamine → Monomethyl ethanolamine

$$\xrightarrow{CH_3} (CH_3)_2N.CH_2.CH_2.OH$$

Dimethylethanolamine

$$\xrightarrow{CH_3} (CH_3)_3N.CH_2.CH_2.OH$$

Choline

8.2 METABOLISM OF AMMONIA

NH_3 is highly toxic in nature and it cannot be stored in the body for a long period. The NH_3 formed during deamination can be detoxified or excreted as NH_3 or urea. Further, NH_3 can be utilised for the amination of keto acids to form their corresponding amino acids.

8.2.1 DETOXIFICATION OF AMMONIA

The NH_3 can be detoxified via the glutamine pathway or excreted as NH_3 or as urea via the ornithine cycle.

Glutamine pathway When a large amount of ammonia is formed, it is necessary to prevent its accumulation in the cells, as it is highly toxic. This is accomplished by detoxification; ammonia couples with glutamic acid to form glutamine. Synthesis of glutamine occurs in extrarenal tissues. The reaction requires energy and involves participation of ATP.

$$\text{Glutamic acid} + NH_3 + \text{ATP} \longrightarrow \text{Glutamine} + \text{ADP} + H_3PO_4$$

Glutamine thus formed passes from the tissues via blood to the kidneys where it is hydrolysed to glutamic acid by glutaminase.

$$\text{Glutamine} \xrightarrow{+H_2O} \text{Glutamic acid} + NH_3$$

$$\begin{array}{l} COOH \\ | \\ CHNH_2 \\ | \\ CH_2 \\ | \\ CH_2 \\ | \\ COOH \\ \text{Glutamic acid} \end{array} + NH_3 \underset{\text{Glutaminase}}{\overset{\text{Glutamine synthetase} \atop \text{ATP} \curvearrowright \text{ADP + P}}{\rightleftharpoons}} \begin{array}{l} COOH \\ | \\ CHNH_2 \\ | \\ CH_2 \\ | \\ CH_2 \\ | \\ CONH_2 \\ \text{Glutamine} \end{array} + H_2O$$

The ammonia so liberated is utilised to replace sodium and potassium of glomerular filtrate and is excreted as ammonium chloride or phosphate.

Formation of urea (Ornithine cycle) Urea is a nitrogenous compound formed by the combination of two molecules of ammonia and one molecule of carbon dioxide with the elimination of a water molecule.

$$CO_2 + 2NH_3 \longrightarrow \underset{\text{Urea}}{C(=O)(NH_2)_2} + H_2O$$

Urea is produced by a cyclic process called ornithine cycle, first elucidated by Kreb and Henseleit, based on intensive studies of metabolism in rat liver slices; their work has been supplemented by other investigators using modern tracer techniques.

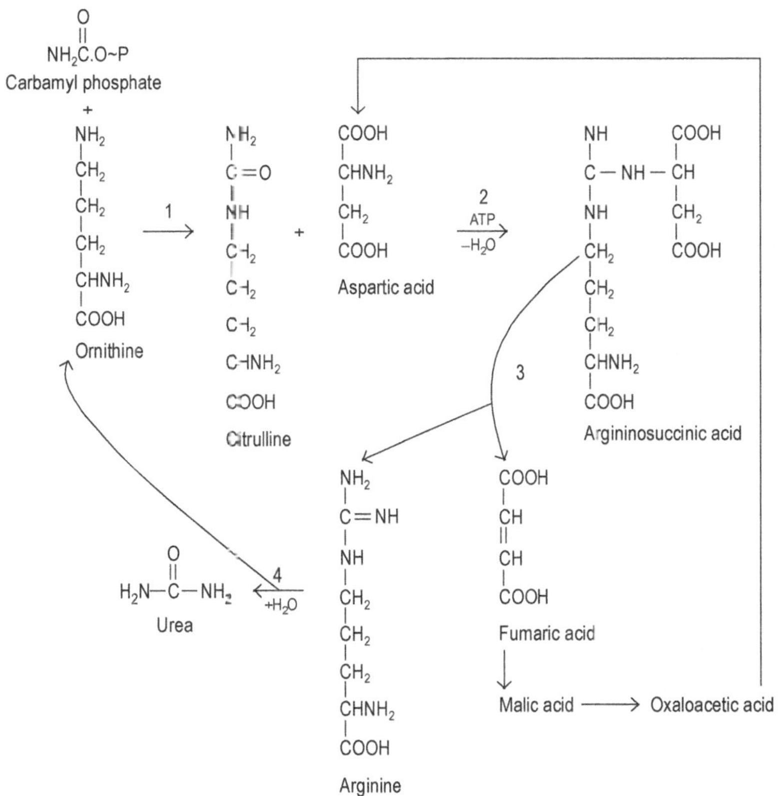

Fig. 8.1 Ornithine cycle
1. Ornithine transcarbamylase 2. Arginino succinate synthetase 3. Arginosuccinase 4. Arginase

In the ornithine cycle, ammonia formed by deamination combines with carbon dioxide to form carbamyl phosphate in the presence of the enzyme **carbamyl phosphate synthetase** and certain cofactors such as magnesium ions and acetyl glutamate. Two molecules of ATP are utilised in this process.

$$CO_2 + NH_3 + 2ATP \xrightarrow[\text{N acetyl glutamate, } Mg^{++}]{\text{Carbamyl phosphate synthetase}} \underset{\text{Carbamyl phosphate}}{NH_2.\overset{\overset{O}{||}}{C}.O{\sim}P} + 2ADP + Pi$$

The carbamyl group is then transferred to ornithine in the presence of the enzyme **transcarbamylase** to form citrulline. Citrulline couples with aspartic acid (arising through the transamination of oxaloacetic acid, from citric acid cycle) to form argininosuccinic acid; one molecule of ATP is utilised to derive the energy for coupling citrulline with aspartic acid. The argininosuccinic acid then splits into arginine and fumaric acid. Arginine, in the presence of a co-enzyme **arginase**, yields urea and ornithine (Figure 8.1).

Urea is eliminated as a nitrogenous waste while ornithine once again combines with NH_3 and CO_2 and repeats the cycle. The fumaric acid formed during the ornithine cycle is converted into malic acid from which oxaloacetic acid is produced for the citric acid cycle. Three molecules of ATP are spent to convert the toxic ammonia into a molecule of urea.

8.3 METABOLISM OF CARBON SKELETON

After deamination, the carbon skeleton left behind consists of only carbon, hydrogen and oxygen and is called as α keto acid.The α keto acids are converted into their respective amino acids either by transamination or transdeamination. Each α keto acid follows a special metabolic pathway. It follows either the glycogenic pathway or the ketogenic pathway and yield energy-rich compounds. Some of the keto acids are both glycogenic and ketogenic.

Glycine, alanine, serine, threonine, valine, aspartic acid, glutamic acid, cysteine, cystine, methionine, proline, hydroxyproline, arginine, and histidine are glycogenic amino acids.

The glycogenic pathway is shown below:

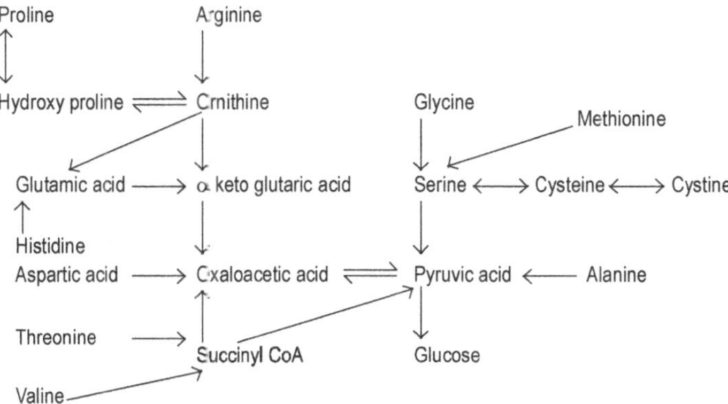

Leucine follows ketogenic pathway. It is converted to acetate or acetoacetic acid, the intermediates formed during fatty acid metabolism. Hence, leucine is called ketogenic amino acid.

Leucine ⟶ α keto isocoproic acid ⟶ Acetyl CoA ⟶ Fatty acids

Some of the amino acids like isoleucine, lysine, phenylalanine, tyrosine and tryptophan are both glycogenic and ketogenic. The metabolic pathway of these amino acids is given below:

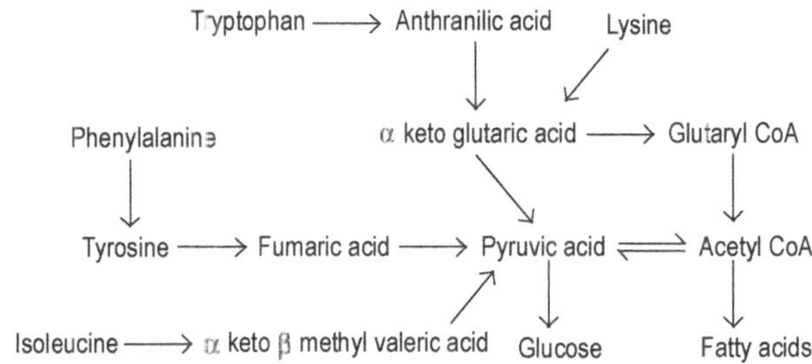

Fate of sulphur There are some amino acids such as cystine, methionine, cysteine which contain sulphur in their constitution. After deamination of these amino acids, sulphur is removed and is usually oxidised to sulphates. These sulphates may be converted to inorganic sulphates such as potassium sulphate or conjugated sulphates in the detoxification mechanism.

Fate of nucleic acids When nucleic acids are completely hydrolysed, H_3PO_4, simple monosaccharides and nitrogen bases such as purine and pyrimidine are formed. The monosaccharides may be involved in carbohydrate metabolism and H_3PO_4 may be utilised in the synthesis of phospholipids or inorganic phosphates. The purines such as adenine and guanine are oxidised to uric acid and eliminated in the urine. But the metabolic fate of pyrimidine is not yet completely known.

8.4 ENERGY PRODUCTION IN GLUTAMIC ACID OXIDATION

Energy production

i. Glutamic acid $\rightarrow$ α-ketoglutaric acid: $NADH + H^+$ is produced which on oxidation through respiratory chain yields	3 ATP
ii. α-ketoglutarate $\rightarrow$ succinyl CoA : $NADH + H^+$ oxidation	3 ATP
iii. Succinyl CoA $\rightarrow$ succinate : substrate phosphorylation	1 ATP
iv. Succinate $\rightarrow$ fumarate: $FADH_2$ oxidation	2 ATP
v. Malate $\rightarrow$ oxaloacetate: $NADH+ H^+$ (Oxaloacetate $\rightarrow$ Pyruvate + CO_2)	3 ATP
vi. Pyruvate $\rightarrow$ Acetyl CoA $\rightarrow$ CO_2 and H_2O)	15 ATP
	27 ATP

Energy expenditure

i. Two molecules of NH_3 are utilised for the synthesis of one molecule of urea. Glutamic acid contributes only one of the two nitrogens of urea, therefore, the energy expenditure for glutamic acid metabolism is	2 ATP
Net gain	25

Thus, 25 ATP are produced in the oxidation of one molecule of glutamic acid.

8.5 BIOSYNTHESIS OF PROTEINS

Proteins in the body are constantly broken down and synthesised. Synthesis of plasma proteins, tissue proteins, enzymes and hormones takes place constantly. The sites of protein synthesis within the cells are ribosomes which are present in the cytoplasm.

The following steps are involved in the biosynthesis of proteins.

i. Transcription
ii. Attachment of mRNA with 30S ribosomes and formation of polyribosome
iii. Transfer of amino acids to the site of protein synthesis.
   a. Activation of amino acids
   b. Attachment of activated amino acid to tRNA
iv. Initiation of protein synthesis
v. Elongation of polypeptide chain
vi. Termination and release of polypeptide chain
vii. Modification of released polypeptide

8.5.1 TRANSCRIPTION

The process of protein synthesis is started by the uncoiling of strands of DNA molecule. One strand of DNA molecule acts as a template for the formation of mRNA. In the presence of **RNA polymerase** mRNA is formed according to the triplet codes of DNA by transcription process. Thus, **transcription** is the enzymatic process catalysed by the enzyme RNA polymerase, whereby the genetic information contained in one strand of DNA is used to specify a complementary sequence of bases in a mRNA chain. As soon as the mRNA is formed, it leaves the nucleus and reaches the cytoplasm where it attaches with the 30S subunit of the ribosomes. The mRNA carries the triplet codons for the synthesis of the proteins.

8.5.2 ATTACHMENT OF mRNA WITH 30S RIBOSOMES AND FORMATION OF POLYRIBOSOME

In prokaryotic cells it has been observed that before the process of protein synthesis the ribosomes occur in dissociated and inactive state. The mRNA binds with the 30S ribosomal subunit in the presence of IF_2 protein factor. Soon the N-formylmethionine–tRNA (F-met tRNA) comes from the cytoplasmic amino acid pool binds with the first triplet codon (5′–AUG→3′) of the mRNA to initiate the process of protein synthesis and to form the initiation complex. The formation of initiation complex is aided by the GTP (guanosine triphosphate) and three protein factors (F_1, F_2, and F_3). After the formation of initiation complex the 30S ribosomal subunit unites with 50S ribosomal subunit to form the 70S ribosome. The union of ribosomal subunits occurs in the presence of Mg^{++} ions, F_1, and F_2 factors. The codes of mRNA are read not just by one ribosome but many ribosomes move and read the codes of the mRNA. When many ribosomes bind with the mRNA, the formation of polysome or polyribosome occurs.

8.5.3 TRANSFER OF AMINO ACIDS TO THE SITE OF PROTEIN SYNTHESIS

The amino acids are activated and transferred from the intracellular amino acid pool to the active ribosomes by the tRNA.

(a) Activation of amino acids

Each of the 20 amino acids occur in the cytoplasm in an inactive state. Each amino acid before its attachment with its specific tRNA is activated by a specific activating enzyme known as the **aminoacyl synthetase** and ATP. The free amino acid react with ATP, resulting in the production of amino acyl adenylate and pyrophosphate:

$$\underset{\text{Amino acid}}{\text{AA}} + \text{ATP} + \underset{\substack{\text{Aminoacyl}\\\text{synthetase}}}{\text{Enzyme}} \longrightarrow \underset{\substack{\text{Aminoacyl}\\\text{adenylate}\\\text{enzyme complex}}}{\text{AA} \sim \text{AMP} - \text{Enzyme}} + \underset{\text{Pyrophosphate}}{\text{PP}}$$

The reaction product aminoacyl adenylate is bound to the enzyme in the form of a monocovalent complex. This aminoacyl adenylate enzyme complex then esterifies to specific tRNA molecule. The cell has atleast 20 aminoacyl synthetase enzymes for the 20 amino acids. Each enzyme is specific and it attaches with the specific amino acid without any error.

(b) Attachment of activated amino acid to tRNA The aminoacyl adenylate remains bounded with enzyme until it collides with the specific tRNA molecule and its synthetase is recognised by dihydrouridine (DHU) loop of specific tRNA. Then, amino acid residue of aminoacyl adenylate is transferred to amino acid attachment site of tRNA where its carboxyl group forms bond or linkage with the 3–OH group of the ribose of the terminal adenosine at CCA end of tRNA. As a result, AMP and enzyme are released and a final product aminoacyl tRNA is formed.

$$\underset{\text{Aminoacyl adenylate enzyme complex}}{\text{AA} - \text{AMP} - \text{Enzyme}} + \text{tRNA} \longrightarrow \underset{\text{Aminoacyl-tRNA}}{\text{AA} - \text{tRNA}} + \text{AMP} + \text{Enzyme}$$

The aminoacyl-tRNA moves towards the site of protein synthesis, i.e. ribosomes with mRNA.

8.5.4 INITIATION OF PROTEIN SYNTHESIS

The initiation of protein synthesis in the bacterium *Escherichia coli (E.coli)* involves the formation of 70S complex. The mRNA always has first triplet codon as AUG at its beginning (i.e. 5′ end). The AUG codons are the codes for the amino acid methionine. The methionine remains formylated and it has a very important role in initiating the process of protein synthesis. In every type of protein the formyl methionine occupies the first place in the molecule and when the protein molecule is completely synthesised, the formyl methionine often detaches from the newly synthesised protein molecule by the activity of a hydrolytic enzyme.

Since in protein synthesis the peptide chain always grows in a sequence from the free terminal amino ($-NH_2$) group towards the carboxyl (–COOH) end, the function of F-met tRNA) is to ensure that proteins are synthesised in that direction. In the F-met tRNA, the amino ($-NH_2$) group is blocked by the formyl group leaving only the –COOH group available to react with the $-NH_2$ group of the second amino acid (AA_2).

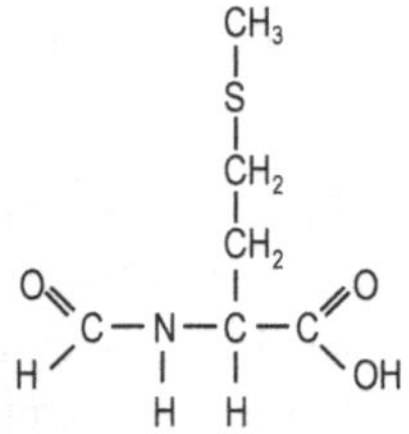

$$H-\overset{O}{\overset{\|}{C}}-\underset{H}{N}-\underset{H}{\overset{CH_2-CH_2-S-CH_3}{C}}-\overset{O}{\overset{\|}{C}}-OH$$

N-formyl methionine

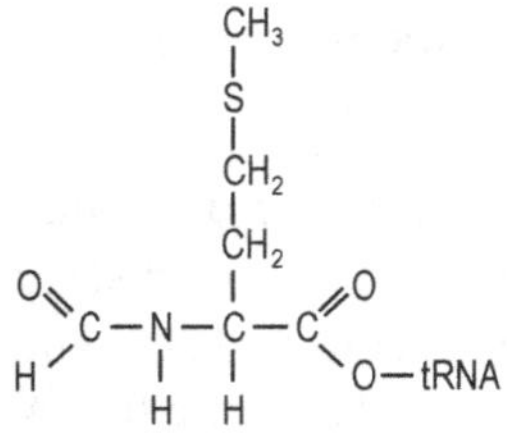

$$H-\overset{O}{\overset{\|}{C}}-\underset{H}{N}-\underset{H}{\overset{CH_2-CH_2-S-CH_3}{C}}-\overset{O}{\overset{\|}{C}}-O-tRNA$$

N-formyl methionine tRNA

In this way, the synthesis of protein chain follows the correct sequence.

8.5.5 ELONGATION OF POLYPEPTIDE CHAIN

With the formation of functional 70S ribosome (i.e. 70S-mRNA-F-met tRNA), the elongation of polypeptide chain is brought about by the regular addition of amino acids and relative movements of ribosome and mRNA in the presence of GTP molecules, so that a new triplet codon remains available for new aminoacyl tRNA at the decoding or 'A' site of ribosome in each step. Thus, F-met tRNA must move from decoding site ('A' site) to peptidyl site or 'P' site, before the second aminoacyl-tRNA (i.e. AA_2 tRNA) can bind to the next triplet codon occurring at decoding or 'A' site of ribosome. The aminoacyl-tRNA (AA_2 tRNA) binds with the codon of 'A' site in the presence of GTP and two proteins, called transfer factors (Tu and Ts), which remain associated with ribosome. During this binding process, a complex is formed between GTP, the transfer factors and all aminoacyl-tRNA (viz., AA_2 tRNA), which ultimately deposits aminoacyl-tRNA at the 'A' site of ribosome with the release of transfer factors—GDP complex and inorganic phosphate.

In the next step, due to relative movement of ribosome and mRNA in the presence of EFG factor or translocase and GTP molecule, the next triplet codon becomes available for next aminoacyl-tRNA (viz. AA_2 tRNA) at the 'A' site of ribosome. At this stage, F-met tRNA occurs at exit or 'E' site, while AA_2 tRNA occurs at peptidyl or 'P' site. Now, an enzyme known as transferase I removes tRNA from formyl-methionine (F-met or AA_1) and flips the formyl-methionine to the aminoacyl-tRNA (AA_2-tRNA) bound at the peptidyl or 'P' site. The 'G' factor is supposed to release the discharged or deacylated tRNA from the ribosome. The G factor is found to function in the presence of an acidic and contractile protein in the 50S ribosomal subunit.

It follows the next stage of elongation process which involves the synthesis of a peptide bond by a reaction (peptidyl transferse reaction) between the free amino group of incoming amino acid (i.e. AA_2) and the carboxyl group of the first amino acid (AA_1), which is esterified to tRNA. The enzyme which catalyses this reaction is called **peptidyl transferase** (or peptide synthetase) and is an integral part of the 50S subunit. The energy for peptide bond synthesis is derived from cleavage of the ester link between an amino acid and its tRNA.

Thus, during the elongation of polypeptide chain, each charged or loaded tRNA (aminoacyl-tRNA) enters the decoding or 'A' site, moves to the condensing or 'P' site, transfers its amino acid to the carboxyl end of the polypeptide, moves to the exit site where the polypeptide chain is transferred to the adjacent tRNA on the condensing site, and tRNA is then released from the ribosome (Figure 8.2). This sequence of events involved in elongation must take place very rapidly since it has been calculated that in *E.coli* growing under optimal conditions, a polypeptide chain of about 40 amino acids can be produced in 20 seconds.

8.5.6 TERMINATION AND RELEASE OF POLYPEPTIDE CHAIN

The synthesis of a polypeptide chain is concluded when a given ribosome in a polysome encounters a terminator codon, which is a genetic signal encoded in the mRNA, which specifies that the C-terminal amino acid of a polypeptide has been added to the chain. A terminator codon is not recognized by the anticodons of any of the normally occurring, aminocyl- tRNAs, and its presence in the decoding or aminoacyl site precludes the addition of any further amino acids to the chain. In *E.coli*, its phages and in eukaryotes, the RNA triplet UAA, UGA and UAG all function as terminator codons.

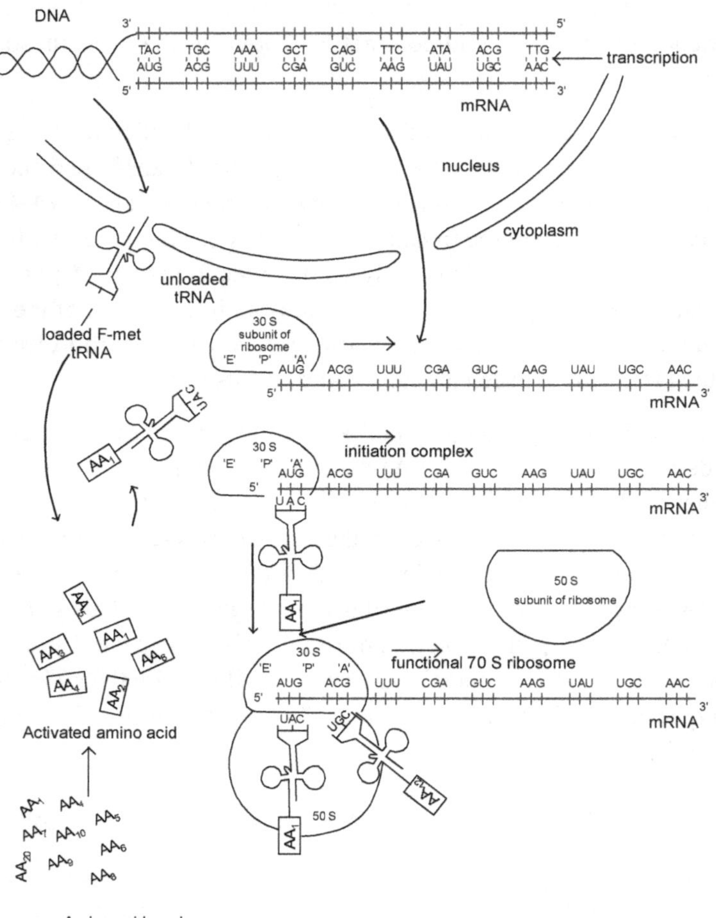
DNA
TAC TGC AAA GCT CAG TTC ATA ACG TTG
AUG ACG UUU CGA GUC AAG UAU UGC AAC
transcription
mRNA
nucleus
cytoplasm
unloaded tRNA
loaded F-met tRNA
30 S subunit of ribosome
'E' 'P' 'A'
AUG ACG UUU CGA GUC AAG UAU UGC AAC
mRNA
initiation complex
50 S subunit of ribosome
functional 70 S ribosome
Activated amino acid
Amino acid pool

Contd.

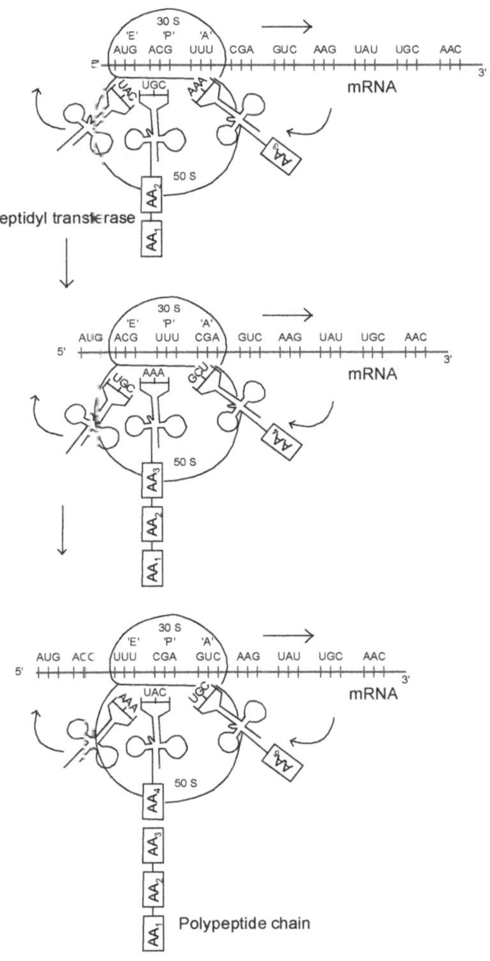

Fig. 8.2 Biosynthesis of proteins

When a terminator codon moves into an aminoacyl site, the following events takes place. The terminator first interacts with one of the two specific protein factors called **releasing factors** (R_1 and R_2). The R_1 is specific for UAG and UAA and R_2 for codons UAA and UAG. This complex of releasing factor terminator codon and ribosome effectively blocks further chain elongation. With the aminoacyl site so clogged, the completed polypeptide remains esterified to the final tRNA occupying the peptidyl site. This linkage is then broken by hydrolysis in a reaction mediated by still another protein factor, and both a free tRNA molecule and a complete polypeptide are released from the ribosome. The ribosome then dissociates into its large and small subunits, an event that may be mediated by F_3 protein factor. The dissociated subunits are now free to form new initiation complexes and participate in another round of polypeptide synthesis.

8.5.7 MODIFICATION OF RELEASED POLYPEPTIDE

The released polypeptide chain contains the formylated methionine at its one end. An enzyme **deformylase** removes the formyl group of methionine. The **exopeptidase** enzyme may remove some amino acids from N-terminal end or the C-terminal end of polypeptide chain. At this stage, the polypeptide (protein) possesses its primary and probably its secondary structures. The linear sequence of amino acids forms the primary structure; at least some portion of many proteins has a secondary structure in the form of an alpha helix. The protein chain may then fold back upon itself forming internal bonds (including strong disulphide bonds) which stabilise its tertiary structure into a precisely and often intricately folded pattern. Two or more tertiary structures may unite into a functional quaternary structure. For example, hemoglobin consists of four polypeptide chains, two identical α-chains and two identical β-chains. A protein does not become an active enzyme until it has assumed its tertiary or quaternary pattern.

8.6 REGULATION OF PROTEIN SYNTHESIS

Living cells have accurate mechanisms for regulating the synthesis of their proteins. Protein synthesis in prokaryotes is regulated primarily at the level of transcription of DNA to yield mRNA. In 1961, Jacob and

Monod have proposed a hypothesis to explain the regulation of protein synthesis at the level of transcription. This hypothesis is called the **Operon hypothesis** for which they were awarded the Nobel Prize in 1965. The **operon** may be defined as a genetic unit which comprises the operator gene and the adjacent group of structural genes that are under its control. The activity of the operator gene in turn depends on the regulator gene, which acts through a repressor.

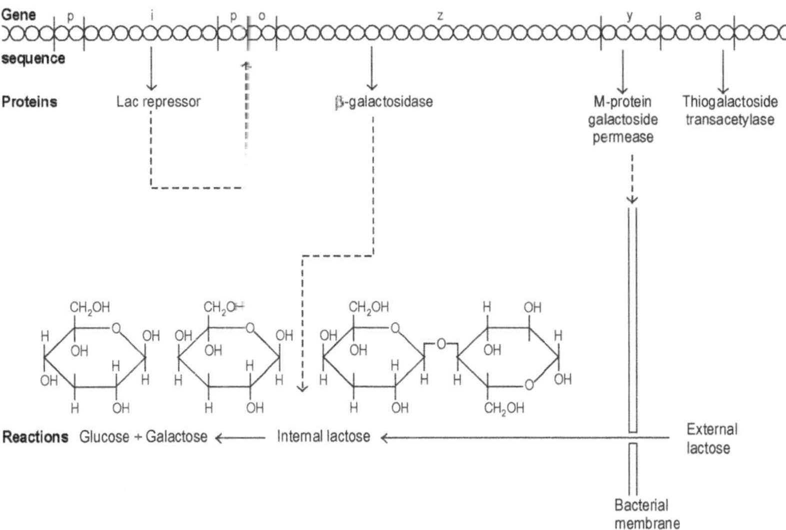

Fig. 8.3 The lac operon of *E.coli*

i–regulator gene, p–promotor site, o–operator gene, z–the structural gene for β-galactosidase enzyme protein, y–the gene responsible for synthesis of β galactoside permease enzyme which is required for the permeation of galactose into the cell, a–the gene for β-galactoside transacetylase enzyme protein.

One of the best examples of operon is the lactose *(lac)* operon *E.coli,* which comprises in addition to the operator gene o and the promoter gene p, three structural genes z, y and a (Figure 8.3). These three genes are responsible for the production of three enzymes called **β-galactosidase, β-galactoside permease, β-galactoside transacetylase** respectively, which control the metabolism of lactose. At a remote portion of the chromosome is located another gene called the regulator gene i. This has a controlling effect on the operator gene. It synthesises large protein molecules called repressors, which combine with and block the function of the operator gene, which in turn will result in blocking the function of all the structural genes in the operon unit. There is another portion of the operon 'p' the promoter site, to which the **RNA polymerase** binds. Blockage of this site by combination with repressor molecule will block the functioning of the adjacent operator gene and thus block the functioning of the structural genes.

An enzyme inducer (like lactose or even its analog, isopropylthiogalactoside or IPTG) can combine with the repressor molecule and thus prevent its combination with the operator gene. The operator gene and the structural genes hence become functional and mRNA for the functional proteins is synthesised.

The promoter of the lac operon contains two active components that regulate transcription. One is the RNA polymerase binding site, and the second is the cyclic AMP receptor protein (CRP) or catabolite activator protein (CAP) binding site. The CAP site functions to prevent transcription of the lac operon when sufficient glucose is present. The CAP–cyclic AMP complex acts as a positive regulator.

In eukaryotes, regulation of protein synthesis is much more complex. Although every cell of a vertebrate contains the entire genome of the organisms, only a fraction of its structural genes are expressed in any given cell type. Nearly all the cells of higher animals contain the basic enzymes required for the central metabolic pathways. However, different cell types such as muscle, brain and liver have characteristically different structures and biological functions, each dependent upon distinctive sets of specialised proteins. Moreover, the biosynthesis of different sets of specialised proteins during the orderly differentiation and growth of higher organisms must also be accurately programmed with respect to time and sequence of their appearance. Set of regulatory genes operate, transforming given areas of an egg cell into predictable structures in the adult organisms.

The initiation of transcription is a major regulation point for both prokaryotic and eukaryotic gene expression. Although some of the same regulatory mechanisms are used in both systems, there is a fundamental difference in the mechanism of transcription in eukaryotes and bacteria.

The inherent activity of promoters and transcriptional machinery *in vivo* in the absence of regulatory sequences can be defined as the transcriptional ground state. In bacteria, RNA polymerase generally has access to every promoter and can bind and initiate transcription in the absence of activators or repressors. The transcriptional ground state is therefore, non-restrictive. In contrast, eukaryotic promoters are generally inactive *in vivo* in the absence of regulatory sequences; that is, the transcriptional ground state is restrictive. This fundamental difference gives rise to at least four important features that distinguish the regulation of gene expression in eukaryotes from that in bacteria.

i. Access to eukaryotic promoters is restricted by the structure of chromatin, and activation of transcription is associated with multiple changes in chromatin structure in the transcribed region.

ii. Although both positive and negative regulatory elements are found in eukaryotic cells, positive mechanisms predominate. Since the transcriptional ground state is restrictive, eukaryotic gene requires activation.

iii. Eukaryotic cells have larger, more complex multimeric regulatory proteins than do bacteria.

iv. Transcription in the eukaryotic nucleus is separated in both space and time from translation in the cytoplasm.

In eukaryotes, hormones also regulate gene expression. Steroids or thyroid hormones interact directly with intracellular receptors that are regulatory proteins; binding of the hormone has either positive or negative effects on the transcription of genes targeted by the hormone. Nonsteroid hormones bind to cell surface receptors, triggering a signaling pathway that can lead to phosphorylation of a regulatory protein, affecting its activity.

REVIEW QUESTIONS

1. Discuss the general catabolic pathway of amino acids.
2. With suitable examples explain deamination and transamination process.
3. Describe ornithine cycle.

4. Describe the biosynthesis of proteins.
5. Write short notes on:
    i. Transdeamination
    ii. Deamination
    iii. Transamination
    iv. Decarboxylation
    v. Transmethylation
    vi. Lac operon hypothesis

9

NUCLEIC ACIDS

Nucleic acids are biopolymers of high molecular weight with mononucleotides as their repeating units. Friedrich Miescher (1871) isolated a substance from the nuclei of the pus cells and named it as nuclein. Later, it was found that the nuclein had acid properties and hence it was renamed as nucleic acid. The nucleic acid contains carbon, hydrogen, oxygen, nitrogen and phosphorus. There are two kinds of nucleic acids—deoxyribonucleic acid (DNA) and ribonucleic acid (RNA). DNA is found mainly in the chromatin of the cell nucleus, whereas, most of the RNA is present in the cell cytoplasm and a little in the nucleolus.

9.1 BIOLOGICAL FUNCTIONS OF NUCLEIC ACIDS

Nucleic acids play a vital role in the biosynthesis of cellular proteins. Ribonucleic acids are involved in the actual process of synthesis of protein molecules. The soluble RNA (sRNA) acts as the acceptor for specific amino acid units and transfers the amino acid to the polyribosome which is the site of protein biosynthesis. Messenger RNA (mRNA) carries the genetic information stored in the DNA of the nucleus (genetic code) to the RNA strand in the ribosome. The specific arrangement of the nitrogenous bases in the genetic code determines the sequence of amino acids in a specific protein molecule to be synthesised. Ribosomal RNA (rRNA) is associated with the ribosomes in the cytoplasm. They play an unspecified role in the biosynthesis of protein. The concentration of RNA is increased in organs which show high energy intake due to treatment with thyroxine, insulin and cortisone. The concentration of RNA is diminished in high fat diets and in states of deficiency of thiamine and vitamin B_{12}. According to the current concept, RNA is speculated to be associated with memory storage function in the brain.

Viruses which invade plant and animal cells contain RNA whereas, bacteriophages which are viruses infecting bacterial cells contain DNA. Viruses, after invading living cells, multiply by the synthesis of the nucleic acid and the protein of which they are composed, thus altering the metabolism of the host cell.

DNA participates in two important roles in the life of the cells:

i. During cell division, it undergoes replication and transmits as exact copy to each daughter cell.

ii. It provides genetic information and directs synthesis of a specific protein.

Nucleic acids are concerned with mutation and carcinogenesis of living cells. Agents like X-rays, ultraviolet light, nitrogen and mustard, which are considered mutagenic and carcinogenic produce abnormal types of nucleic acids in the cells.

9.2 MOLECULAR STRUCTURE OF NUCLEIC ACIDS

A nucleic acid may be visualised as a polymer of a nucleotide monomer. In other words, it may be considered as a polynucleotide.

Upon hydrolysis, the two nucleic acids yield three components—phosphoric acid, a pentose sugar and nitrogenous bases.

An important property of the pentoses is their capacity to form esters with phosphoric acid. In this reaction, the OH groups of the pentose, especially those at C3 and C5, are involved forming a 3', 5'-phospho diester bond between adjacent pentose residues. This bond, in fact, is an integral part of the structure of nucleic acids.

9.2.3 NITROGENOUS BASES

Two types of nitrogenous bases are found in all nucleic acids. These are derivatives of pyrimidine and that of purine.

Pyrimidine derivatives These are derived from their parent heterocyclic compound pyrimidine. The common pyrimidine derivatives found in nucleic acids are uracil, thymine and cytosine (Figure 9.1).

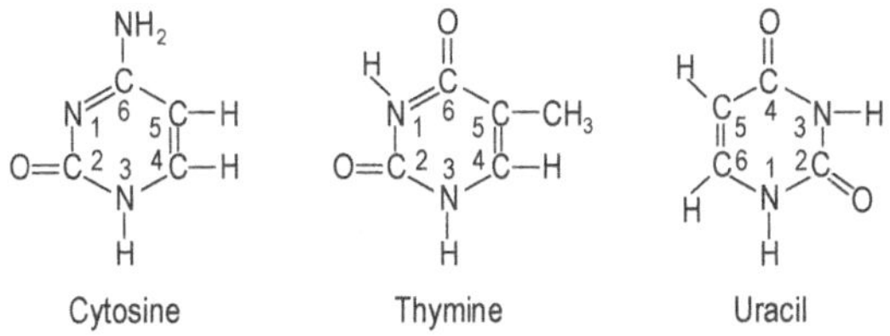

Cytosine Thymine Uracil

Fig. 9.1 Pyrimidine

Purine derivatives The prevalent purine derivatives found in nucleic acids are adenine and guanine (Figure 9.2).

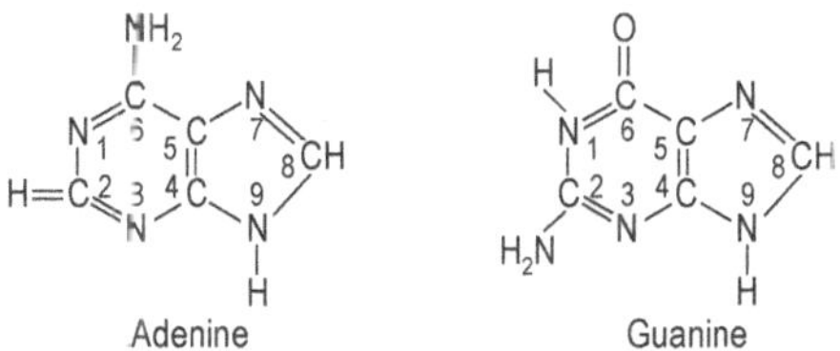

Fig. 9.2 Purine

Tautomeric forms of nitrogenous bases Compounds that exist in two structural isomeric forms which are mutually interconvertible and exist in dynamic equilibrium are called **tautomers** and the phenomenon is termed as **tautomerism**.

Tautomerisation In a normal molecule of DNA, the purine adenine (A) is linked to the pyrimidine thymine (T) by two bonds, while the purine guanine (G) is linked to pyrimidine cytosine (C).

in the distribution of hydrogen atoms (tautomeric shift). Due to tautomerization the amino (–NH_2) group of cytosine and adenine is converted into imino (–NH) group and likewise keto group (C=O) of thymine and guanine is converted to enol group (–OH) (Figure 9.3).

In its rare or tautomeric state, a nitrogenous base cannot pair to its normal partner. Rather a tautomeric purine adenine pairs with the normal cytosine and tautomeric guanine with thymine. Similarly, tautomeric thymine pairs with normal guanine and cytosine with adenine. Such pairs of nitrogenous bases are known as forbidden base pairs or unusual base pairs (Figure 9.4).

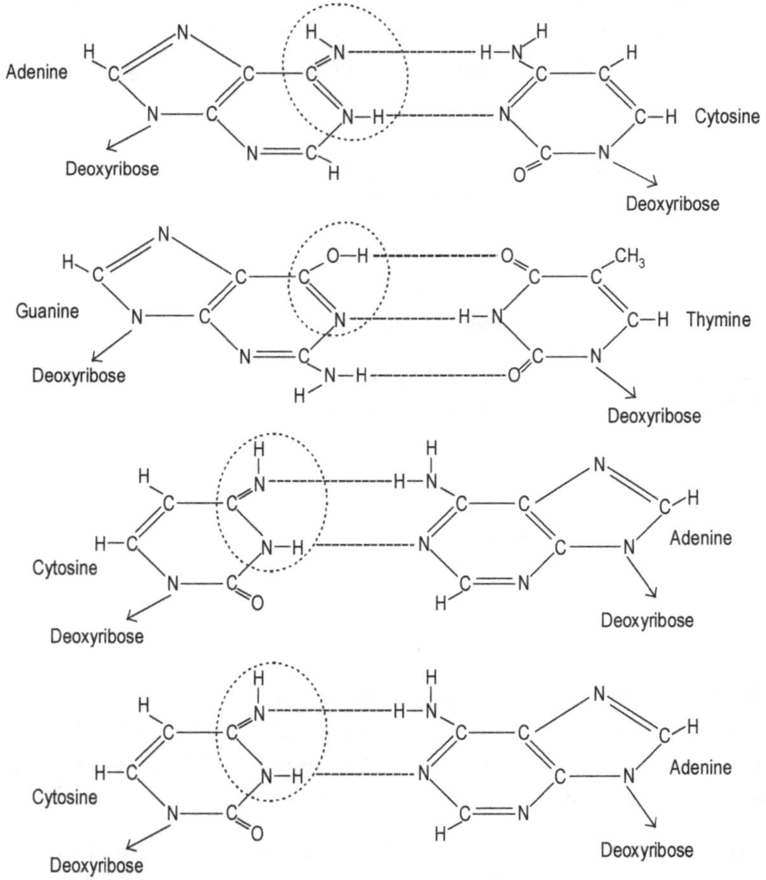

Fig. 9.4 Forbidden base pairs resulting from tautomerisation

Nucleosides containing purines are

i. Adenosine—Adenine and ribose
ii. Guanosine—Guanine and ribose

Nucleosides containing pyrimidines are

i. Uridine—Uracil and ribose
ii. Cytidine—Cytosine and ribose
iii. Thymidine—Thymine and ribose

9.2.5 NUCLEOTIDES

Each nucleotide unit consists of a phosphoric acid, a pentose and a nitrogenous base, which may be either a purine or a pyrimidine. Out of the four mononucleotides, two contain each a purine base and two each a pyrimidine base. Nucleotides are phosphoric esters of nucleosides (Figure 9.6).

The nucleotides are named according to the purines and pyrimidines contained in them. Nucleotides containing purines are

i. Adenylic acid (adenine nucleotide)
ii. Guanylic acid (Guanine nucleotide)

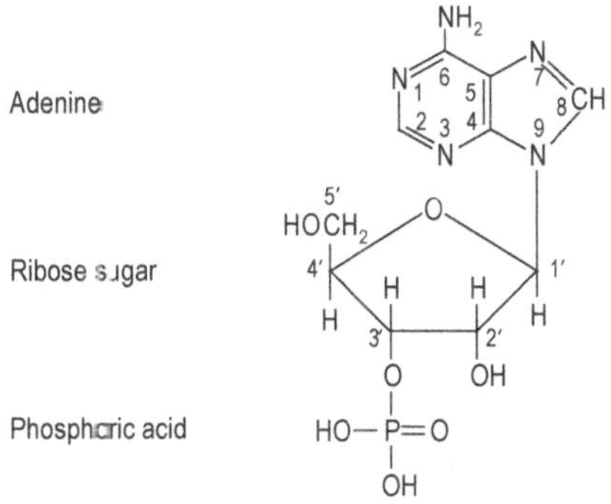

(a) Adenylic acid

Contd.

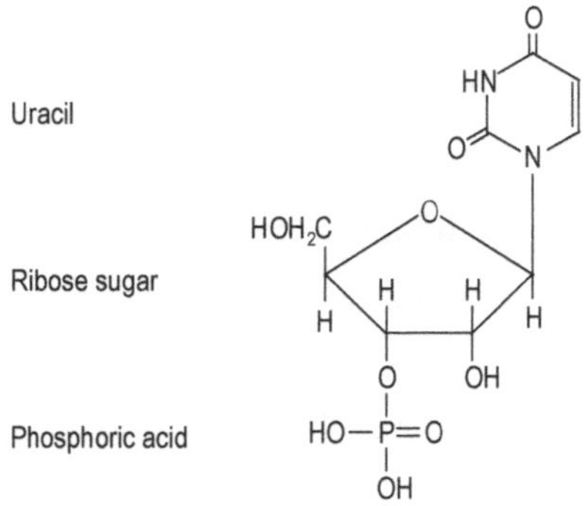

Uracil

Ribose sugar

Phosphoric acid

(e) Uridylic acid

Fig. 9.6 Nucleotides

Nucleotides containing pyrimidines are

i. Uridylic acid (Uracil nucleotide)

ii. Cytidylic acid(Cytosine nucleotide)

iii. Thymidylic acid (Thymine nucleotide)

Besides the nucleotides which form the integral components of the nucleic acids, the following nucleotides exist in free state in the tissues.

i. Adenylic acid also known as adenosine monophosphate (adenosine-3, monophosphate) or AMP. It is important for its metabolic role in the activation of phosphorylase.

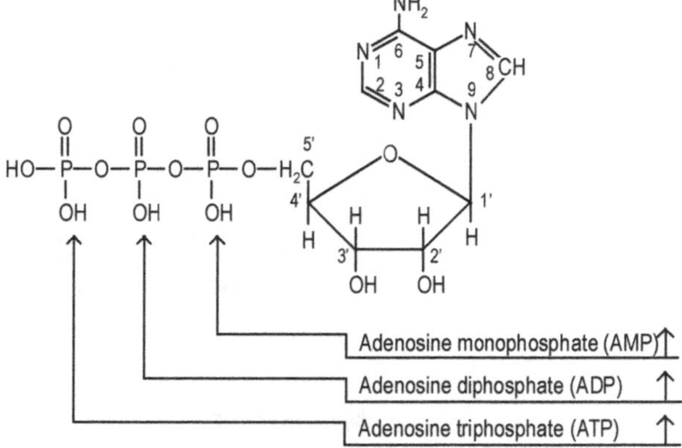

Fig. 9.7 AMP, ADP, ATP

viii. Cytosine derivatives: Cytidine triphosphate (CTP) is a high-energy phosphate compound. It reacts with phosphoryl choline to form cytidine diphosphate choline (CDP-choline), which combines with the diglyceride to form phospholipid.

9.3 DEOXYRIBONUCLEIC ACID (DNA)

DNA is a polynucleotide. The constituent units are coupled by means of 3′-5′ phosphodiester bonds. The nature, properties and function of the nucleic acids depend on the exact order of the purine and pyrimidine bases in the molecule. This sequence of specific bases is termed the primary structure. The purine and pyrimidine bases of DNA carry genetic information whereas the sugar and phosphate groups perform a structural role. The structure of a small segment of DNA is shown in the Figure 9.9.

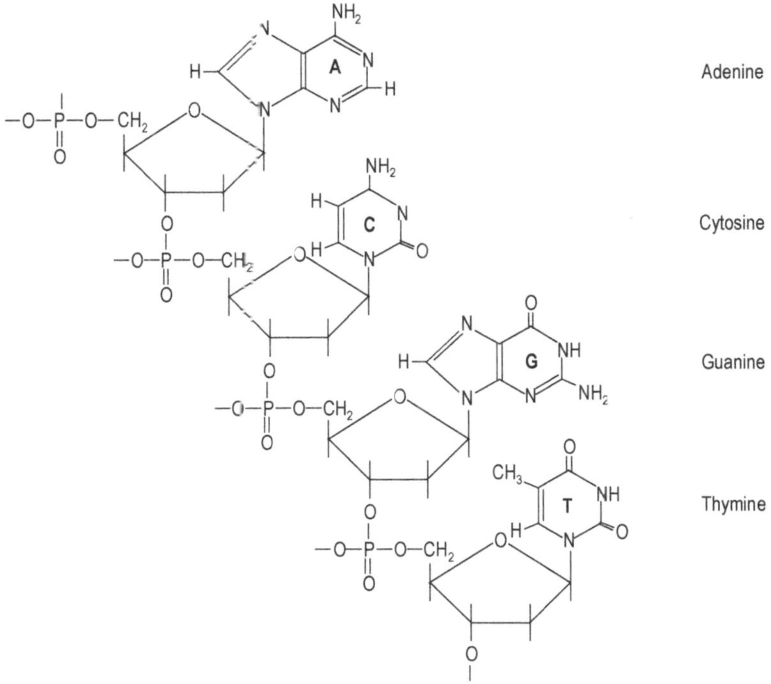

Adenine

Cytosine

Guanine

Thymine

Fig. 9.9 A small segment of DNA chain

According to this model, DNA consists of two helical polynucleotide chains, which are coiled around a common axis in the form of a right-hand double helix. The diameter of the helix is about 20A° and each chain makes a complete turn every 34 A°. The bases are 3.4 A° apart along the helix axis and are related by a rotation of 36°. Each turn of the helix thus contains 10 nucleotide residues. The phosphate and deoxyribose units are found on the periphery of the helix, whereas, the purine and pyrimidine bases occur in the centre. The two complementary chains do not run in the same direction with respect to their internucleoside linkages, but are rather antiparallel. The two chains are held together by hydrogen bonds between pairs of bases. Adenine always pairs with thymine by two hydrogen bonds and guanine with cytosine by three hydrogen bonds (Figure 9.11). This specific positioning of the bases is called base complementarity. The individual hydrogen bonds are weak in nature but, as in the case of proteins, a large number of them involved in the DNA molecule confer stability to it. The stability of the DNA molecule is primarily a consequence of vander Waals forces between the planes of stacked bases. An important feature of the double helix is the specificity of the pairing of bases. Pairing always occurs between adenine and thymine and between guanine and cytosine. This is due to steric and hydrogen bonding forces.

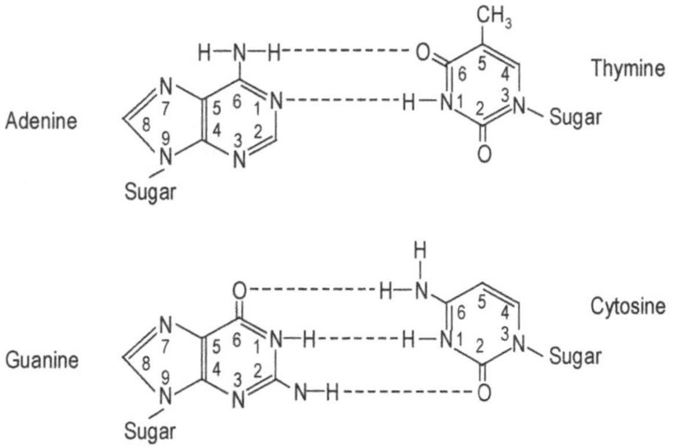

Fig. 9.11 The pairing of purine and pyrimidine bases

In recent years, many evidences have been collected in favour of the Watson and Crick model of DNA structure. They include:

the RNA are formed. Thus like the DNA, RNA is also made up of nucleotides which are called ribonucleotides. These are as follows:

Adenosine-3′monophosphate (Adenylic acid)

Guanosine-3′ monophosphate (Guanylic acid)

Cytosine-3′ monophosphate (Cytidylic acid)

Uridine-3′ monophosphate (Uridylic acid)

These 3′ nucleotides also occur as 5′ nucleotides. In RNA polynucleotide, these four types of nucleotides are linked together by 3′-5′ phosphodiester bond. A small segment of RNA chain is shown in Figure 9.12.

9.4.1 TYPES OF RNA

Depending on the site of occurrence and functions, four types of RNA are described—messenger RNA (mRNA), ribosomal RNA (rRNA), transfer RNA (tRNA) or soluble RNA (sRNA) and viral RNA.

Messenger RNA or nuclear RNA mRNA is synthesised inside the nucleus as a complementary strand to DNA and carries genetic information from chromosomal DNA to the cytoplasm for the synthesis of proteins. It was for this reason that, it was named messenger RNA (mRNA) by Jacob and Monod in 1961. It constitutes about 10% of the total RNA present in the cell. It has the following characteristics:

i. It is formed as a complementary strand to one of the two strands of a DNA.
ii. It, therefore, carries the same sequence of base sequence as found in that particular segment of DNA from which it is copied, except that the thymine of DNA which is substituted by uracil in mRNA. mRNA, therefore, contains the same information as coded in that part of DNA.
iii. After synthesis, it immediately diffuses out of the nucleus into the cytoplasm, where it is deposited on certain number of ribosomes.
iv. Here, mRNA acts as a template for protein synthesis.

a. Monocistronic mRNA molecule contains the codons of a single cistron which codes for one complete molecule of protein.

b. Polycistronic mRNA molecule contains the codons for more than one cistron which may lie close together. This type of mRNA synthesises more than one protein chain. For example, mRNA molecule which governs the metabolism of histidine, codes for the synthesis of ten specific enzymes.

In eukaryotes mRNA is synthesised as heterogeneous nuclear RNA (Hn-RNA) inside the nucleus. These molecules are much bigger in size than mRNA molecules. A sequence of about 200 nucleotides of polyadenylic acid (poly-A) is added to the 3′ end of Hn-RNA. Immediately the 5′ end of Hn-RNA starts disintegrating to release poly-A–mRNA. These poly-A–mRNA molecules diffuse out into the cytoplasm.

Ribosomal RNA (rRNA) It occurs in ribosomes, which are nucleoproteins. Inside the ribosomes of eukaryotic cells rRNA occurs in the form of particles of three different dimensions. These are designated as 28S, 18S and 5S. The 28S and 5S molecules occur in large subunit (60S subunit) of ribosome, whereas 18S molecule is present in the small subunit (40S subunit) of ribosomes.

Base composition of rRNA rRNA differs in base content from tRNA and mRNA. It is relatively rich in guanine and cytosine.

Biogenesis of rRNA rRNA although present in ribosomes, is formed inside the nucleus. DNA associated with the nucleolus is responsible for coding rRNA. This part of DNA is known as nucleolar organiser.

The two types of rRNA (28S and 18S) are transcribed from the nucleolar DNA as a single elongated unit of 45S. Inside the nucleolus, the 45S RNA is methylated and complexed with protein. Finally, via a number of steps it is cleaved into 32S and 18S segments. The 18S RNA gets associated with basic proteins to form the small subunit of ribosome. The 32S segment is further severed and finally changes to 28S RNA. This 28S RNA gets associated with proteins and forms the large subunit of ribosome. These units then come out into the cytoplasm.

on itself (Cloverleaf model), and the two arms are coiled over one another (Figure 9.13). Some of the bases of the two arms of tRNA molecule exhibit intramolecular base pairing. The 3′ end of the polynucleotide chains ends in CCA base sequence. This represents site for the attachment of activated amino acids. The end of the chain terminates with guanine base. The bend in the chain of each tRNA molecule contains a definite sequence of three nitrogenous bases, which constitute the anticodon. It recognises the codon on mRNA.

Four different regions or special sites can be recognised in the molecule of tRNA. These are:

i. **Amino acid attachment site**—It occurs at the 3′ end of the tRNA chains and has –OH group which combines with the specific amino acid in the presence of ATP forming aminoacyl-tRNA. It is common to all the tRNA molecules.

ii. **Recognition site**—It contains a specific base sequence, which dictates the attachment of correct amino acid to the tRNA molecule. It matches with the amino acid activating enzyme through which attachment of amino acid to tRNA takes place.

iii. **Anticodon or codon recognition site**—This site has three unpaired bases whose sequence iscomplementary with a codon (triplet) in mRNA. Therefore, it determines the pairing of tRNA with the specific codon (triplet) of mRNA. It is, therefore, the most specific regions of tRNA molecule.

iv. **Ribosome recognition site**—This helps in the attachment of tRNA to the ribosome. This site is common to all the molecules of tRNA. In addition to the usual bases of RNA (cytosine, guanine, adenine and uracil), each tRNA molecule consists of several unusual bases. Some of them are pseudouridine, inosinic acid, methyl guanine, methyl aminopurine, etc.

The presence of these rare nucleotides (unusual) does not affect the pairing of tRNA with mRNA. These probably prevent intermolecular base pairing in the open tRNA loop or help in the recognition of aminoacyl-tRNA synthetase enzyme.

tRNA molecules occur in both active and inactive forms. The inactive molecules of tRNA lack the CCA sequence of nitrogenous bases at 3′

and protein. This hypothesis proposes that (1) RNA molecule first catalysed their own replication and developed a number of enzyme activities. (2) In the next stage, RNA molecules began to synthesise proteins. These proteins became superior enzymes because their 20 side chains are more versatile than the four bases of RNA. (3) Finally, DNA was formed by reverse transcription of RNA. DNA replaced RNA as the genetic material because its double helix is a more stable and reliable store of genetic information than the single-stranded RNA. At this point, RNA retained only the functions of information carrier (mRNA) and adapter (tRNA) in protein synthesis. It also remained as components of ribosomes (rRNAs) and other assemblies involved in gene expression. Thus, the present mechanism of information transfer from gene to protein began when RNA alone was responsible for all activities of life in the world.

REVIEW QUESTIONS

1. Explain the molecular structure of DNA.
2. Explain the structure and functions of RNA.
3. Write short notes on:
   - i. Purines
   - ii. Pyrimidines
   - iii. Tautomerism
   - iv. Nucleotides
   - v. Nucleosides
   - vi. mRNA
   - vii. tRNA
   - viii. RNA world hypothesis

10

ENZYMES

Life is an intricate meshwork involving a perfect coordination of a vast number of chemical reactions. In living cells the enzymes catalyse these reactions. Kuhne (1878) coined the term 'enzymes' to designate the 'biological catalysts' that had previously been called ferments. The word enzyme means 'in yeast'. This was referred to denote one of the most noteworthy reactions wherein the production of ethyl alcohol and carbon dioxide through the agency of an enzyme, **zymase**, present in yeast takes place.

10.1 PROPERTIES OF ENZYMES

Enzymes are biocatalysts. They promote chemical reactions; whether they partake in the chemical reaction is not known. But enzymes retain their identity at the end of the reactions as in the beginning. All enzymes are proteins and exhibit all the properties of proteins. Enzymes are colloidal in nature and are soluble in water, salt solutions and glycerine. They are precipitated by protein precipitating agents such as alcohol, ammonium sulphate, and alkaloid reagents. Chemical analysis of purified crystalline enzymes gives a composition typical of proteins. They contain C, H, N and S. Extremely small quantities of enzymes are sufficient to bring about measurable reaction.

10.1.1 SPECIFICITY

One characteristic feature which distinguishes an enzyme from a catalyst is its specificity. Four kinds of specificity are described. They are:

Absolute specificity When enzymes catalyse only one particular reaction, they are said to exhibit absolute specificity. e.g. **urease** acts only on urea.

$$H_2N-\overset{\overset{O}{\|}}{C}-NH_2 + H_2O \xrightarrow{\text{Urease}} 2NH_3 + CO_2$$

Group specificity Some enzymes are capable of catalysing the reaction of a structurally related group of compounds. e.g. lactic dehydrogenase (LDH) catalyses the interconversion of pyruvic and lactic acid and also of a number of other structurally related compounds.

$$\underset{\text{Pyruvic acid}}{CH_3.CO.COOH} + NADH + H \xrightarrow{\text{Lactic dehydrogenase}} \underset{\text{Lactic acid}}{CH_3.CHOH.COOH} + NAD$$

Optical specificity (Stereo specificity) The most striking aspect of specificity of enzymes is that a particular enzyme will react with only one of the two optical isomers. **D-amino acid oxidase** oxidises the D-amino acids only to the corresponding keto acids.

$$\underset{\text{D-amino acid}}{H_2N-\overset{\overset{R}{|}}{\underset{\underset{COOH}{|}}{C}}-H} + H_2O \xrightarrow[\text{Oxidase}]{\text{D-amino acid}} \underset{\text{Keto acid}}{\overset{\overset{R}{|}}{\underset{\underset{COOH}{|}}{C}}=O} + NH_3 + H_2$$

Geometrical specificity Some enzymes exhibit specificity towards the cis and trans forms. e.g. **fumarase** catalyses the interconversion of fumaric acid and malic acid.

$$\underset{\text{Malic acid}}{\begin{matrix} H & & COOH \\ & C & \\ & \| & \\ & C & \\ H & & COOH \text{ 'cis'} \end{matrix}} \xrightarrow{\text{Fumarase}} \underset{\text{Fumaric acid}}{\begin{matrix} H & & COOH \\ & C & \\ & \| & \\ & C & \\ \text{'trans' } HOOC & & H \end{matrix}}$$

10.1.2 REVERSIBILITY

Reversibility of action is seen in certain enzymes. This can be illustrated by the following example. In the liver, excess of glucose in the bloodstream is converted to glycogen by an enzyme called **glycogenase**, and stored. When there is deficiency of glucose in the blood, the same enzyme is capable of reconverting insoluble glycogen into soluble glucose, which then enters the bloodstream.

$$\text{Glucose} \underset{}{\overset{\text{Glycogenase}}{\rightleftharpoons}} \text{Glycogen}$$

10.1.3 PROENZYMES

Often enzymes are produced in an inactive form called proenzymes or zymogen. For example, the gastric glands secrete inactive pepsinogen and this is converted into active pepsin after the addition of HCl. Trypsin is secreted in an inactive form called trypsinogen which is activated by enterokinase. The substances which activate the inactive enzyme are called activators. Activation usually involves cleavage of the peptide bond.

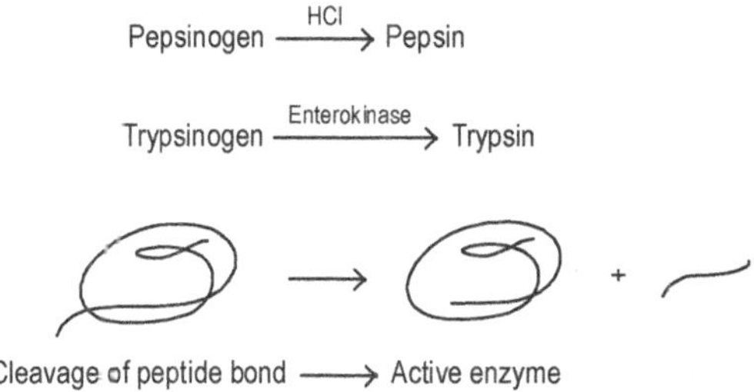

10.1.4 ANTI-ENZYMES

Most intestinal worms secrete anti-enzymes which guard the animal against the action of digestive enzymes of their host. e.g. anti-trypsin, anti-pepsin, anti-amylase.

10.1.5 FACTORS INFLUENCING ENZYME ACTION (ENZYME KINETICS)

The enzyme activity is influenced by temperature, pH, enzyme concentration and substrate concentration.

Effect of temperature The rate of enzyme action increases with increase of temperature. The temperature at which the enzyme action is at its maximum is called optimum temperature. After exceeding the optimum temperature, the enzymes, being protein in nature, get denatured; as a result, the rate of reaction decreases constantly and with further temperature rise, falls to zero (Figure10.1). The optimum temperature varies from enzyme to enzyme.

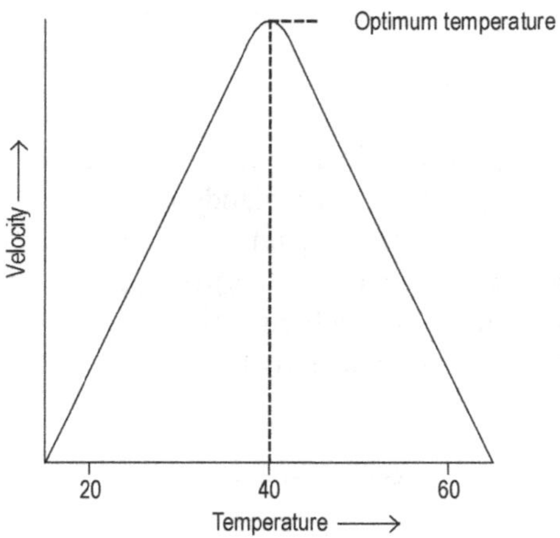

Fig. 10.1 Effect of temperature on the reaction rate of an enzyme-catalysed reaction

Effect of pH (hydrogen ion concentration) There is a particular range of pH over which the activity of an enzyme in general is maximum. The pH at which the enzyme activity is high is called optimum pH. Increase in H^+ ion concentration beyond this peak point will destroy the enzymes because of their protein nature, and bring down the enzyme action. The activity of the enzyme pepsin is high at acidic pH 1.2, whereas trypsin requires alkaline pH 8.2. It is also noteworthy that enzymes of the same type but of different origin may have different optimum pH. For example, the optimum pH for intestinal sucrase is 6.8, while yeast sucrase requires the optimum pH 4.5. Further, the same enzyme acting on different substrates may require different optimum pH. For example, the optimum pH required for pepsin to act on albumin is 1.4, while for the same enzyme to act on gelatin, the optimum pH required is 3.3. Excessive acidity or alkalinity causes the destruction of enzymes. The effect of pH on the activity of various enzymes is shown in Figure 10.2.

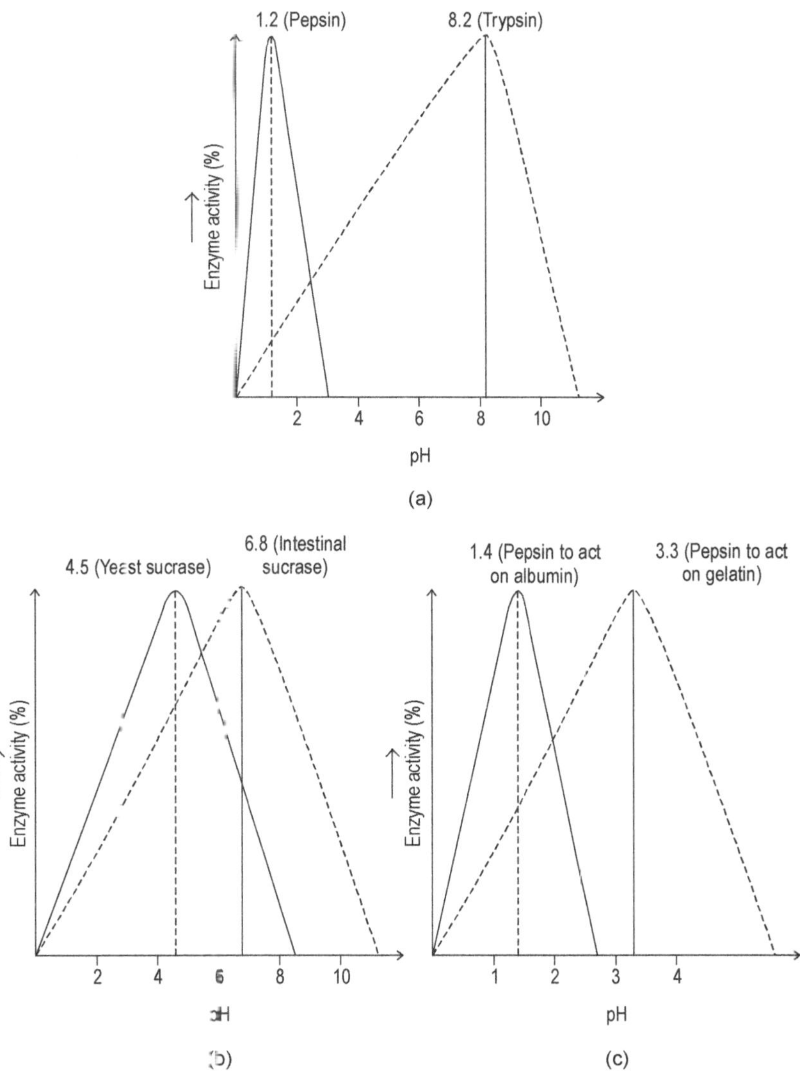

Fig. 10.2 (a) Effect of pH on the activity of pepsin and trypsin (b) Effect of pH on the activity of yeast sucrase and intestinal sucrase (c) Effect of pH on the activity of pepsin to act on albumin and to act on gelatin

Effect of enzyme concentration When other factors influencing the enzyme reaction remain constant, the velocity of enzyme action increases proportionately with the concentration of enzyme (Figure 10.3). The relationship between the enzyme concentration and velocity of reaction is of great practical importance, because it can be used to detect the strength of a particular enzyme in a mixture of proteins. This relation can be expressed as

$$v \propto E$$

velocity enzyme concentration

or

$$v = K \times E$$

(K= constant)

Therefore $E = v/K$

v can be expressed as $v = \dfrac{\text{amount of product}}{\text{time of reaction}}$

Therefore $E = \dfrac{\text{amount of product}}{\text{time of reaction} \times K}$

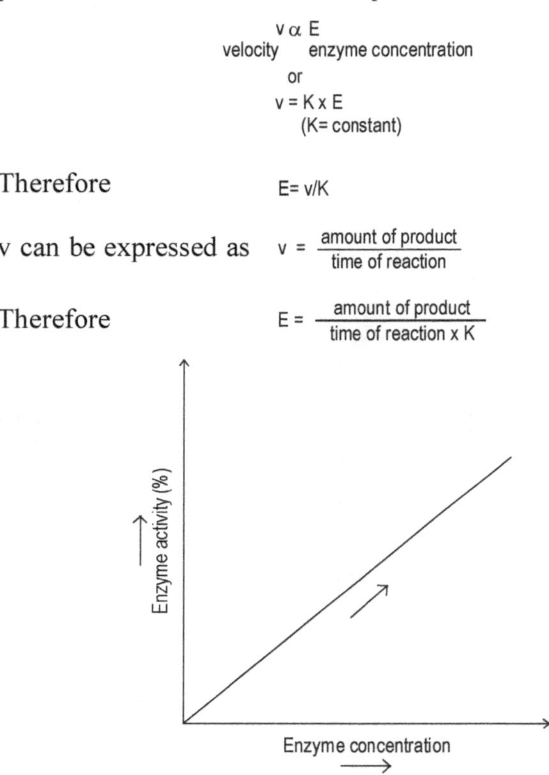

Fig. 10.3 Effect of enzyme concentration on the rate of an enzyme-catalysed reaction

Effect of substrate concentration Keeping the concentration of enzyme constant, the rate of reaction increases proportionately with the concentration of the substrate upto a point, beyond which increase in substrate concentration will not cause further increase in the rate of reaction. This is due to the increasing saturation of the active sites of the enzymes with the substrate molecules, which block further action.

The curve is first linear and later assumes a hyperbolic pattern (Figure 10.4).

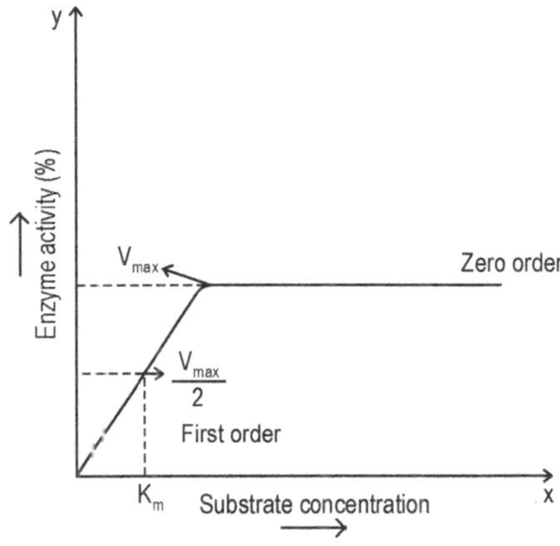

Fig. 10.4 Effect of substrate concentration on the rate of an enzyme-catalysed reaction

10.2 MECHANISM OF ENZYME ACTION

Enzyme forms an intermediate complex with the substrate. In 1880 Wurtz observed that after addition of the soluble proteinase papain to the insoluble protein fibrin, repeated washing of the fibrin did not stop the proteolysis. He then concluded that the papain had formed a substance or complex with the fibrin.

The compound on which the enzyme acts is called the substrate. Enzymes being proteins, have on their surface, reactive sites into which the substrates can fit. All enzymatic reactions involve a temporary union of the substrate with the enzymes. The substrate just fits on the reactive sites, to form the enzyme–substrate complex, which is also referred to as the Michaelis complex. The combination weakens the structure of the substrate and the bond in the substrate is strained to the point of rupture. When the bond ruptures, the reaction products are liberated into the solution and the original enzyme molecule is regenerated (Figure 10.5). The enzyme molecule can now catalyse fresh substrate molecules and repeat the process. The reaction can be written as follows:

$$\text{Enzyme} + \text{Substrate} \longrightarrow \text{Enzyme–substrate complex (Michaelis complex)}$$

$$\text{Enzyme substrate complex} \longrightarrow \text{Enzyme} + \text{Products}$$

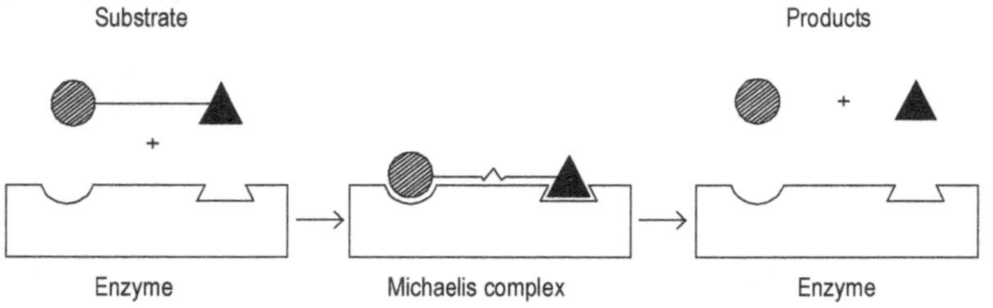

Fig. 10.5 Mechanism of enzyme action

10.2.1 LOCK AND KEY HYPOTHESIS

In 1898, Emil Fischer proposed the lock and key model to explain the mechanism of enzyme action. According to this model, the union between the substrate and the enzyme takes place at the active site more or less in a manner in which a key fits a lock and results in the formation of an enzyme–substrate complex as shown in figure 10.6.

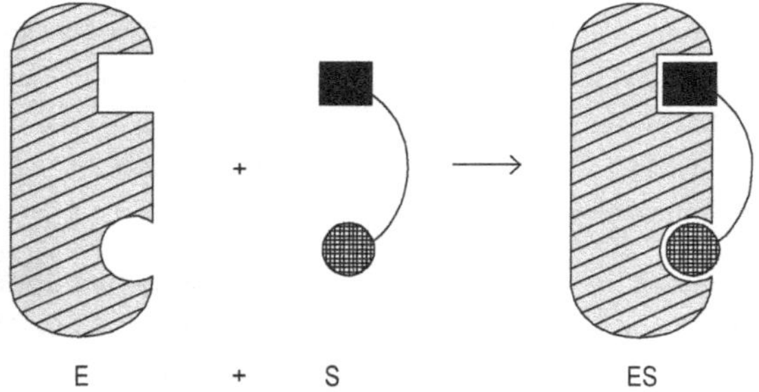

Fig. 10.6 Formation of enzyme–substrate complex

Active sites are usually represented by free hydroxyl groups of serine, sulphydryl groups of cysteine, imidazole groups of histidine or phenolic groups of tyrosine.

10.2.2 KOSHLAND'S INDUCED FIT MODEL

In order to explain the enzyme properties more efficiently, Koshland in 1958, modified the Fischer's model. An important feature of Fischer's model is the rigidity of the active site. Koshland presumed that the enzyme molecule does not retain its original shape and structure. But the contact of the substrate induces some configurational or geometrical changes in the active site of the enzyme molecule. Consequently, the enzyme molecule is made to completely fit the configuration and active centres of the substrate.

To explain the theory, a hypothetical illustration may be given (Figure 10.7). The hydrophobic and charged groups both are involved in substrate binding. A phosphoserine (–P) and the –SH group of cysteine residue are involved in catalysis. Other amino acid residues not involved in either substrate binding or catalysis are lysine (LYS) and methionine (MET). In the absence of substrate, the substrate binding and catalytic groups are far apart. But the proximity of the substrate induces a conformational change in the enzyme molecule aligning the groups for both substrate binding and catalysis. Simultaneously, the spatial orientation of other regions is also changed so that the lysine and methionine are now much closer.

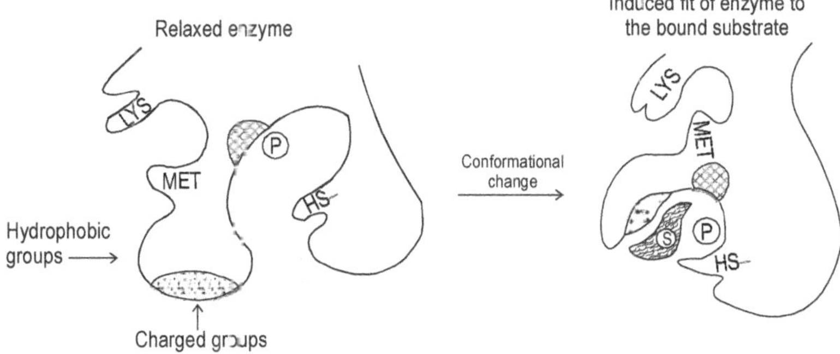

Fig. 10.7 Conformational changes brought about by induced fit in an enzyme molecule

As to the sequence of events during the conformational changes, three possibilities exist. The enzyme may first undergo a conformational change (A), then bind substrate (B). An alternative pathway is that the substrate may first be bound (C) and then a conformational change (D) may occur. Thirdly, both the processes may occur simultaneously (E) with further isomerisation (F) to the final conformation (Figure 10.8).

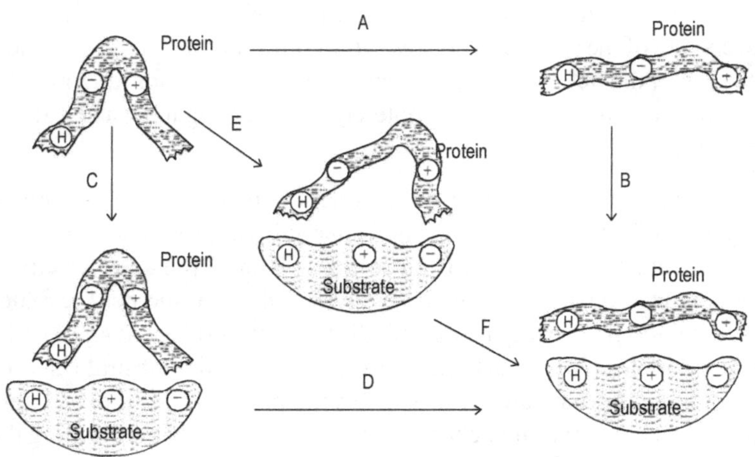

Fig. 10.8 Alternative pathways for a substrate-induced conformational change

The active sites exert a binding force on the substrate molecule both by hydrophilic and hydrophobic groups. In most of the cases, enzyme–substrate complex formation occurs by multiple bonding with the substrate. This may involve formation of covalent bonds, hydrogen bonds or electrostatic bonds.

10.3 MICHAELIS–MENTEN HYPOTHESIS

Michaelis and Menten, while studying the hydrolysis of sucrose catalysed by the enzyme invertase, proposed this theory. This theory is based on the following assumptions.

i. Only a single substrate is involved and a single product is formed.

ii. The process proceeds essentially to completion.

iii. The concentration of the substrate is much greater than that of the enzyme in the system.

iv. An intermediate enzyme–substrate complex is formed.

v. The rate of decomposition of the substrate is proportional to the concentration of the enzyme–substrate complex.

This theory postulates that the enzyme (E) forms a weakly bonded complex (ES) with the substrate (S). This enzyme–substrate complex, on hydrolysis, decomposes to yield the reaction product (P) and the free enzyme (E). These reactions may be symbolically represented as follows:

$$E + S \longrightarrow ES \longrightarrow E + P$$

The following symbols may be used for deriving Michaelis–Menten equation:

E_t - total concentration of enzyme

S - total concentration of substrate

ES - concentration of ES Complex

E_t–ES - concentration of free enzyme

The velocity of reaction is proportional to the concentration of the enzyme–substrate complex.

$$v = K \times (ES) \qquad (1)$$

The maximum reaction rate V_{max} will occur at a point where the total enzyme (E_t) is bound to the substrate. Then the maximum concentration of ES will be equal to the total enzyme concentration, E_t. Thus,

$$V_{max} = K \times (E_t) \qquad (2)$$

Dividing equation(1) by (2), we get

$$\frac{v}{V_{max}} = \frac{(ES)}{(E_t)} \qquad (3)$$

Equilibrium constant for dissociation of ES is K_m

$$K_m = \frac{(E_t - ES) \times S}{(ES)} \qquad (4)$$

$$K_m \times (ES) = (E_t - ES) \times (S)$$

$$K_m \times (ES) = (E_t) \times (S) - (ES) \times (S)$$

$$K_m \times (ES) + (ES) \times (S) = (E_t) \times (S)$$

$$ES\,(K_m + S) = (E_t) \times (S)$$

$$\frac{ES}{E_t} = \frac{S}{K_m + S} \qquad (5)$$

Substituting $\frac{v}{V_{max}}$ for $\frac{(ES)}{(E_t)}$ (as shown in the equation (3))

$$\frac{v}{V_{max}} = \frac{S}{K_m + S}$$

$$v = \frac{V_{max} \times (S)}{K_m + S} \qquad (6)$$

This equation is called Michaelis–Menten equation. This can be used to calculate K_m, after experimentally determining the velocity of reaction at various substrate concentrations. This equilibrium constant, K_m is usually called Michaelis constant. K_m is the Michaelis constant which represents the substrate concentration at half maximal velocity. The greater the concentration of ES complex, the lower is the concentration of free enzyme and consequently the lower is the value of K_m.

Three important derivations follow the hypothesis:

i. When the substrate concentration is so low as to produce half maximal velocity, i.e. K_m value, the velocity v depends upon the substrate concentration S and increases as S increases. This may be represented as $v \approx K(S)$.

   When $S < K_m$, $v = K(S)$.

ii. When the substrate concentration S is much greater than K_m, the velocity v also increases till it reaches the maximal velocity V_{max}. This may be represented as $v = V_{max}$.

   When $S > K_m$, $v = V_{max}$

iii. When the substrate concentration S is equal to K_m value, the observed velocity is equal to half the maximal velocity ($V_{max}/2$). This may be represented as $v = V_{max}/2$.

When $S = K_m$, $v = V_{max}/2$.

Since K_m value represents the substrate concentration when the velocity is half maximal, it gives an indication as to how much substrate should be used.

10.4 LINEWEAVER–BURK EQUATION

Taking the reciprocal of Michaelis–Menten equation, the following equation is obtained.

$$\frac{1}{v} = \frac{K_m + S}{V_{max} \times S}$$

$$\frac{1}{v} = \frac{K_m}{V_{max}} \times \frac{1}{S} + \frac{1}{V_{max}}$$

This is known as Lineweaver–Burk equation. This equation is of the form, $y = mx+b$; if one considers the variables to be $1/v$ and $1/S$ and plots the graph against these two variables, a straight line is obtained (Figure 10.9). The slope of this line corresponds to K_m/V_{max} and the $1/v$ intercept corresponds to $1/V_{max}$. Since V_{max} can be determined from the intercept, the K_m may also be calculated.

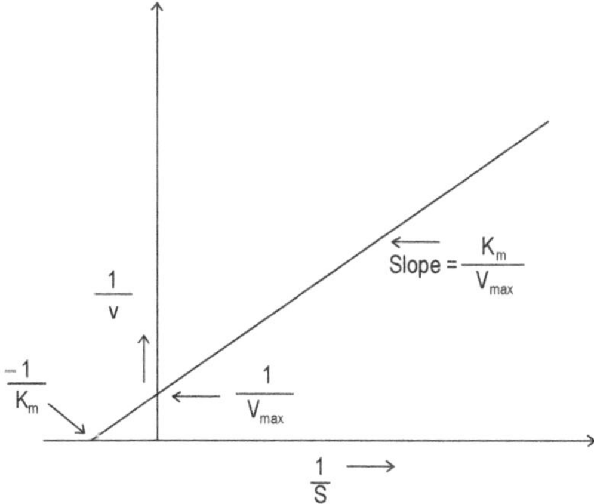

Fig. 10.9 Lineweaver–Burk plot showing determination of K_m value of an enzyme

10.5 COENZYMES

Many enzymes require the presence of certain non-protein compounds which help in accelerating the enzyme action. In contrast to enzymes, these are heat stable, are of smaller molecular weight and are therefore dialysable. These are called coenzymes or prosthetic groups. If the activator is firmly attached to the enzyme protein, it is called prosthetic group. Conjugated proteins are examples of this kind. On the other hand, if the non-protein compounds are not firmly attached to the enzyme protein, but exist in free state in the solution and contact the enzyme protein only at the instant of enzyme action, they are called coenzymes.

The entire enzyme system consisting of the enzyme protein and the coenzyme or prosthetic group is called the *holoenzyme* and the protein portion is sometimes called *apoenzyme* (Figure 10.10).

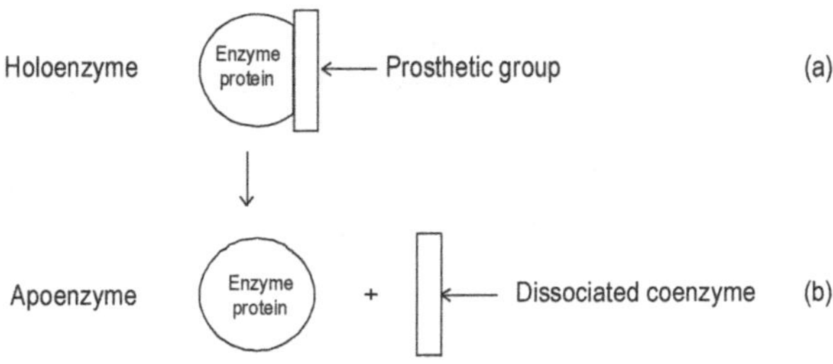

Fig. 10.10 (a) Holoenzyme (b) Apoenzyme and dissociated coenzyme

Coenzymes are necessary for enzyme action because they accelerate enzyme action. Most of them are derivatives of vitamin B-complex. Many of the coenzymes act as intermediate carriers of hydrogen atoms in the biological oxidation–reduction reactions.

10.5.1 MECHANISM OF COENZYME ACTION

Coenzymes accelerate the enzymatic reaction by rupturing the bond in the substrate and acting as acceptors for one of the cleavage products. The mechanism of coenzyme action is shown in Figure 10.11.

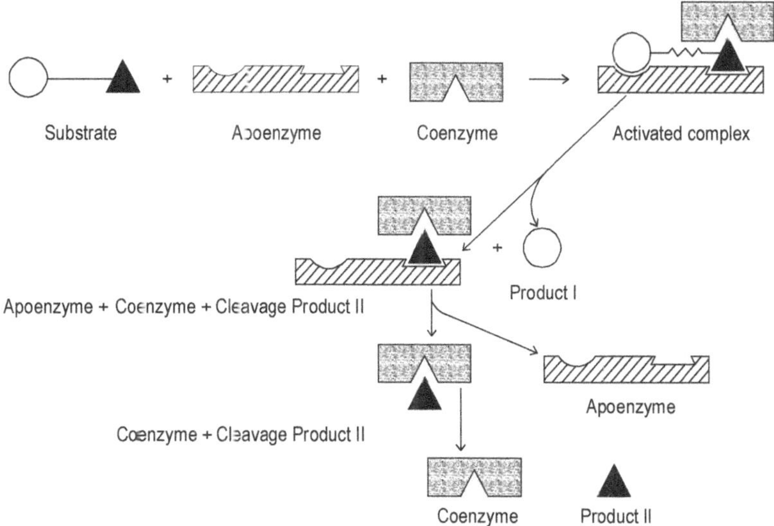

Fig. 10.11 Mechanism of coenzyme action

i. The substrate combines with the apoenzyme to form activated complex in the presence of the coenzyme.

ii. The bond in the substrate is strained and ruptures, when one of the cleavage products now dissociates from the apoenzyme.

iii. The other cleavage product now dissociates from the apoenzyme and the apoenzyme is now free for fresh reaction.

iv. The cleavage product attached to the coenzyme is next released from the surface of the coenzyme.

Now both apoenzyme and coenzyme are regenerated to their original forms and are ready for fresh enzyme and coenzyme action. In this way, one enzyme system can catalyse many substrate molecules. A prosthetic group also acts in a similar fashion, with the difference that the prosthetic group remains firmly attached to the surface of the apoenzyme. Some common coenzymes are given below:

i. Thiamine pyrophosphate (TPP) and lipothiamide pyrophosphate (LTPP), are derivatives of thiamine. They act as cocarboxylase in carbohydrate metabolism.

ii. Diphosphopyridine nucleotide (DPN) or NAD (Nicotinamide adenine dinucleotide) and nicotinamide adenine dinucleotide phosphate (NADP) are derivatives of nicotinic acid. They act as coenzymes of dehydrogenase and function as hydrogen acceptors and donors in biological oxidation–reduction reactions.

Ethyl alcohol → Acetaldehyde

$$H-\overset{H}{\underset{H}{C}}-\overset{H}{\underset{H}{C}}-OH \xrightarrow[\text{has NAD as coenzyme}]{\text{Alcohol dehydrogenase}} H-\overset{H}{\underset{H}{C}}-\overset{H}{C}-O + NADH + H^{+}$$

iii. Flavin mononucleotide (FMN) and flavin adenine dinucleotide (FAD) are derivatives of riboflavin and act as carriers of hydrogen atoms from one substrate to another in biological oxidation reaction.

iv. Pyridoxol phosphate is the derivative of pyridoxine. It is the coenzyme of decarboxylases and transaminases.

v. Biotin acts as a coenzyme for carboxylase. e.g. Carboxylation of acetyl CoA to malonyl CoA is the initial reaction involving synthesis of fatty acids.

vi. Coenzyme A is the derivative of pantothenic acid. It is the coenzyme for condensing enzymes and functions in the transfer of acyl groups in the fatty acid oxidation reactions.

vii. Tetrahydrofolic acid is the derivative of folic acid. It acts as coenzyme in reactions involving the transfer and utilisation of the formyl groups.

viii. Vitamin B_{12} is the cofactor in transmethylation reaction.

ix. Heme is the prosthetic group of catalases, peroxidases and cytochrome oxidase.

x. Uridine derivatives namely uridine diphosphate glucose (UDPG) and uridine diphosphate galactose (UDPGal) act as coenzymes in reactions involving epimerisation of galactose and glucose. Uridine diphosphate glucuronic acid (UDPgluc) serves as a source of active glucuronide in the formation of bilirubin glucuronide.

xi. ATP and related compounds act as coenzymes in transphosphorylation reactions.

xii. Specific inorganic ion activators are necessary for certain reactions. Mn, Mg, and Zn are necessary for the action of various peptide-splitting enzymes.

10.6 ISOENZYMES

Many enzymes occur in more than one molecular form in the same species, in the same tissue or even in the same cell. In such cases the different forms of the enzyme catalyse the same reaction but since they possess different kinetic properties and different amino acid compositions they can be separated by appropriate techniques such as electrophoresis. Such multiple forms of the enzymes are called isoenzymes or isozymes. Over a hundred enzymes are now known to be of isozymic nature and consequently occur in two or more molecular forms.

Lactic dehydrogenase (LDH) for example, is an enzyme which exists in five possible forms in various organs of most vertebrates. LDH catalyses the reversible oxidation–reduction reaction:

$$\text{Lactate} - \text{NAD}^+ \longrightarrow \text{Pyruvate} + \text{NADH} + \text{H}^+$$

Two basically different types of LDH are found:

i. **Heart LDH** This predominates in the heart and is active at low levels of pyruvate. This has four identical subunits called H subunits (H for heart).

ii. **Muscle LDH** This is characteristic of many skeletal muscles and maintains its activity in much higher concentrations of pyruvate. This also has four identical subunits called M subunits (M for muscle) which are enzymatically inactive.

The two types of subunits H and M, have the same molecular weight (35,000) but differ in amino acid composition and in immunological properties. The two subunits are produced by two separate genes. LDH can be formed from H and M subunits to yield a pure H tetramer and a pure M tetramer. Combinations of H and M subunits will, however, produce 3 additional types of hybrid enzymes, thus making the total number of possible forms as five (Figure 10.12). This is confirmed by the fact that when the two subunits are mixed in equal proportions, a sequence of 5 bands is obtained by electrophoresis. The various isozyme forms of LDH differ significantly in the maximum activities V_{max}, in the Michaelis constant K_m for their substrates, especially for pyruvate and in the degree of their allosteric inhibition by pyruvate.

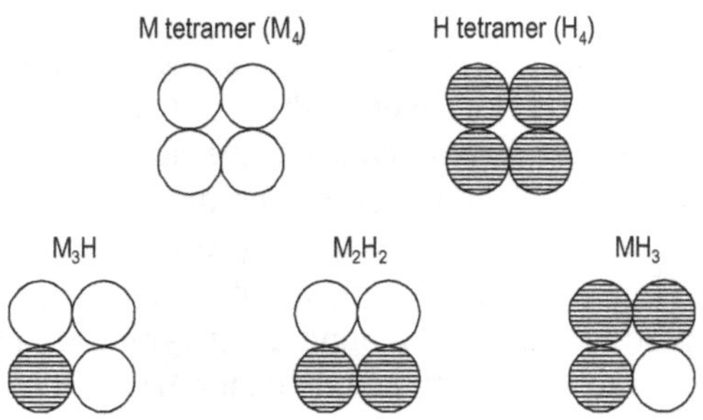

Fig. 10.12 Isoenzymes

Other examples of isozymes are malic dehydrogenase (MDH), hexokinase, esterase and glycol dehydrogenase.

10.7 ENZYME REGULATION

In each enzyme system there is at least one enzyme, the *pacemaker* that sets the rate of the overall sequence. Such pacemaker enzymes not only have a catalytic function but are also capable of increasing or decreasing their catalytic activity in response to certain signals. By the action of such pacemaker enzymes, the rate of each metabolic sequence is constantly adjusted to changes in the cell's demand for energy and for the building- block molecules required in cell growth and repair.

In cell metabolism, groups of enzymes work together in sequential chains or systems to carry out a given metabolic process, such as the conversion of glucose into lactic acid in skeletal muscle or the synthesis of an amino acid from simpler precursors. In such enzyme systems, the reaction product of the first enzyme becomes the substrate of the next and so on. Multienzyme systems may have as many as 15 or more enzymes acting in a specific sequence.

In most multienzyme systems, the first enzyme of the sequence is the pacemaker enzyme. The other enzymes in the sequence, which are usually present in amounts providing a large excess of catalytic activity, simply follow the pacemaker. They can promote their reactions only as fast as their substrates are made available from preceding steps.

Such pacemaker enzymes whose activity is modulated through various types of molecular signals are called regulatory enzymes. There are two major classes of regulatory enzymes: *allosteric* or *noncovalently regulated enzymes* and *covalently regulated enzymes*.

10.7.1 ALLOSTERIC ENZYMES

Allosteric enzymes are regulated by non-covalent binding of modulator molecules. They may be stimulated or activated by their modulators. The modulators that stimulate allosteric enzymes are called stimulatory or positive modulators, whereas, the compounds that inhibit enzymes are called enzyme inhibitors or negative modulators. Enzyme inhibition may be classified under three groups (i) competitive (ii) noncompetitive and (iii) allosteric.

Competitive inhibition Competitive inhibition may be defined as one in which the substrate as well as the inhibitor compete for the same site on the enzyme molecule. In such inhibitions, both enzyme-substrate (ES) and enzyme-inhibitor (EI) complexes are formed during the course of reaction (Figure 10.13). If both the substances are added to the enzyme at the same time, the quantity of ES or EI formed will depend upon the relative affinity of the enzyme for the substrate and inhibitor. If the affinity of the enzyme towards the substrate is greater, the amount of ES formed will be more. On the other hand, if the affinity of the enzyme towards the inhibitor is greater, the amount of EI formed will be more. Secondly, formation of ES and EI also depends on the relative concentration of the substrate and inhibitor. If the inhibitor concentration is kept constant and substrate concentration is increased gradually, it is observed that at low substrate concentrations, the amount of ES formed is smaller and EI is greater; at higher substrate concentration, ES formed is greater. Competitive inhibitions are reversible in nature.

Studies of the chemical structures of competitive inhibitors have revealed that the structure of the inhibitor very closely resembles the substrate structure. The following are some examples of competitive inhibitors resembling their substrate.

Substrate	**Competitive inhibitor**
Succinate	Malonate
P-aminobenzoic acid	Sulphanilamide
Thiamine	Pyrithiamide
Pyridoxine	Deoxy pyridoxine
L-Histidine	D-Histidine
NAD, NADP	ADP, ATP

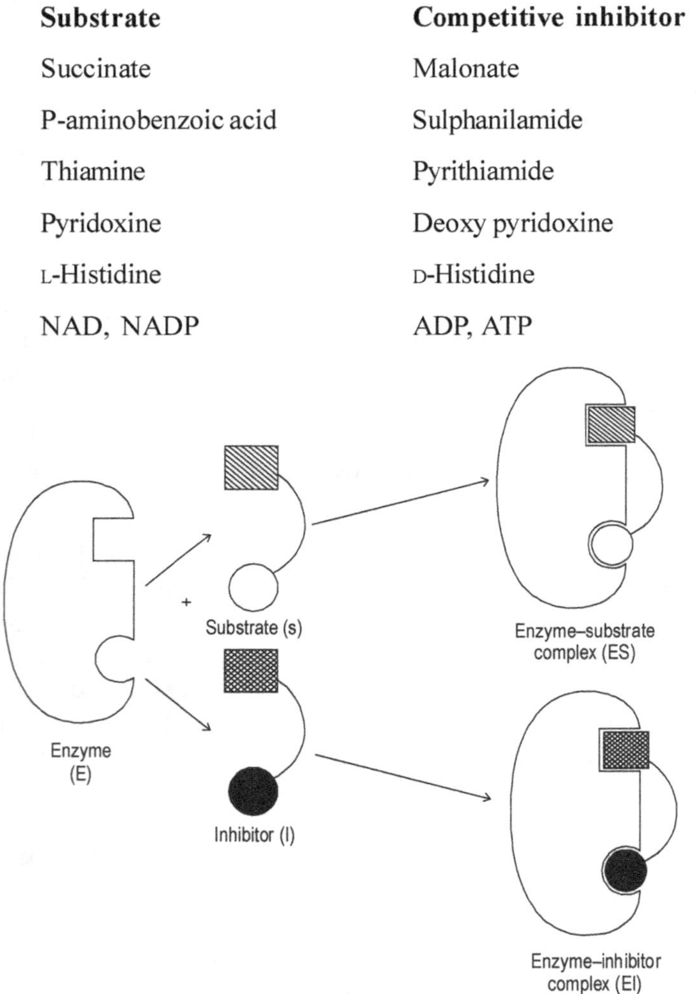

Fig. 10.13 Formation of enzyme–competitive inhibition substrate and enzyme–inhibitor complex

Noncompetitive Inhibition Noncompetitive inhibition is defined as one in which the substrate and inhibitor do not exhibit mutual competition. The sites of attachment of substrate and noncompetitive inhibitor are different on the enzyme molecules (Figure 10.14).

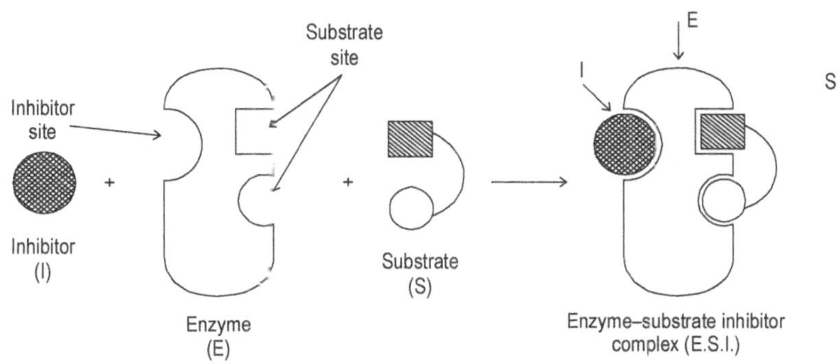

Fig. 10.14 Noncompetitive inhibition

Noncompetitive inhibitors are mostly structurally dissimilar to the substrate. If the inhibitor can be removed from its site of attachment preserving the activity of the enzyme, it is known as *reversible noncompetitive inhibition.* On the other hand, if it cannot be removed at all, or only removed after the loss of enzymic activity, it is known as *irreversible noncompetitive inhibition.* Both reversible and irreversible noncompetitive inhibitions have been found to exhibit similar reaction kinetics. In such inhibitions, the rate of combination of the substrate with the enzyme may not be altered but the rate of its dissociation is definitely reduced.

The double-reciprocal plot of enzyme rate data offers an easy way of determining whether an enzyme inhibitor is competitive or noncompetitive. Figure 10.15a shows a set of double-reciprocal plots obtained in the absence of inhibitor and with two different concentrations of a competitive inhibitor. Competitive inhibitors yield lines with a common intercept on the 1/v axis but with different slopes. This indicates that the V_{max} is unchanged by the presence of competitive inhibitor.

In noncompetitive inhibition similar plots of the rate data yield lines having a common intercept on the 1/s axis, indicating that K_m for the substrate is not altered by a noncompetitive inhibitor but V_{max} decreases (Figure 10.15b).

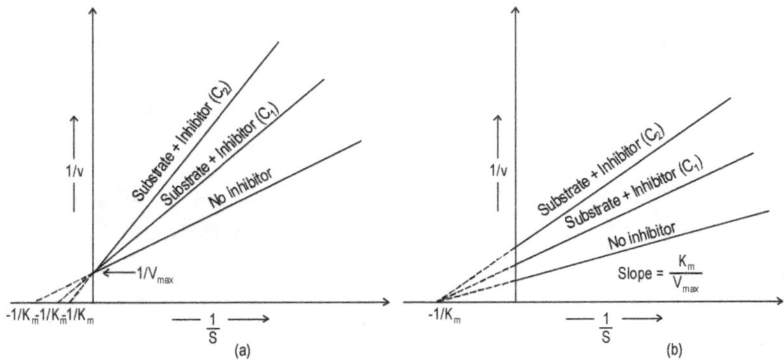

Fig. 10.15 Lineweaver–Burk plots for (a) Competitive inhibition (b) Noncompetitive inhibition

Allosteric Inhibition In mixed inhibitions, the inhibitor molecule attached to the enzyme prevents the breakdown of the enzyme–substrate complex (ES) and also at the same time, it interferes with the binding of the substrate to the enzyme. Thus, such inhibitors exhibit partially competitive and partially noncompetitive kinetics. The sites other than the active sites, lying on different regions in the enzyme molecule where an inhibitor molecule may get attached is known as allosteric site (Figure 10.16). An inhibitor molecule on this site affects

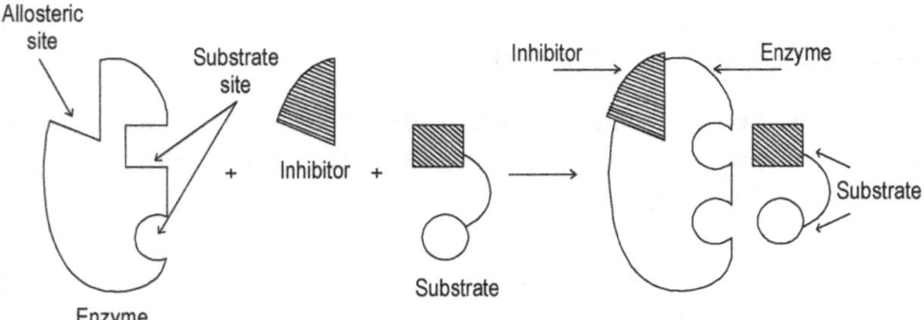

Fig. 10.16 Allosteric inhibition

the conformation around the active site. The conformational changes occurring on the active site affect the affinity of the active site towards substrate. It results in reduced enzyme–substrate complex formation, and reduced reaction rate. Similarly, if the substrate molecules have pre-occupied the active site, it may affect conformation around the allosteric site thereby reducing enzyme–inhibitor complex formation, and thus reduces the extent of inhibition.

End Product Inhibition (Feed-back inhibition) During end product inhibition, the end product molecule occupies the allosteric site on the enzyme, thus checking the reaction when the end product molecule dissociates from the allosteric site, the enzyme immediately returns to its unmodified level of activity.

10.7.2 COVALENT MODIFICATION

Another important class of regulatory enzymes is modulating through interconversion of their active and inactive forms by covalent modification of the enzyme molecule. An important example is the regulatory enzyme **glycogen phosphorylase** of muscle and liver, which catalyses the reaction.

$$(\text{Glucose})_n + \text{Phosphate} \longrightarrow (\text{Glucose})_{n-1} + \text{Glucose-1-phosphate}$$

The glucose -1- phosphate so formed can then be broken down to lactic acid in the muscle or free glucose in the liver. Glycogen phosphorylase occurs in two forms, the active phosphorylase *a* and the relatively inactive form phosphorylase *b*. Phosphorylase *a* has two polypeptide chain subunits, each with one specific serine residue in its sequence that is phosphorylated at its hydroxyl group. These serine phosphate residues are required for maximum activity of the enzyme. The phosphate groups can be hydrolytically removed from phosphorylase *a* by an enzyme called phosphorylase phosphatase.

$$\underset{\text{(more active)}}{\text{Phosphorylase } a + 2H_2O} \xrightarrow{\text{Phosphorylase phosphatase}} \underset{\text{(inactive)}}{\text{Phosphorylase } b + 2\text{ Pi}}$$

In this reaction, phosphorylase *a* is converted to phosphorylase *b* which is much less active than phosphorylase *a* in catalysing glycogen breakdown. Thus, the active form of glycogen phosphorylase is converted into the relatively inactive form by the cleavage of two covalent bonds between phosphoric acid and two specific serine residues in the enzyme.

Phosphorylase *b* can in turn be reactivated, i.e. covalently transformed back into active phosphorylase *a*, by another enzyme *phosphorylase kinase*, which catalyses transfer of phosphate groups from ATP to the hydroxyl groups of the specific serine residues in phosphorylase *b*.

$$\underset{\text{(inactive)}}{2\text{ATP} + \text{Phosphorylase}\,b} \xrightarrow{\text{Phosphorylase kinase}} \underset{\text{(active)}}{2\text{ADP} + \text{Phosphorylase}\,a}$$

Thus, the breakdown of glycogen in skeletal muscles and the liver is regulated through variations in the ratio of the active and inactive forms of the enzymes. These two forms differ in their quaternary structure and consequently change in catalytic action.

10.8 CONCEPT OF ENERGY ACTIVATION

Enzymes are biological catalysts. A catalyst is any organic or inorganic substance that accelerates the rate of chemical reaction by lowering its energy of activation. The greater the energy possessed by reactants, the less reactive they are. In other words, substances possess the energy of activation. Accordingly, reactants must first acquire enough energy to pass over the energy activation barrier, during which the reactants are placed into a reactive stage, which is favourable for the

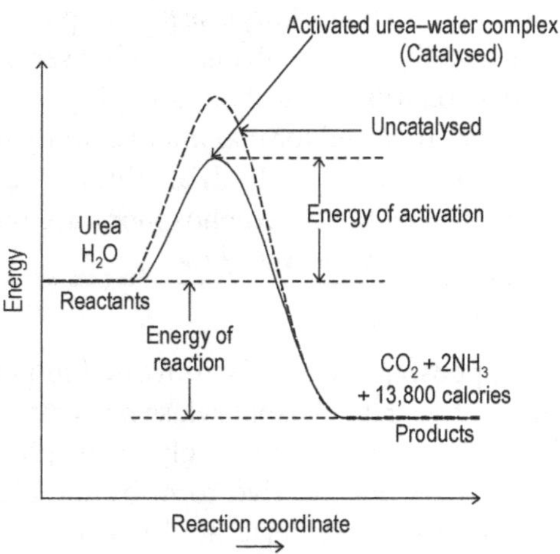

Fig. 10.17 Energy changes in catalysed and uncatalysed reactions

reaction. During hydrolysis of urea, in order to react, urea and water must have sufficient energy to form an activated complex. At room temperature very few molecules have sufficient energy to form activated complex, hence the rate of reaction is extremely slow even though the formation of products is energetically very favourable. Enzyme catalysis lowers down the energy of activation; as a result, the reaction proceeds even at low temperatures generally possessed by living systems (Figure 10.17). All enzymatic reactions are characterised by lower activation energies than are necessary for the corresponding non-enzymatic reactions.

10.9 CLASSIFICATION OF ENZYMES

Enzymes may be classified into six major groups. Each of the groups has many subgroups.

10.9.1 OXIDOREDUCTASES

They catalyse oxidation and reduction reactions between two substances. These enzymes are concerned with reactions of biological oxidation and involve transfer of hydrogen atoms or electrons from one substrate to another. e.g oxidases, dehydrogenases, peroxidases, and catalase.

$$\underset{\text{Alcohol}}{CH_3CH_2OH + NAD} \xrightarrow{\text{Alcohol dehydrogenase}} \underset{\text{Acetaldehyde}}{CH_3CHO} + NADH + H^+$$

10.9.2 TRANSFERASES

They bring about transfer of certain groups from one organic compound to another and include the following:

Transaminases They transfer amino groups of certain amino acids to certain ketoacids.

$$\underset{\text{Glutamic acid}}{\begin{array}{c} COOH \\ | \\ CH_2 \\ | \\ CH_2 \\ | \\ CHNH_2 \\ | \\ COOH \end{array}} + \underset{\text{Oxalacetic acid}}{\begin{array}{c} COOH \\ | \\ CH_2 \\ | \\ C{=}O \\ | \\ COOH \end{array}} \xrightarrow{\text{GOT}} \underset{\alpha \text{ keto glutaric acid}}{\begin{array}{c} COOH \\ | \\ CH_2 \\ | \\ CH_2 \\ | \\ C{=}O \\ | \\ COOH \end{array}} + \underset{\text{Aspartic acid}}{\begin{array}{c} COOH \\ | \\ CH_2 \\ | \\ CHNH_2 \\ | \\ COOH \end{array}}$$

e.g. Glutamate oxaloacetate transaminase (GOT) catalyses the transfer of NH_3 from glutamic acid to oxaloacetic acid, resulting in the formation of α keto glutaric acid and aspartic acid.

Transacylases They transfer acyl and acetyl groups.

Transpeptidases They transfer aminoacids or peptides from one peptide to another.

10.9.3 HYDROLASES

These enzymes catalyse hydrolysis of substrates having large molecular weight. They bring about hydrolysis, that is, they add water to the substrates and simultaneously break them. Some examples are:

i. *Carbohydrases* split carbohydrates.

ii. *Esterases* split ester linkages.

iii. *Proteases* split peptide linkages.

iv. *Cholinesterase* split acetylcholine

$$\underset{\text{Acetylcholine}}{CH_3CO-\text{choline}} + H_2O \xrightarrow{\text{Cholinesterase}} \underset{\text{Acetic acid}}{CH_3COOH} + \text{Choline}$$

10.9.4 LYASES

These enzymes catalyse the addition or removal of some chemical group of a substrate by mechanism other than hydrolysis, oxidation, or reduction. They differ from transferases in that the chemical group liberated is not merely transferred from one substrate to another but is liberated in the free state. Depending on the nature of the chemical groups added or removed they are further classified as:

Dehydratases These enzymes catalyse the reversible removal and addition of a molecule of water from or to their respective substrates. e.g.

i. Enolase

$$\underset{\text{2-phosphoglycerate}}{C_3H_6O_4-P} \underset{+H_2O}{\overset{-H_2O}{\rightleftharpoons}}_{\text{Enolase}} \underset{\text{Phosphoenol pyruvate}}{C_3H_4O_3-P}$$

ii. Fumarase

$$C_4H_4O_4 \xleftarrow[\text{Fumarase}]{-H_2O} C_4H_6O_5$$

$$\text{Fumaric acid} \xrightarrow[+H_2O]{} \text{Malic acid}$$

Desulphydrases These enzymes catalyse the removal of H_2S with pyridoxol phosphate as the coenzyme. e.g. Cysteine desulphydrase.

$$\underset{\text{Cysteine}}{HSCH_2-\overset{\overset{\Large H}{|}}{\underset{\underset{\Large NH_2}{|}}{C}}-COOH} \xrightarrow[+H_2O]{-H_2S} \underset{\text{Pyruvic acid}}{CH_3C{=}OCOOH} + \underset{\text{Ammonia}}{NH_3}$$

Decarboxylase This enzyme catalyses the removal of CO_2 from amino acids to form the corresponding amines and from keto acids to form aldehydes.

$$\underset{\text{Pyruvate}}{CH_3CO\ COOH} \xrightarrow{\text{Pyruvate carboxylase}} \underset{\text{Acetaldehyde}}{CH_3CHO + CO_2}$$

10.9.5 ISOMERASES

These are enzymes, which catalyse the interconversion of isomers. e.g. Triose phosphate isomerase.

$$\text{3-Phosphoglyceraldehyde} \xrightarrow{\text{Triose phosphate isomerase}} \text{Dihydroxyacetone phosphate}$$

10.9.6 LIGASES

These are enzymes which catalyse the linking together of two compounds, coupled with activation with ATP or GTP.

$$\text{Fatty acid} + ATP + CoA \longrightarrow \text{Acyl CoA} + AMP + pp$$

10.10 NOMENCLATURE

Enzymes are customarily named by adding the suffix 'ase' to the name of the substrate on which it acts:

Substrate	Enzyme
Protein	Protease
Lipid	Lipase
Sucrose	Sucrase
Maltose	Maltase
Lactose	Lactase

Some enzymes were given names which were not directly related to the substrate. e.g. pepsin, trypsin, chymotrypsin etc.

Another system of nomenclature of the enzymes was based upon the type of reaction catalysed by the enzyme. The suffix 'ase' is added to the type of reaction catalysed by the enzyme. e.g. hydrolases, isomerases, transaminases, etc. However, this system does not indicate the type of substrate acted upon.

In 1961, the International Union of Biochemists (IUB) has recommended a numerical system of classification whereby each enzyme can be identified by 4 numbers- e.g. 1.1.1.1.–Alcohol dehydrogenase. The first number indicating the number of the above 6 classes to which it belongs; the next number representing the subclass; the third figure indicating further subclass (sub-subclass); and the fourth number indicating the serial number of the enzyme in that sub-subclass. The method is cumbersome for routine use, but is very useful in pinpointing a single enzyme without ambiguity.

REVIEW QUESTIONS

1. Discuss the general properties of enzymes.
2. Discuss the mechanism of enzyme action with suitable illustration.
3. What are coenzymes? Discuss the mechanism of coenzyme action. Give examples.
4. Explain the classification of enzymes according to IUB system giving examples for each class.
5. Describe different types of enzyme inhibition.

6. Give an account of isoenzymes.
7. Discuss the theories of enzyme action.
8. Derive Michealis–Menten hypothesis.
9. Give an account of regulatory enzymes.
10. Write short notes on:
    i. isoenzymes
    ii. coenzymes
    iii. competitive inhibition
    iv. noncompetitive inhibition
    v. covalently regulated enzymes

11

HIGH-ENERGY COMPOUNDS

Some compounds like adenosine diphosphate (ADP) and adenosine triphosphate (ATP), are called 'high-energy' or 'energy-rich' compounds, or 'energy-carriers', because they exhibit a large decrease in 'free energy' when they undergo hydrolysis.

These compounds are unique in having one or more high-energy phosphate bonds. The formation of such bonds requires large input of energy. The bonds also release large amount of energy when they are broken down. Thus, the conversion of adenosine monophosphate (AMP) to ADP, and ADP to ATP requires not only additional phosphate groups but also large input of energy. The energy is derived in living cells from the various exergonic (energy-yielding) reactions. e.g. the oxidation of organic food materials. The respiration actually functions primarily to create high-energy bonds in ATP, which may be said to be the chief products of respiration.

$$\text{Organic food} \xrightarrow[\text{decomposition}]{\text{Exergonic}} \text{Simple molecules} + \text{Energy}$$

$$\text{ADP} + \text{Phosphate} + \text{Energy} \rightleftharpoons \text{ATP}$$

ATP is formed in the mitochondria of living cells. The energy in this ubiquitous fuel is used to 'drive' many energy-requiring processes of the cell. While acting as the donor of energy, ATP is split to ADP and phosphate with the release of large amount of usable energy. The estimated amount of energy available from the conversion of ATP into ADP and phosphate is about 9000 calories per mole of ATP (hence, free energy change or $\Delta F=9000$ cal/mole).

11.1 TYPES OF ENERGY-RICH COMPOUNDS

Energy-rich compounds include adenosine triphosphate, adenosine diphosphate, acetyl phosphate and 1,3 diphosphoglyceric acid. The enolic phosphate includes phosphoenolpyruvic acid. Guanidinium phosphates include creatine phosphate and arginine phosphate. The examples of thioesters are acetyl coenzyme A, S-adenosylmethionine and S-acetyl glutathione. The position of the high-energy bond in each of the compounds is given as follows:

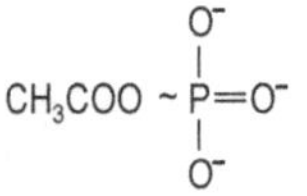

$CH_3COO \sim PO_3^{2-}$ ($CH_3COO \sim P(=O)(O^-)_2$)

Acetyl phosphate

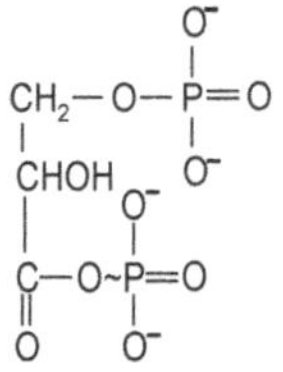

$CH_2{-}O{-}P(=O)(O^-)_2$ / $CHOH$ / $C(=O){-}O{\sim}P(=O)(O^-)_2$

1, 3-Diphosphoglyceric acid

$CH_2{=}C{-}COOH$ / $O \sim P(=O)(O^-)_2$

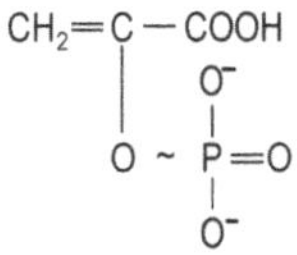

Phosphoenolypruvic acid

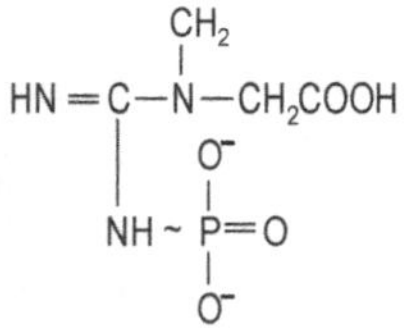

$HN{=}C{-}N(CH_3){-}CH_2COOH$ / $NH \sim P(=O)(O^-)_2$

Creatine phosphate

$CH_3CO \sim S.CoA$

Acetyl coenzyme A

$R.CH_2CO \sim S.CoA$

Acyl coenzyme A

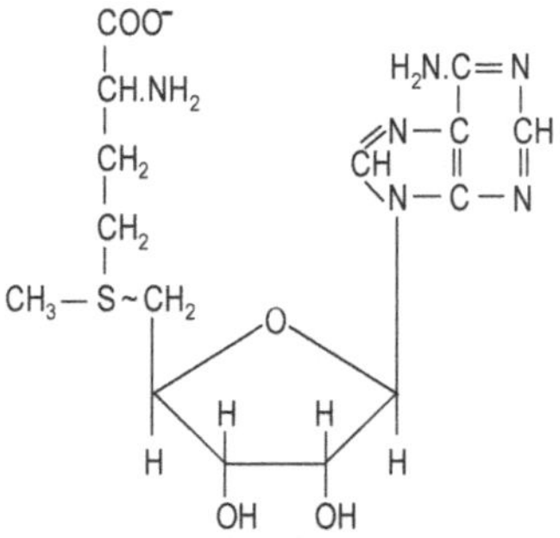

$COO^-{-}CH.NH_2{-}CH_2{-}CH_2{-}S(CH_3) \sim CH_2$–ribose (H, H, OH, OH)–adenine ($H_2N.C{=}N$, $N{-}C$, CH, $N{-}C{-}N$)

S-Adenosyl methionine

makes actomyosin supple and elastic and able to contract at will. When ATP disappears or separates from actomyosin, for example, in extreme fatigue or during *rigor mortis* after death, muscles become rigid and stiff.

During a contraction–extension cycle, the ATP of actomyosin–ATP complex yields its energy. To prepare a muscle for a new contraction-extension cycle, new energy must be supplied in the form of ATP. ATP is supplied to a contracting muscle by at least three important sources:

1. Phosphagens that make up about 0.5% of muscle and furnish a reserve of high-energy phosphates for ADP, when ADP is converted to ATP in muscular contraction.

$$\begin{matrix}\text{Phosphocreatine} \\ \text{or} \\ \text{Phosphoarginine}\end{matrix} + \text{ADP} \rightleftharpoons \begin{matrix}\text{Creatine} \\ \text{or} \\ \text{Arginine}\end{matrix} + \text{ATP}$$

2. Glycolysis or anaerobic oxidation of the muscle glycogen which is synthesised from blood glucose.
3. Aerobic oxidation of substrates in the muscle mitochondria.

Thus, it may be stated that under anaerobic conditions, when the oxygen supply to the muscle is not adequate, muscle can contract by using up its store of phosphocreatine. This happens during sustained exercise. This store is replenished during exercise by the glycolysis of glucose. In man and other vertebrates the phosphagen is creatine phosphate and in nonchordates, except echinodermata, the common phosphagen is phosphoarginine.

Non-muscular movements In all types of non-muscular movements, ATP appears to be the common source of energy for mechanical work. However, it is not clear how the chemical energy of ATP is translated into the mechanical energy of motion. The movements of various cell parts and the processes like active transport, pinocytosis, cyclosis, etc. are ATP-dependent. Among other intracellular movements are the precise movements of chromosomes during cell division. In addition to such intracellular movements, groups of cells and indeed whole tissues and organs undergo numerous types of motions associated with the growth and development. Thus, ATP energises all types of mechanical cell functions.

Synthesis of cellular components ATP is required for the formation of certain important cellular constituents like nucleic acids, proteins, polysaccharides and fats.

ATP is also necessary for the synthesis of enzymes, hormones and urea and for the synthesis of certain protective agents like irritants and poisons.

REVIEW QUESTIONS

1. Discuss the role of ATP in the biological systems.
2. Write notes on:
   i. High-energy compounds
   ii. ATP
   iii. Creatine phosphate

12

VITAMINS

Vitamins are organic compounds, which are needed in small quantities in the diet and are essential for the normal functioning of an organism. The protective substances present in milk were named 'accessory factors' by Hopkins. In 1911, Funk isolated a crystalline substance from rice polishing, which could prevent polyneuritis in pigeons. It was chemically an amine and was vital to life and hence he named it 'vitamine'. However, only a few vitamins have amines and hence the letter 'e' is dropped and the term vitamins was retained to designate the accessory food factors that are necessary for the health and growth of animals and whose absence lead to the deficiency diseases.

Vitamins are classified into two groups (a) Fat-soluble vitamins and (b) Water-soluble vitamins.

FAT-SOLUBLE VITAMINS

Fat-soluble vitamins include vitamins A, D, E and K.

12.1 VITAMIN A (RETINOL)

Vitamin A is derived from certain pigments called carotenoids, which are widely distributed in nature. They are known as provitamin A.

Provitamin A include four compounds: α-carotene, β-carotene, γ-carotene and cryptoxanthine.

Chemistry and structure The vitamin has a characteristic ring structure called the **β-ionone ring** attached to a long hydrocarbon side chain ending in an alcohol group and is chemically known as 'retinol'. Its aldehyde form is known as 'retinal'. The provitamins, carotenes have one or more beta ionone rings connected by a long hydrocarbon chain. In case of β-carotene, the two rings at either end are β-ionone rings and it is capable of forming two vitamin A molecules (Figure 12.1).

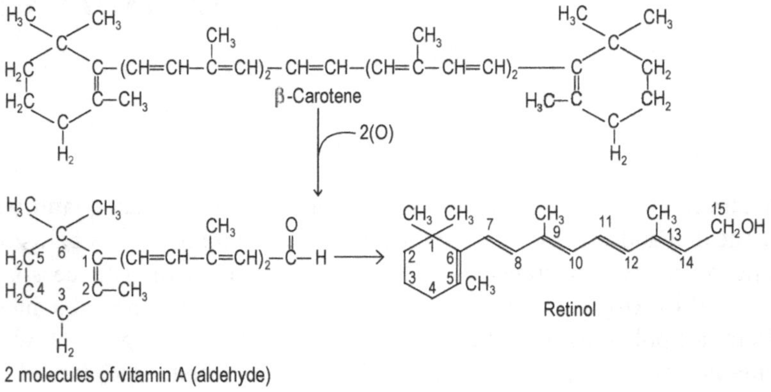

$$H_3C\text{–}C(CH_3)\ \text{ring}\text{–}C\text{–}(CH{=}CH\text{–}\overset{CH_3}{C}{=}CH)_2\text{–}CH{=}CH\text{–}(CH{=}\overset{CH_3}{C}\text{–}CH{=}CH)_2\text{–}C\ \text{ring}$$

β-Carotene

↓ 2(O)

$$\text{ring}\text{–}C\text{–}(CH{=}CH\text{–}\overset{CH_3}{C}{=}CH)_2\text{–}\overset{O}{\overset{\|}{C}}\text{–}H \longrightarrow \text{Retinol} \ (\ldots CH_2OH)$$

2 molecules of vitamin A (aldehyde)

Retinol

Fig. 12.1 β-carotene, retinal and retinol

α and γ carotenes possess only one β-ionone ring and hence give rise to only one vitamin A molecule. The vitamin derived as above is called A_1 and is the most predominant form. It is the vitamin present in marine fish liver. Fresh water fish liver contain vitamin A_2, which has an additional double bond in the β-ionone ring.

Sources Vitamin A occurs only in animal tissues but its precursors, carotenoid pigments, occur extensively in the vegetables. Cod liver oil and other fish liver oils, animal liver, milk and milk products and eggs contain vitamin A. The carotenoid pigments are present in carrots, sweet potato, green vegetables like spinach and amaranth.

Deficiency diseases Retinol is necessary for growth and reproduction. Retinol plays an important role in physiological mechanism of vision. It plays an important role in the construction of normal bone. Deficiency causes defective endochordal bone formation and decreased osteoblastic activity.

Vitamin A is necessary for the functional integrity of the epithelial cells of the skin and the mucous membranes. Deficiency results in keratinisation of the cells. The skin becomes dry, scaly and rough, resulting in the formation of follicular hyperkeratosis. Vitamin A deficiency causes a widespread atrophy of the mucous membrane of living cells leading to keratinisation. This can occur in the mucous membrane of the eyes, lacrimal glands, respiratory tract, gastrointestinal tract, and genitourinary tract. The conjunctiva is mostly affected. Keratinisation of conjunctiva and lacrimal glands leads to dryness of the eyes, photophobia, and follicular conjunctivitis. This condition is called **xerophthalmia**. In severe cases, the corneal epithelium becomes inflamed resulting in **keratomalacia**, which may lead ultimately to blindness due to secondary infection and perforation of the cornea. The mucous membrane of nose, nasal sinuses, pharynx and tracheobronchial tract may get keratinised resulting in impairment of the local resistance to infection.

Vitamin A has a specific role to play in the physiological mechanisms involved in vision. An early sign of vitamin A deficiency in man is inability to see objects in dim light, especially after exposure to high light. This condition is called **night blindness** or **nyctalopia**. The mechanism, which concerns the active participation of vitamin A to maintain normal vision, is described by Wald as a cyclic phenomenon, which involves the constant splitting and resynthesis of light sensitive vitamin A containing pigment, rhodopsin present in the retina. This cyclic mechanism is known as **Wald's visual cycle** (Figure 12.2).

There are two types of receptor cells in the retina. These are the cones and rods. Cones are specialised for colour and the vision in bright light and rods are specialised for dim light vision. When light rays fall on these receptors, chemical changes take place. Vitamin A plays an important role in the photochemical process. The rod cells contain a photosensitive pigment called rhodopsin or visual purple which is a conjugated protein containing opsin a protein, and retinene the prosthetic group. Similarly, the cones contain a photosensitive pigment iodopsin, which is also called visual violet. This is also a conjugated protein containing protein photopsin and retinene. When light strikes the retina, rhodopsin is converted to unstable orange coloured products lumirhodopsin and metarhodopsin. Finally, it is split to form a yellow coloured mixture of opsin and trans-retinene termed retinal which is inactive vitamin aldehyde. Trans-retinene (visual yellow) is reduced to trans-vitamin A by retinene reductase and NADH. Trans vitamin A, which is inactive, passes into the blood and is carried to the liver, where it is converted to its cis isomer. The cis-vitamin A thus formed in the liver enters circulation again and reaches the retina, where it is oxidised to cis-retinene by the reverse action of retinene reductase (oxidation) and NAD. Cis-retinene couples with opsin to form the active visual purple rhodopsin in the retina. Resynthesis of rhodopsin occurs in dim light or darkness. If there is a deficiency of vitamin A, the rate of resynthesis is delayed and the individual continues to have difficulty in seeing objects in darkness. This is the defect in night blindness (nyctalopia) and is the result of deficient intake of vitamin A.

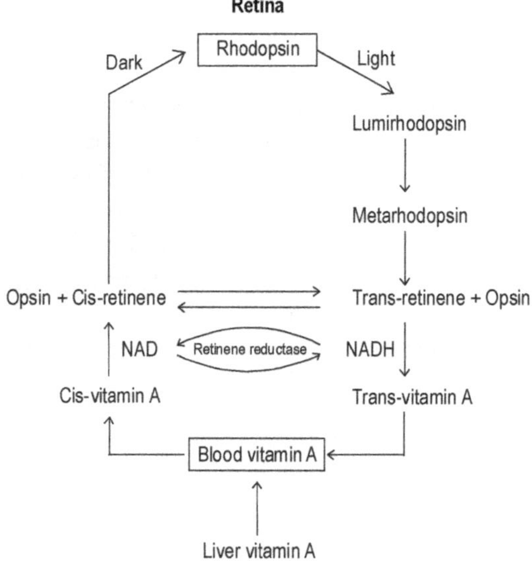

Fig. 12.2 Wald's visual cycle

12.2 VITAMIN D

Vitamin D is known as **antirachitic vitamin**. The active vitamin D was identified to be calciferol in the irradiated food.

Chemistry and structure The naturally occurring provitamins are inactive. They are sterols possessing similar structural characteristics. They have the same empirical formula and are isomers.

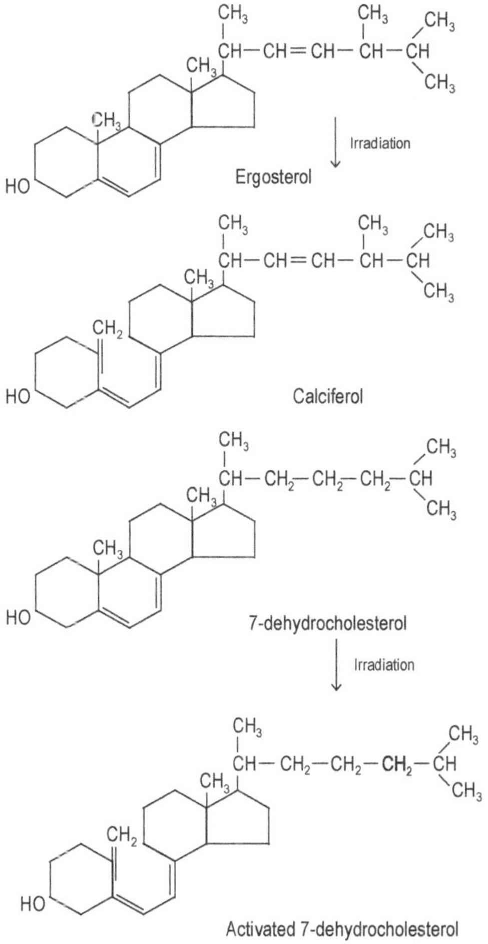

Ergosterol

Irradiation

Calciferol

7-dehydrocholesterol

Irradiation

Activated 7-dehydrocholesterol

Fig. 12.3 Structure of vitamin D

The two major types of vitamin D are (1) calciferol or vitamin D_2 obtained by irradiating the plant sterol, ergosterol and (2) vitamin D_3 (activated vitamin D_3) formed by irradiating the animal sterol, 7-dehydrocholesterol (Figure 12.3).

Sources The rich sources of vitamin D are cod liver oil and other fish liver oils. Other important sources are egg yolk and animal liver. Very little vitamin D is present in the milk. It can be synthesised in the skin itself from 7-dehydrocholesterol by solar irradiation.

Deficiency diseases Vitamin D is known as antirachitic vitamin and its prolonged deficiency leads to **rickets**. In the growing children, the disease is associated with bow-legs and enlarged joints. Another disease associated with vitamin D deficiency is **osteomalacia**. In this disease, the bones become softer than rickets, and the Ca : P ratio is changed and calcium loss is increased. The softness of the bones results in deformities.

12.3 VITAMIN E (TOCOPHEROL)

Vitamin E is chemically known as **tocopherol**. It possesses antisterility property.

Chemistry and structure The structure of vitamin E depicts a tocol nucleus; α, β, γ and δ tocopherols are naturally occurring tocopherols (Figure 12.4). They differ from one another in their number and position of methyl groups in the first ring of tocol nucleus. α tocopherol is most active.

Sources Tocopherols occur abundantly in plants. All green plants, lettuce and alfalfa are rich sources of the vitamin. Vegetable oils like wheat germ oil and seed germ oil are particularly rich sources. Milk, egg and meat are good sources.

Deficiency diseases Vitamin E has specific function based on its antioxidant activity. Vitamin E has specific effect on selenium metabolism. Vitamin E protects cell membranes particularly their unsaturated fatty acid constituents from oxidative damage against peroxidation. Tocopherolactone, a metabolite of tocopherol is necessary for the biosynthesis of coenzymes, which is a component in electron

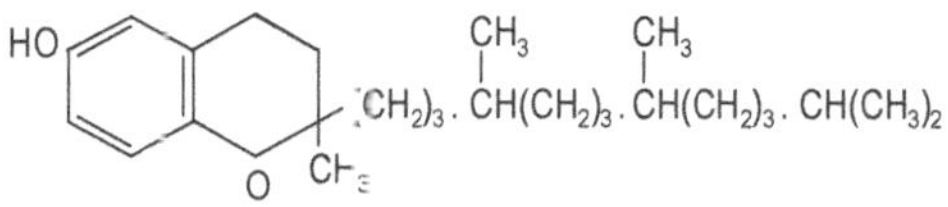

Tocol nucleus

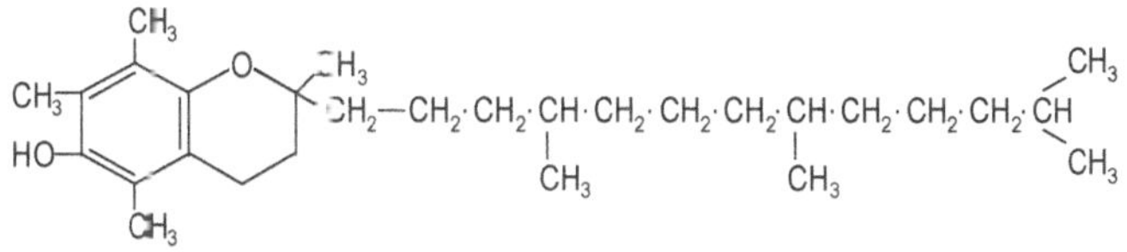

α-tocopherol

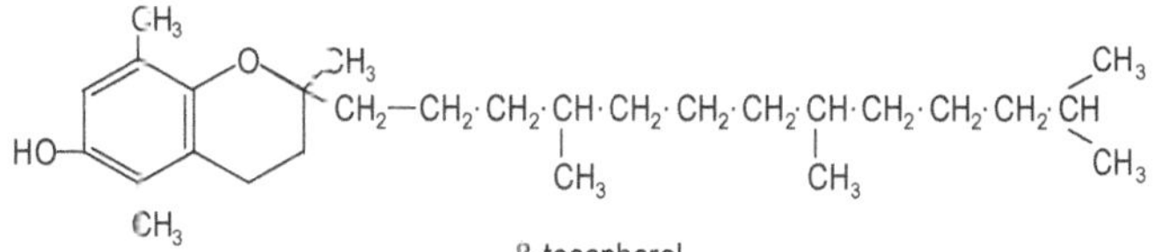

β-tocopherol

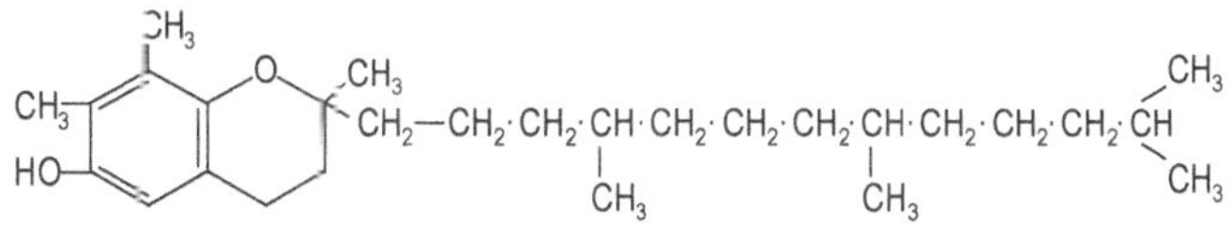

γ-tocopherol

Fig. 12.4 Structure of Vitamin E

transport chain. Deficiency in female rats causes intrauterine death of the foetus and in male rats degeneration of the gereminal epithelium, resulting in complete sterility. Deficiency of vitamin E may also result in muscular dystrophy characterised by degenerative changes in skeletal muscles leading to necrosis, inflammation and fibrosis accompanied by weakness or paralysis.

12.4 VITAMIN K

Vitamin K is commonly known as **antihaemorrhage factor** and chemically it is naphthoquinone. The important vitamins of this group are vitamins K_1, K_2 and menadione. Vitamin K_1 has been identified to be 2-methyl-3-phytyl-1, 4-naphthoquinone. Vitamin K_2 has been obtained from *Bacillus brevis* and has been identified to be 2-methyl-3-difarnesyl-1, 4- naphthoquinone (Figure 12.5).

Sources Food sources are principally derived from plants. Green leafy vegetables, cabbage, tomatoes and oat shoots are rich sources of vitamin K. Vitamin K is the product of metabolism of most bacteria and present in large amounts in putrified fish.

Deficiency diseases Vitamin K promotes the biosynthesis of prothrombin in the liver tissue. Vitamin K facilitates the process of blood coagulation. Therefore, deficiency of this vitamin results in delay in coagulation of blood and blood flows profusely even from very minor wounds.

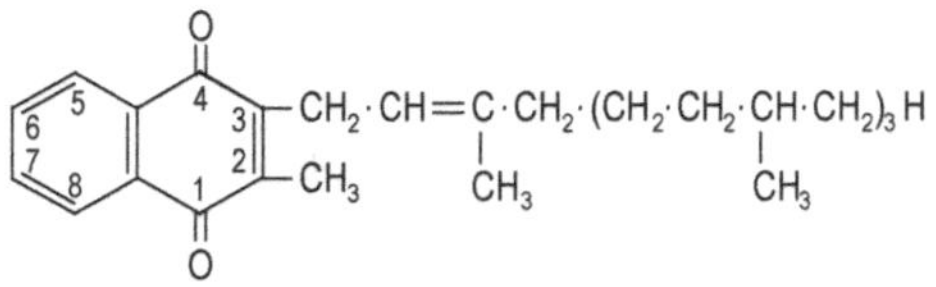

Vitamin K_1 (2-methyl-3-phytyl-1, 4-naphthoquinone)

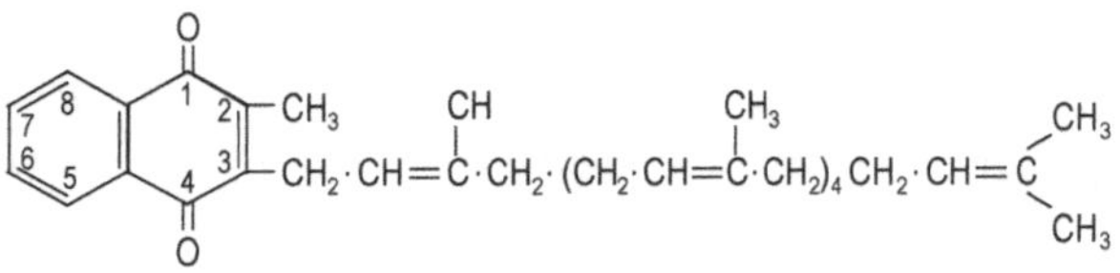

Vitamin K_2 (2-methyl-3-difarnesyl, 4-naphthoquinone)

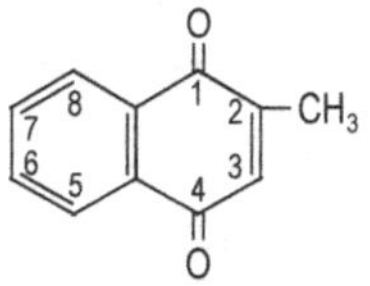

Menadione (2-methyl-1, 4-naphthoquinone)

Fig. 12.5 Structure of vitamin K

WATER-SOLUBLE VITAMINS

The water soluble vitamins include vitamin B-complex and vitamin C.

12.5 VITAMIN B-COMPLEX

Vitamin B-complex comprises a large number of compounds which form coenzymes or prosthetic groups of different enzymes. Structurally and functionally they differ from each other. Most of them are synthesised by the microorganism of the intestine.

12.5.1 THIAMINE

This is also known as vitamin B_1 or anti-beriberi factor, aneurin or anti-neuretic factor.

Chemistry and structure Thiamine possesses a pyrimidine ring attached to a thiazole nucleus (Figure 12.6). Intestinal microorganisms can synthesise this vitamin, whereas human body is unable to do so.

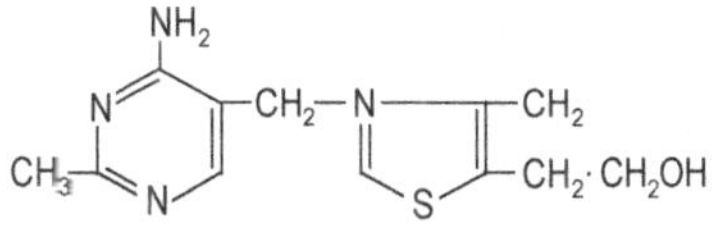

Fig. 12.6 Thiamine

Sources Thiamine is found to be present in whole grains, legumes, nuts, yeast, eggs, fish, pork, beef, liver, heart, kidney and many other vegetables. Milk and fresh fruits also contain appreciable amounts of thiamine.

Deficiency diseases In birds, polyneuritis develops and they become unable to fly or even to stand. Rats develop bradycardia. Polyneuritis is believed to develop due to raised pyruvic acid levels.

In man, deficiency of thiamine causes a neurological disease, **beriberi**. In this disease there occurs polyneuritis, muscular atrophy, cardiovascular changes and oedema. Other symptoms of the disease are weakness, loss of appetite, headache, dizziness and insomnia. Beriberi in infants results when they are restricted only to milk diet. The symptoms include constipation, weakness, oedema, and enlargement of heart, cyanosis and irregular pulse.

12.5.2 RIBOFLAVIN

Riboflavin is also known as vitamin B_2.

Chemistry and structure Riboflavin is chemically 6,7-dimethyl-9-D-1-ribityl-isoalloxazine. It contains a sugar alcohol, ribitol, which is derived from ribose sugar. The attachment occurs between nitrogen at 9′ position and carbon at 1′ position (Figure 12.7).

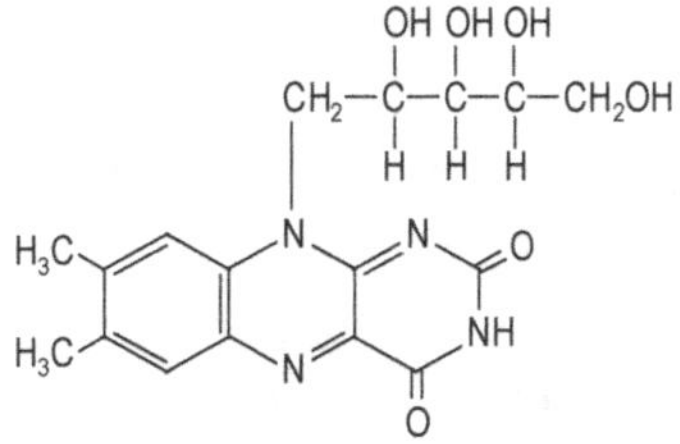

Fig. 12.7 Riboflavin

Sources Riboflavin occurs widely in nature and the most important source is milk. Besides milk, the other important sources are yeast, egg, liver, leafy vegetables, whole grains and nuts.

Deficiency diseases Riboflavin is involved in the metabolism of proteins, fats, carbohydrates and nucleic acids by forming a part of the flavoproteins. Deficiency of this vitamin retards growth. Characteristic lesions appear in the mouth region and face is affected. Lips appear red and shiny with characteristic painful fissures at the angles of the mouth called **cheilosis**. Painful **glossitis** with swollen and abnormally red tongue having flattened papillae occurs in some cases. Eyes are affected in case of severe deficiencies leading to photophobia, opacity and ulceration of cornea and sometimes cataract formation.

12.5.3 NIACIN (NICOTINIC ACID)

This vitamin is also known as anti-pellagra factor.

Chemistry and structure Niacinamide or nicotinamide is the acid amide of nicotinic acid. It derives its name from nicotine from which it can be prepared by oxidation. Nicotinic acid contains a pyridine nucleus, possessing a carboxylic group at third position (Figure 12.8a). In nicotinamide, the carboxylic group is replaced by amide group (Figure 12.8b). It forms two phosphorylated compounds, nicotinamide adenine dinucleotide (NAD) and nicotinamide adenine dinucleotide phosphate (NADP), which function as coenzymes for hydrogen transfer enzymes (dehydrogenases) in metabolic reactions.

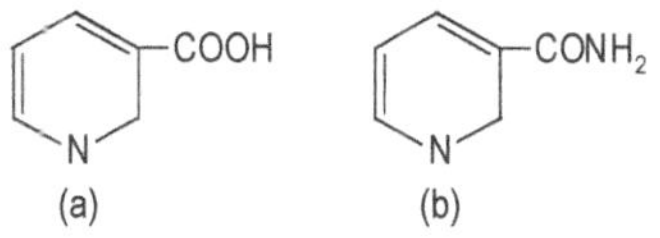

Fig. 12.8 (a) Nicotinic acid (b) Nicotinamide

Sources Yeast, liver, milk, tomatoes and green leafy vegetables are good sources of niacin. Wheat and unpolished rice also contain this vitamin in considerable quantities.

Deficiency diseases Niacin in the form of NAD and NADP plays an important role in the metabolic reactions. Deficiency of niacin causes **pellagra**. Administration of niacin cures the condition. Hence, it is also called as pellagra-preventing factor or anti-pellagra factor. Symptoms include dermatitis, diarrhoea and dementia, and these are exhibited as skin, gastrointestinal and cerebral manifestations.

12.5.4 PYRIDOXINE (VITAMIN B_6)

Pyridoxine refers to a group of naturally occurring pyridine derivatives.

Chemistry and structure Pyridoxine is a derivative of pyridine. The naturally occurring pyridoxine includes pyridoxol, pyridoxal and pyridoxamine (Figure 12.9 a, b and c). Pyridoxine is highly soluble in water. It is destroyed by exposure to light but it is resistant to heat. The primary alcoholic group of pyridoxine is oxidised to aldehydic group in pyridoxal and to methylamino group in pyridoxamine. Pyridoxine, pyridoxal as well as pyridoxamine exist in the tissues as pyridoxine phosphate, pyridoxal phosphate and pyridoxamine phosphate.

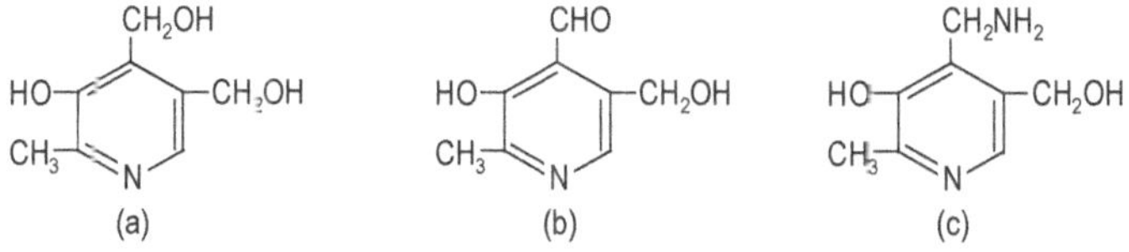

Fig. 12.9 (a) Pyridoxine (b) Pyridoxal (c) Pyridoxamine

Sources The rich sources of pyridoxine are egg yolk, meat, fish, milk, yeast, and germinating cereal grains. High concentration occurs in royal jelly.

Deficiency diseases Prolonged deficiency leads to fall in hemoglobin content, mental depression and confusion. It also causes vomitting, diarrhoea, abdominal distension and convulsions.

12.5.5 PANTOTHENIC ACID

Pantothenic acid is essential for nutrition.

Chemistry and structure The structure consists of an alanine chain in peptide linkage with a dihydroxy-dimethyl butyric acid (Figure 12.10). It is viscous and soluble in water. It is heat labile. The free acid is destroyed by an acid or alkali. The calcium and sodium salts of pantothenic acid are more stable.

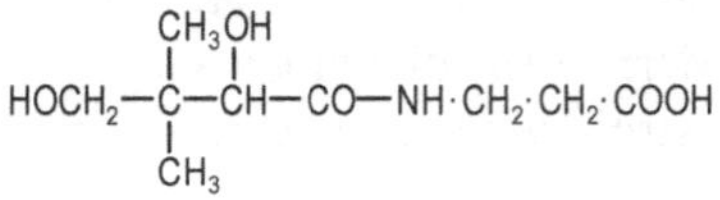

$$HOCH_2-\overset{\overset{CH_3}{|}}{\underset{\underset{CH_3}{|}}{C}}-\overset{\overset{OH}{|}}{CH}-CO-NH\cdot CH_2\cdot CH_2\cdot COOH$$

Fig. 12.10 Pantothenic acid

Sources The important sources of pantothenic acid are liver, kidney, eggs and milk. Among the plant sources are molasses, peas, cabbage, sweet potatoes, yeast, potatoes and tomatoes.

Deficiency diseases The metabolic functions of pantothenic acid are due to its co-enzyme derivative CoA. CoA gains further importance after its conversion to acetyl CoA. In man, there are no deficiency manifestations; in animals, deficiency causes changes in skin leading to dermatitis and alopecia (loss of hair), degenerative changes in the nervous system, gastrointestinal disorders and fatty livers.

12.5.6 LIPOIC ACID

Lipoic acid is also known as α lipoic acid, pyruvate oxidation factor (POF) or protogen. It is a crystalline, insoluble substance and has been identified to be 6,8 dimercapto-n-caprylic acid (Figure 12.11).

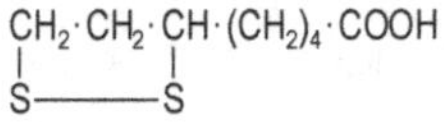

$$\underset{\underset{S}{|}}{CH_2}\cdot CH_2\cdot \underset{\underset{S}{|}}{CH}\cdot (CH_2)_4\cdot COOH$$

(with the two S atoms joined: S——S)

Fig. 12.11 Lipoic acid

Lipoic acid is involved in the oxidative decarboxylation of α keto acids such as pyruvic acid and α ketoglutaric acid. It acts in these reactions as an oxidising agent. It can be easily oxidised and reduced in the biological systems.

12.5.7 *p*-AMINOBENZOIC ACID

p-Aminobenzoic acid (PABA) (Figure 12.12) is a white crystalline substance, which is slightly soluble in cold water but soluble in hot water.

Chemistry and structure

NH_2

COOH

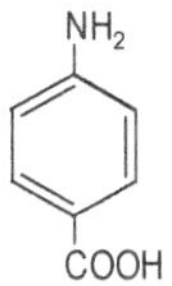

Fig. 12.12 p-Aminobenzoic acid

Deficiency diseases Deficiency of PABA for longer periods led to graying of hairs in man and administration of PABA led to darkening of such hairs.

12.5.8 BIOTIN

This vitamin is also known as coenzyme R or vitamin H. Chemically this form has been identified to be N-biotinyllysine (Figure 12.13).

Chemistry and structure

```
     O
     ‖
    C
NH     NH
|      |
CH  —  CH
|      |
CH2    CH·(CH2)4·COOH
    S
```

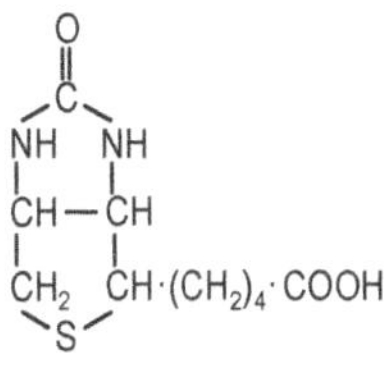

Fig. 12.13 Biotin

Sources The important sources of biotin are liver, kidney, milk, royal jelly, molasses and to some extent other vegetables.

Deficiency diseases Biotin is involved in the carboxylation reactions. Deficiency of this vitamin, result in transient dermatitis in man which is followed by anorexia, muscular pain and hyperesthesia.

12.5.9 FOLIC ACID

Folic acid is also known as pteroylglutamic acid (PGA). This vitamin is abundantly present in the green leaves and grass and hence the name folic acid (folium denotes leaf). Folic acid is made up of glutamic acid, p-aminobenzoic acid (PABA) and pteridine (Figure 12.14). The portion of folic acid molecule, which contains pteridine and p-aminobenzoic acid, is known as pteroic acid. In folic acid group of vitamins, the number of glutamic acid molecules differs and it may be one, three, or six. These compounds may be converted into PGA by the action of vitamin B_c conjugase, which is found in most of the animal tissues. The reduced formylated derivative of folic acid is called folinic acid.

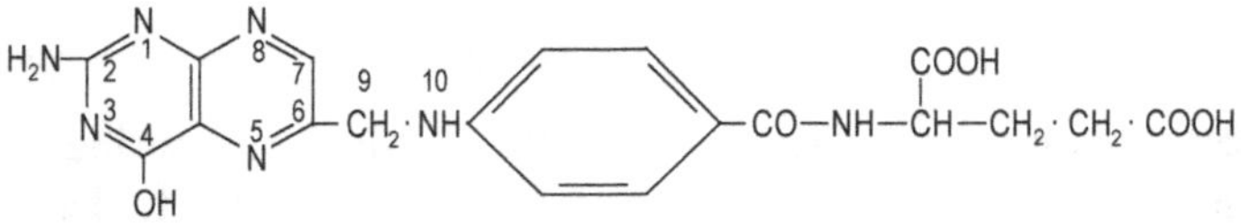

Fig. 12.14 Folic acid

Sources Green leafy vegetables, cauliflower, yeast, liver and kidney are good sources.

Deficiency diseases Folic acid deficiency in man causes megaloblastic anaemia, glossitis and disorder of gastrointestinal tract.

12.5.10 INOSITOL

Inositol is also known as muscle sugar and has been identified to be hexahydroxycyclohexane (Figure 12.15).

Chemistry and structure

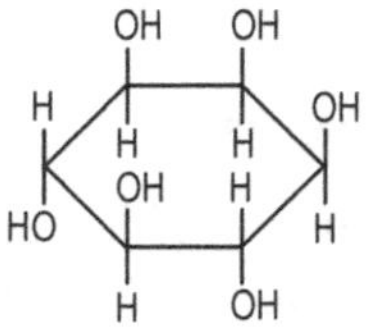

Fig. 12.15 Inositol

Sources Animal tissues such as muscles, brain, red blood cells and eye contain inositol. It is also widely distributed in the plants e.g. fruits, vegetables, whole grains and yeast. Milk also contains inositol.

Deficiency diseases Inositol has been shown to exert a curative effect on fatty liver in the rats. In mice, its deficiency causes retardation in growth and alopecia. In human beings, its nutritional significance is not known definitely. Its lipotropic action is now generally agreed. In the presence of inositol, blood cholesterol level is not increased as expected on high cholesterol diets in the experimental animals. It has been reported to act as antiketogenic compound.

12.5.11 CHOLINE

Choline is chemically trimethyl-hydroxyethyl-ammonium hydroxide and forms constituent of lecithin and acetylcholine (Figure 12.16).

Chemistry and structure

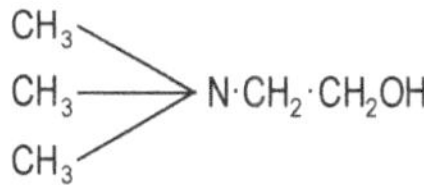

Fig. 12.16 Choline

Sources The important sources of choline are meat, egg yolk , pancreatic tissue, bread, cereals and vegetables. It is widely distributed.

Deficiency diseases Deficiency of choline is accompanied by development of fatty-liver. Liver cirrhosis is often observed in the rats. No definite deficiency symptom of choline alone has been shown in human.

12.5.12 VITAMIN B_{12}

Vitamin B_{12} is also known as **antipernicious anaemia factor**. Vitamin B_{12} contains cobalt cyanide and amino groups in its molecular structure and is chemically known as **cyanocobalamine**. Its structure is very complex and contains 63 carbon atoms.

Chemistry and structure A molecule of vitamin B_{12} contains an atom of cobalt in the trivalent state. Three species of vitamin B_{12} are known namely B_{12} a, B_{12} b, B_{12} c. In B_{12} a, cobalt is bound to cyanide but in B_{12} $v = V_{max}$ b, and B_{12} c it is bound to hydroxyl group and nitrite group respectively (Figure 12.17).

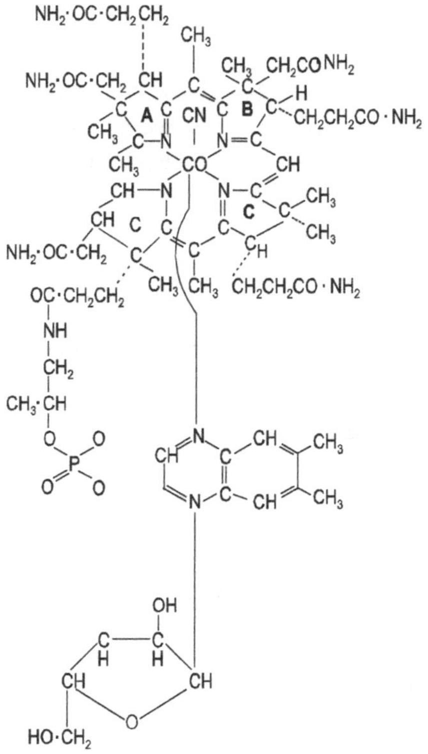

Fig. 12.17 Cyanocobalamine (vitamin B_{12})

Sources The most important source of vitamin B_{12} is liver. Other sources are milk, meat, eggs and fish. Microorganisms of the intestine can also synthesise these vitamins.

Deficiency diseases Vitamin B_{12} is essentially required for normal hematopoiesis (formation of blood) and erythrocyte maturation. This is involved in the metabolism of glycine, serine, methionine, choline, and methyl groups. These favour biosynthesis of thymine, choline, methionine and proteins. Vitamin B_{12} is also involved in the enzymic conversion of methylmalonyl CoA to succinyl CoA. In humans, its deficiency causes **pernicious anaemia**. Deficiency of vitamin B_{12}

and folic acid results in accumulation of megaloblasts in bone marrow, leading to the development of macrocytic type of anaemia. These abnormalities are corrected by administration of vitamin B_{12} and folic acid.

12.6 VITAMIN C (ASCORBIC ACID)

Vitamin C is a hexose derivative (Figure 12.18).

Chemistry and structure

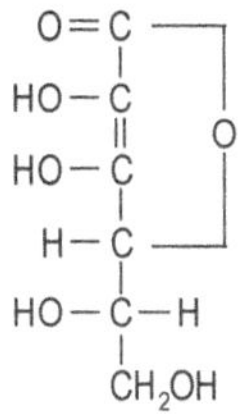

```
  O=C ─────┐
    |      |
 HO─C      |
    ‖      O
 HO─C      |
    |      |
  H─C ─────┘
    |
 HO─C─H
    |
   CH2OH
```

Fig. 12.18 L-Ascorbic acid

Sources The richest sources of vitamin C are the citrus fruits such as lemons and oranges. The other rich sources are amla, plums, guava, grapes, strawberries and apples. Lesser important sources are sprouts during germination of seeds, cauliflower, cabbage, tomatoes and potatoes. Vitamin C is also present in milk.

Deficiency diseases Vitamin C is an essential factor responsible for collagen biosynthesis. It is involved in tyrosine metabolism, electron transport chain and acts as coenzyme for cathepsins and esterases. It also helps in the conversion of folic acid into tetrahydrofolic acid and in the control of cholesterol metabolism. Man, monkey and guinea pigs require vitamin C in the diet, as these cannot synthesise this vitamin. Deficiency of this vitamin leads to retardation of growth, swelling, and pain in the joints. Haemorrhage of gums starts, followed by loosening of teeth. Extreme deficiency of ascorbic acid in humans leads to **scurvy**. Further, deficiency of vitamin C impairs neuromuscular coordination.

Table 12.1 Vitamins

S. No.	Vitamin	Physiological Function	Principal food sources	Deficiency Symptoms & Diseases	Requirements for normal body/day
A.	**Fat- Soluble vitamins**				
1.	**Vitamin A** (Antixerophthalmic vitamin) $C_2O\ H_{30}O$	Essential for visual purple regeneration, contain growth factor and maintain the integrity of the epithelial tissues.	Carrot, green vegetables, milk, butter, fish liver (cod and halibut liver oil), eggs, etc.	Retardation of growth, decreased resistance to infection, **Xerophthalmia** (Dryness of cornea, no tear secretion **Keratomalacia** (inflammation of the cornea, ulceration leading to blindness) **Nyctalopia** (night blindness)	15 to 24 mg.
2.	**Vitamin D** (Calciferol or Antirachitic vitamin) $C_{28}H_{44}O$	Essential for normal development of bones and teeth. It regulates calcium and phosphorus metabolism.	Fish liver oil, butter, egg, yolk and milk and also from exposure of skin to ultraviolet radiation.	**Rickets** in children and **Osteomalacia** in adult, bone become soft and muscles become weak.	1.3 mg.

Table 12.1 Vitamins (contd.)

3. **Vitamin E** (α- Tocopherol) Antisterility factor $C_{19}H_{29}N_7O_6$	Very essential for reproduction.	Wheat germ oil, green vegetable, egg, liver, meat.	Loss of fertility in male and female rats; sometimes abortion; defective fat metabolism. Effect not found in man.	
4. **Vitamin K** (Antihaemorrhagic factor) $C_{31}H_{46}O_2$	Essential for the formation of prothrombin and coagulation of blood.	Green leafy vegetables, alfalfa, spinach, liver, eggs, soyabean.	Excessive bleeding, blood clotting period prolonged.	Not perfectly known.
B Water-soluble vitamins				
1. **Vitamin B-complex**				
(a) **Vitamin B_1** Thiamine (anti beriberi factor) $C_{12}H_{17}N_4$	Maintenance of the tone of gastrointestinal tract, promotes appetite. Controls the carbohydrate metabolism.	Yeast, cereals, peas, beans, eggs, liver, pork.	**Polyneuritis** or **beriberi**, heart enlargement, pain in the eyes and headache, reduced motility of the digestive tract.	1.2 to 1.8 mg.
(b) **Vitamin B_2** (Riboflavin) $C_{17}H_{20}N_4O_6$	Essential for growth and as enzyme in certain tissue.	Yeast, milk, liver, eggs, peas, green vegetables, meat, cheese and kidney.	**Dermatitis,** digestive disorders, anaemia, blurred vision, **cheilosis**, soreness of tongue.	2.0 to 2.5 mg.

Table 12.1 Vitamins (contd.)

(c) **Niacin** (Nicotinic acid amide) $C_6H_5NO_2$ (antipellagra factor)	It acts as coenzyme in tissue reaction and control the cellular function.	Meat, liver, peanuts, chicken and yeast.	**Pellagra**, fatty liver, and kidney hemorrhage.	12.0 to 18 mg.
(d) **Vitamin B$_6$** (pyridoxine) $C_8H_{12}NO_3$ HCl	Act as coenzyme in metabolism of tyrosine and other amino acid and act as accelerator in several activities.	Yeast, cereals, milk, liver, wheat, germ, meat and nuts.	Causes **anaemia**, loss of weight in man and animals, retarded growth in rats. *Acrodymia.*	1.7 to 1.9 mg.
(e) **Pantothenic acid** $C_6H_{17}NO_5$	It maintains the nerve, skin health.	Yeast, liver, milk, sweet potatoes, egg, meat,cane, molasses.	Causes **dermatitis**, greying of hairs and anaemia in animals, but deficiency disease not observed in man.	1.0 to 1.2 mg.
(f) **Para amino benzoic acid** (PABA)	Serves as coenzyme for certain enzyme systems. An important constituent of folic acid.	Yeast, liver, brain, germ of cereal and leguminous plants.	No disease known in man, but in rat produced grey hair.	Not known in human nutrition.
(g) **Biotin** $C_{10}H_{16}N_2O_3S$	Essential for carbohydrate metabolism and growth in insect.	Liver, kidney, yeast, tomatoes, fresh vegetables and fruits.	Diarrhoea, **dermatitis**, nervous disorder. 'Spectacle eye'. Deficiency not observed in man.	Not known.

Table 12.1 Vitamins (contd.)

(h) **Folic acid**	Essential for blood formation and growth.	Green leaves, liver, soyabeans, egg yolk and yeast.	**Anaemia** in man, slow growth and anaemia in chick and rat.	Not known in man.
(i) **Inositol** $C_6H_{12}O_6$	Lipotropic agent and prevent the formation of fatty liver.	Grain cereals, wheat bran and yeast.	Baldness in mice, fatty liver in rats, spectacled eye condition.	Not known in human diet.
(j) **Choline**	Essential for the formation of lecithin.	Eggs, liver, kidney, brain, meat, peas, beans, leafy vegetables and germ of cereals.	Fat storage in body; develop fatty livers; kidney damage.	
(k) **Vitamin B_{12}** (Cyanocobalamin) $C_{64}H_{92}N_{14}O_3PCO$	Formation of blood cells and responsible for growth.	Milk, liver, kidney, meat, soyabeans, egg yolk and bacteria.	Causes **pernicious anaemia** in man, slow growth in animals.	-
2. **Vitamin C** (Ascorbic acid) $C_6H_8O_6$	Maintains integrity of the capillary walls and produces cementing materials of teeth and jaw bones.	Fresh vegetables, fruits especially orange, lemons, tomatoes and chillies.	Loosening of teeth, Scurvy: sore gum, loss of weight, fatigue, bleeding of gums.	75mg.

REVIEW QUESTIONS

1. Give an account of fat-soluble vitamins.
2. Explain the chemistry and physiological functions of water-soluble vitamins.
3. Write brief account of the chemistry, functions, sources and deficiency diseases caused by:
   i. Vitamin A
   ii. Calciferol
   iii. Tocopherol
   iv. Thiamine
   v. Cyanocobalamin
   vi. Ascorbic acid

13

HORMONES

Chemically, a hormone may be any kind of organic molecule. Most known hormones are either steroids or peptides with usually high molecular weights. A third group of hormones, which is less common, consists of amino acid derivatives with relatively low molecular weights. Thus, three categories of hormones may be recognised: steroids, peptides and amino acid derivatives (Table 13.1).

STEROID HORMONES

The steroid hormones include the sex hormones and adrenal cortical hormones. The sex hormones are concerned with the sexual processes and the development of secondary characteristics which differentiate males from females. The adrenal cortical hormones perform a variety of important functions related to cell metabolism. Based on the number of carbon atoms present in the molecule, the steroid hormones are named as C_{18}, C_{19}, or C_{21} steroids.

13.1 C_{18} STEROIDS

13.1.1 ESTROGEN

Mammalian ovary contains ovarian follicles and corpus lutea. Hormones produced mainly in the follicles are known as estrogens. Estrogen is a generic term for a substance that induces estrus, which is a cyclic phenomenon of the female reproductive system. The stages and timing differ in various species but, in general, first, a proestrus period occurs, during which the follicle ripens and the organs of reproduction develop. This is followed by estrus, the period of heat, in which the female will receive the male. Ovulation takes place toward the end of the estrus, either spontaneously or as in rabbit, after mating. Then follows a period of retrogression of the accessory reproductive organs and a period of sexual inactivity.

Chemically, the estrogens are derivatives of a C_{18} hydrocarbon estrane (Figure 13.1).

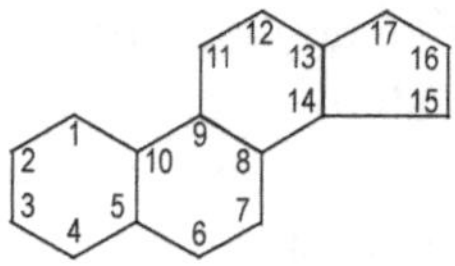

Fig. 13.1 Estrane

The three compounds which exhibit hormonal activity are:

i. β- estradiol ($C_{18}H_{24}O_2$)

ii. Estriol ($C_{18}H_{24}O_3$)

iii. Estrone ($C_{18}H_{22}O_2$)

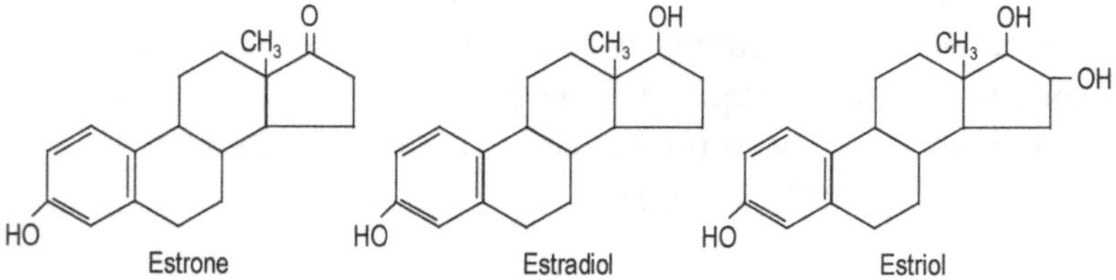

Fig. 13.2 Ovarian hormones

Estrone is the first known member of the sex hormones and was isolated by Butenandt and Doisy independently in 1929 from the urine of the pregnant woman. Later, the estriol and estradiol were isolated.

All these are characterised by the absence of a CH_3 group at carbon 10 and by the aromatic nature of ring A, making the OH group phenolic in character (Figure 13.2). Although ovary is the chief source of estrogens, the testis and the adrenal cortex also produce small quantities of estrogens.

Table 13.1 Vertebrate hormones

Steroid Hormones	Peptide Hormones	Amino acid Derivatives
C_{18} STEROIDS		
Ovarian Hormones	*Hormones of the Pancreas*	*Thyroidal Hormones*
β-estradiol	Insulin	Triodothyronine, T_3
Estriol	Glucagon	Tetraiodothyronine, T_4
Estrone		
C_{19} STEROIDS		
Testicular Hormones from testes	*Hormones of the Hypophysis*	*Adrenal Medullary Hormones*
Testosterone	Adenohypophysis	Adrenalin
Androsterone	Thyrotropin,TSH	Noradrenalin
Dehydroepiandrosterone	Corticotropin, ACTH	
from adrenal gland	Gonadotropin,GTH	
Androst-4-ene-3, 17 dione	FSH	
	LH	
Androst-4-ene-3, 11, 17 trione	LTH	
	Somatotropin,SH	
	Pars intermedia	
	Intermedins, MSH	
	α-MSH	
	β -MSH	
	Neurohypophysis	
	Oxytocin or pitocin	
	Vasopressin or pitressin	

Table 13.1 Vertebrate hormones *(contd...)*

Steroid Hormones	Peptide Hormones	Amino acid Derivatives
C_{21} STEROIDS		
Adrenal Cortical Hormones	*Hormones of the Parathyroid*	
Mineralocorticoids	Parathormone, PTH	
Aldosterone	Calcitonin	
Deoxycorticosterone		
Glucocorticoids		
Cortisone		
Cortisol		
Corticosterone		
Corpus Luteal Hormone	*Hormones of the Gastrointestinal Tract*	
Progesterone		
	Gastrin	
	Secretin	
	Cholecystokinin	
	Pancreozymin	
	Enterogastrone	
	Enterocrinin	
	Hepatocrinin	
	Duicrinin	
	Villikinin	
	Parotin	
	Hormone of the Corpus Luteum	
	Relaxin	

Physiological Functions Estrogens exert specific effects on the growth and development of the female reproductive organs namely, fallopian tubes, uterus, endometrium, cervix, vaginal mucous membrane and female secondary sexual characteristics.

13.1.2 PROGESTERONE

Progesterone is secreted mainly by the corpus luteum of the ovary. It is also formed in the placenta, adrenal cortex and testes. Progesterone makes its appearance on the day of ovulation or one or two days earlier. It exerts its action in the latter half of the menstrual cycle and is concerned mainly with the preparation of the endometrium for implantation of the fertilised ovum for conception. During pregnancy, progesterone is produced by the corpus luteum in the earlier period and by the syncytial cells of the placenta in the later period. The chief excretory product of progesterone in urine is pregnandiol.

Progesterone is a steroid and contains 21 carbon atoms. It has methyl groups at C-10 and C-13 (Figure 13.3).

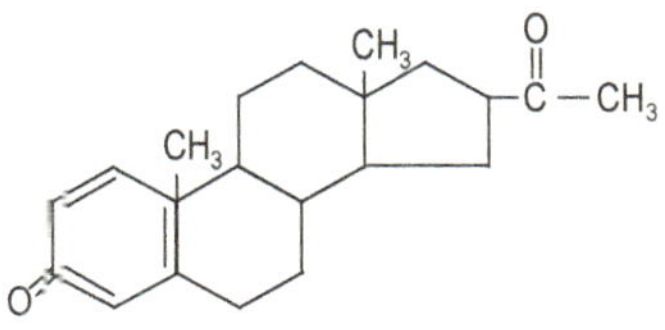

CH_3 CH_3 O C CH_3 O

Fig. 13.3 Progesterone

Progesterone is soluble in all organic solvents except petroleum, ether, acetone and pyridine. It is insoluble in water.

Physiological Functions

i. Progesterone produces characteristic progestational changes in the endometrium during the reproductive period. It prepares the endometrium for the reception and implantation of the fertilised ovum.

ii. It stimulates development and growth of mammary glands and sensitises them for the action of lactogenic hormone.

iii. The suppression of ovulation and menstruation during pregnancy is due to the action of progesterone which, along with estrogen, maintains the quiescent state of the uterus. If pregnancy does not occur, the concentration of progesterone decreases at the end of the menstrual cycle and menstrual bleeding starts, which is associated with shedding of the mucosa of the endometrium.

iv. It is responsible for increasing the BMR during the luteal phase of normal menstrual cycle, that is, from 14 to 28 days. There is a slight rise in basal temperature during this period.

13.2 C_{19} STEROIDS

13.2.1 TESTICULAR HORMONES

These hormones are secreted mainly by the testes, the male reproductive organs and are called as androgens. Chemically, these are derivatives of a C_{19} hydrocarbon, androstane. There are many hormones secreted from testes with androgenic activity (Figure 13.4). The three important hormones are:

a. Testosterone ($C_{19}H_{28}O_2$)

b. Androsterone ($C_{19}H_{30}O_2$)

c. Dehydroepiandrosterone ($C_{19}H_{25}O_2$)

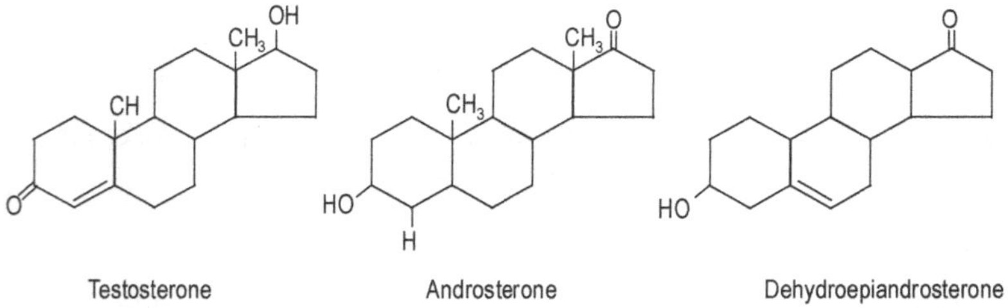

Fig. 13.4 Testicular hormones

Testosterone is the most potent of all these and dehydroepiandrosterone is less active. A few testicular hormones are also produced by the adrenal gland.

Functions

i. Testosterone promotes anabolism of protein by stimulating protein biosynthesis and transport of amino acids into cells.

The protein anabolic effect results in,

a. decreased urinary excretion of nitrogen

b. increase in body weight which is mainly due to increase in skeletal muscle mass and bones.

ii. It produces specific stimulation of the growth and functional activity of prostrate, seminal vesicle, epididymis and vas deferens.

13.3 C_{21} STEROIDS

13.3.1 ADRENAL CORTICAL HORMONES

The adrenals are a pair of glands, located above the kidneys. In mammals each of the two adrenals has two distinct parts namely an outer cortex, and an inner dark-coloured mass called the medulla (Figure 13.5). The cortex is derived from mesodermal glandular tissue and the medulla originates from the cells of neural crest. Both these parts secrete hormones, which differ from each other chemically as well as physiologically. Histologically, the adrenal cortex is made up of three layers:

i. an outer narrow *zona glomerulosa,* the site of biosynthesis of the mineralocorticoid hormones.

ii. a middle, comparatively broader *zona fasiculata*, responsible for the production of glucocorticoid hormones and the adrenal androgens.

iii. an inner narrow *zona reticulosa,* secreting glucocorticoids along with the middle zone.

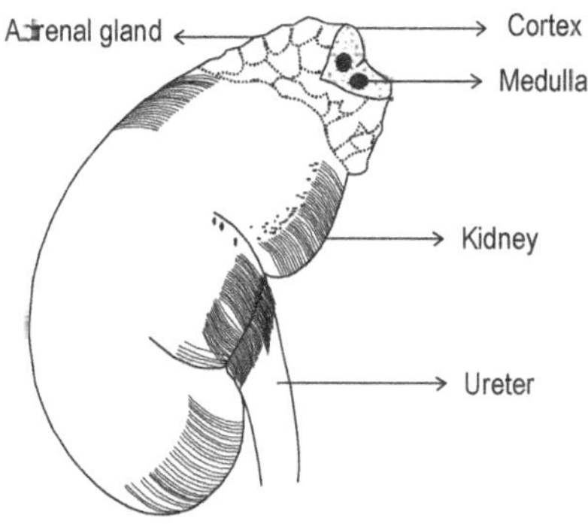

Fig. 13.5 Adrenal gland in man

When there is prolonged stimulation of the adrenal cortex by adrenocorticotropic hormone (ACTH), the middle and inner zones both hypertrophy; but a total lack of ACTH causes these two zones to atrophy almost entirely, leaving the outer zona glomerulosa partially intact. On the other hand, enhanced aldosterone production causes hypertrophy of the zona glomerulosa, while the other two zones remain almost unaffected.

Adrenal cortex secretes some 40–50 closely related C_{21} steroids, collectively called as corticosteroids. The corticosteroids may be grouped under three categories:

Mineralocorticoids concerned primarily with the transport of electrolytes and the distribution of water in tissues. e.g. aldosterone and deoxycorticosterone.

Glucocorticoids concerned primarily with the metabolism of carbohydrates, proteins and fats. e.g. cortisone, cortisol and corticosterone.

Sexcorticoids Adrenal cortex secretes two types of sex hormones—testosterone and estrogens. Androgens of the adrenal cortex are concerned with the development of secondary sexual characteristics and promote protein synthesis. When secreted in normal amounts they do not exert any masculinising effects. Excess secretion before puberty, however, causes precocious development of secondary sexual characters.

Estrogens are secreted from the cortex in very small amounts and may add to the effects of the hormones secreted from ovary.

Chemical nature of corticosteroids

a) Glucocorticoids These include 11-oxysteroids, namely, 11 dehydrocorticosterone, corticosterone, 11-dehydro-17-hydroxycorticosterone (cortisone) and 17-hydroxycorticosterone (cortisol) (Figure 13.6).

11-dehydrocorticosterone

Corticosterone

Cortisone

Cortisol

Fig. 13.6 Glucocorticoids

b) Mineralocorticoids These are further divided into two groups.

i. Deoxy corticosteroid hormone These include 11-deoxy corticosterone and 17-hydroxy-11-deoxycorticosterone (cortisol) (Figure 13.7).

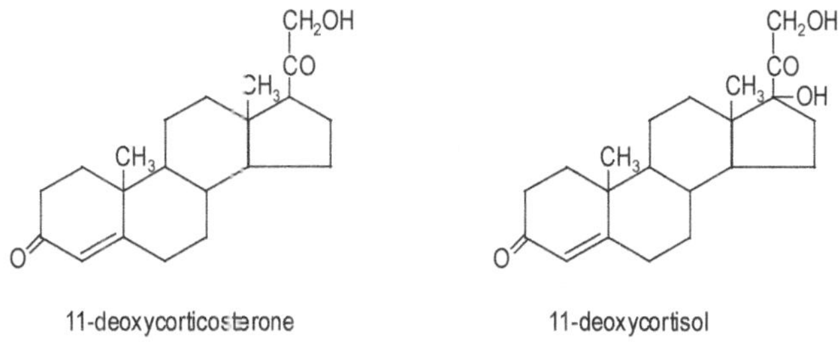

11-deoxycorticosterone

11-deoxycortisol

Fig. 13.7 Mineralocorticoids

ii. Aldosterone This can exist in aldehyde form or in hemiacetal form.

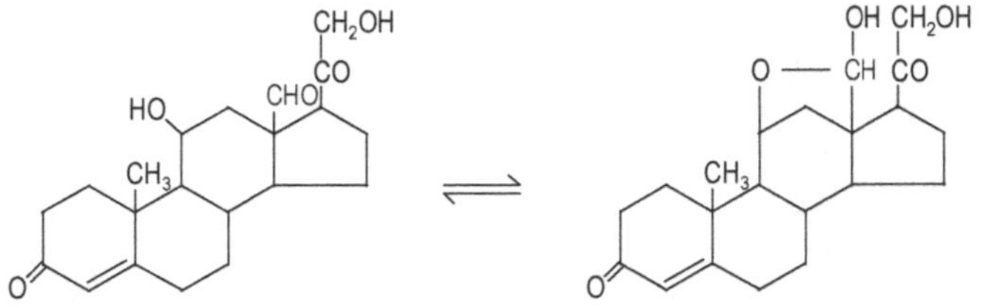

c) **Sexsteroids** These include estrogens, androgens and progesterone. The chemical nature of these hormones have been discussed earlier.

Functions The mineralocorticoid, aldosterone, is chiefly concerned with water–electrolyte balance of the body. It stimulates the reabsorption of Na^+ ion from the kidney tubules and as such regulates NaCl contents of the blood. This also causes excretion of K in the urine. Aldosterone is also more potent in maintaining the life of adrenalectomised animals.

Glucocorticoids, on the contrary, govern many other processes. They perform the following physiological functions:

i. Influence the carbohydrate metabolism, firstly by increasing the release of glucose from the liver and secondly, by promoting the transformation of amino acids to carbohydrates.

ii. Inhibit protein synthesis in muscle tissues.

iii. Control eosinophil cells of the blood.

iv. Regulate lipogenesis.

v. Reduce the osteoid matrix of bone, which results in osteoporosis and heavy loss of calcium from the body.

vi. Decrease immune responses associated with infection and anaphylaxis (immunosuppressive effects).

vii. Cause increased secretion of hydrochloric acid and pepsinogen by the stomach and trypsinogen by the pancreas.

viii. Cause retention of sodium and loss of potassium to some extent. In this respect, it resembles aldosterone in action.

Diseases associated with adrenocortical functions

Hypoadrenocorticism (Addison's Disease) Addison's disease is caused due to chronic insufficiency of the secretion from adrenal cortex. Patients with Addison's disease are highly susceptible to various types of stress. Addisonians are characterised by symptoms like vomitting, diarrhoea, collapse and pyrexia with rigour, mental clouding, hypoglycemia, low blood pressure, dehydration, acidosis, renal failure, progressive loss of weight and pigmentation. The treatment of the disease is carried out by the use of glucocorticoids and mineralocorticoids.

Hyperadrenocorticism This condition refers to hyperfunctioning of adrenal cortex, caused due to tumours of the adrenal cortex, or hyperplasia of cortical tissue initiated by increased secretion of ACTH.

The characteristic findings are:

i. Hyperglycemia and glycosuria.
ii. Retention of sodium and water, resulting in oedema and increased blood volume.
iii. Negative nitrogen balance due to increased gluconeogenesis.
iv. Alkalosis.

Cushing's Syndrome This disease first recognised by Cushing in 1932 is attributed to basophilic tumours of pituitary. Cushing's syndrome is caused by hyperfunction of the adrenal cortex with excessive secretion of glucocorticoids. Excessive steroids cause alterations in carbohydrate and electrolyte metabolism. The syndrome is more frequent in the female sex in the age group of 20–40 years. Patients suffer from obesity with deposition of fat confined to the face, neck, supraclavicular regions and the abdomen. There may be mental depression or psychoses, accompanied by impotency in males and amenorrhoea in females. Effective treatment is generally adrenalectomy.

Conn's syndrome This syndrome is characterised by excessive thirst, polyuria, weakness, hypertension, periodic muscle weakness, cramps and tetany. Physiologically there is an excessive secretion of aldosterone from the adrenal cortex. Tumours of adrenal cortex may lead to the production of excessive amount of aldosterone, which may produce widespread disturbance in the body. The excessive secretion of aldosterone causes sodium retention, low potassium in the blood with alkalosis and secretion of alkaline urine. Treatment generally consists of large doses of potassium, but this provides only temporary relief. Surgical treatment is more effective.

PEPTIDE HORMONES

13.4 HORMONES OF THE PANCREAS

The pancreas is both an exocrine and endocrine gland. It is situated transversely below the stomach between duodenum and spleen. It is a compact and lobulated organ. Generally, in vertebrates, the pancreas is composed of two types of tissues (Figure 13.8).

i. The glandular cells or acinar or acini (exocrine), which make up the bulk of the pancreatic tissues and secrete digestive juices into the duodenum by the pancreatic duct.

ii. The polygonal cells or islets of Langerhans or islet tissue (endocrine), which pour their secretions (i.e. insulin, glucagon and somatostatin) directly into the blood. These were discovered by Langerhans in 1867, hence so named.

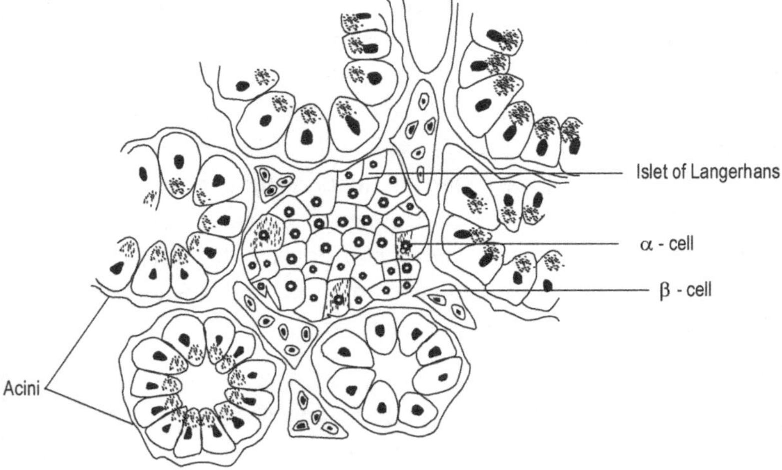

Fig. 13.7 Section of pancreas showing islets of langerhans

The islets of Langerhans in mammals contain different types of cells, α-cells or A cells, β-cells or B cells and D cells. The β-cells secrete a hormone, insulin, the α-cells produce another hormone, glucagon, and D cells secrete somatostatin.

13.4.1 INSULIN

Insulin was first isolated in 1922 from the pancreas of dogs by Banting and Best of the University of Toronto, Canada. They also demonstrated the curative effect of pancreatic extract in dogs with *diabetes mellitus*. Abel and his associates (1926) obtained insulin in crystalline form and also demonstrated its protein nature. It is, in fact, the first hormone to be recognised as a protein. The chemical structure of insulin has been determined by Sanger and his coworkers at Cambridge, England. For achieving this landmark, Sanger was awarded Nobel Prize in 1955. Bovine insulin consists of 51 amino acid residues dispersed in two chains; the two polypeptide chains are held together by cross linkages of two disulphide bonds (Figure 13.9). The acidic chain -A contains 21 residues and the peptide chain -B has 30 residues. It has a molecular weight of 5,733 and is isoelectric at pH 5.4. Human insulin, has a molecular weight of 5808. Insulin is destroyed by alkali but is relatively stable in acid solutions. Reduction of the disulphide bond results in a loss of biological activity.

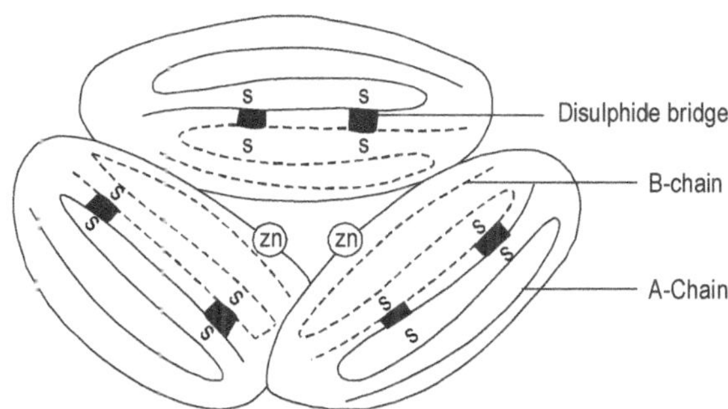

Fig. 13.9 Schematic representation of the three-dimensional structure of insulin

Functions Insulin has a profound influence on carbohydrate metabolism. It facilitates the entry of glucose and other sugars into the cells, by increasing penetration of cell membranes and augmenting phosphorylation of glucose. This results in lowering the sugar content of the blood. Insulin administration promotes protein synthesis by assisting incorporation of amino acids into proteins. This effect is not dependent on glucose utilisation. At the same time, it also acts as an antiproteolytic agent. Synthesis of lipids is also stimulated by administration of insulin.

Insulin also influences the inorganic metabolism, especially that of phosphate and potassium. Insulin administration lowers the blood phosphate level and facilitates absorption of inorganic phosphate by the cells. This phosphate appears within the cells as ATP. A similar mechanism operates in potassium uptake.

Insulin deficiency The deficiency of insulin caused either by inadequate insulin production or by accelerated insulin destruction, leads to *diabetes mellitus* in man. This disease is characterised by:

- an increase in blood sugar (hyperglycemia).
- the appearance of sugar in the urine (glycosuria).
- an increase in concentration of ketone bodies in the blood (ketoanaemia) and in the urine (ketouria).
- the excretion of large quantities of urine (polyuria), frequently at night (nocturia), leading to dehydration.
- the excessive drinking of water (polydipsia) because of constantly feeling thirsty.
- the excessive eating (polyphagia) due to feeling constant hunger.
- the lack of energy (asthenia), which is apparently caused by loss of body protein.

The diabetic person gradually becomes weaker and loses weight due to the failure of glucose utilisation by the body despite a voracious appetite. Diabetic comma ensues, the plasma volume decreases and the kidney fails. This eventually leads to death.

13.4.2 GLUCAGON

Glucagon was first isolated in crystalline form by Behrens and others. This peptide hormone has a molecular weight of 3,485.

Functions Glucagon activates the enzyme adenyl cyclase, which converts ATP to cyclic AMP. The cyclic AMP activates phosphorylase *b* to yield phosphorylase *a*. This promotes glycogenolysis and yields free glucose which increases blood sugar level. Hence, this hormone is also termed as hyperglycemic factor or HG-factor. Glucagon also stimulates gluconeogenesis and affects lipid metabolism by accelerating ketogenesis and inhibiting fatty acid synthesis.

Glucagon has a catabolic action on proteins. Its administration in the body results in excretion of enough nitrogen and phosphorus and in loss of body weight.

13.4.3 SOMATOSTATIN

Somatostatin is also a polypeptide hormone. It contains an interchain disulphide bond. Somatostatin influences the secretion of insulin and glucogen in a complex manner.

13.5 HORMONES OF THE HYPOPHYSIS (PITUITARY GLAND)

Hypophysis is so named because of its location below the brain. Its synonym, pituitary gland, is however misleading as the gland is not concerned with the secretions of mucus or phlegm (*pituita:* phlegm) as was thought previously. This is an unpaired small ovoid gland. Hypophysis is attached to the hypothalamus by a narrow stalk and located in a depression (sella turcica) in the sphenoidal region of the skull.

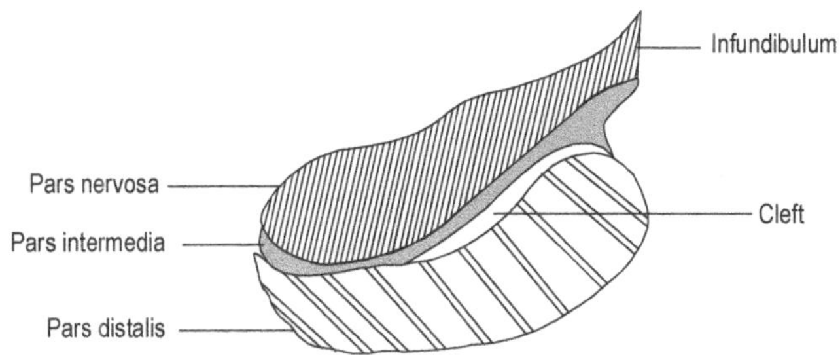

Fig. 13.10 Hypophysis

The hypophysis consists of three lobes: (a) an anterior richly vascular, *pars distalis* or *adenohypophysis* (b) an intermediate relatively avascular *pars intermedia* and (c) a posterior neural lobe, *pars nervosa* or *neurohypophysis* (Figure 13.10).

This gland plays the most dominant role as it secretes hormones that govern the secretion of other endocrine glands (like thyroid, adrenal and gonads) and its secretions have a direct effect on the metabolism of non-endocrine tissues.

ANTERIOR PITUITARY HORMONES

i. The hormones secreted from anterior pituitary or adenohypophysis controls the functional activity and structural integrity of other important endocrine glands namely, thyroid, adrenal and gonads through tropic hormones.
ii. Further, anterior pituitary produces growth hormone, which exerts direct action on certain fundamental aspects of protein, carbohydrate and fat metabolism.

The following are the tropic hormones secreted by the anterior pituitary:

i. Thyrotropic hormone (Thyroid Stimulating Hormone-TSH)
ii. Adrenocorticotropic hormone (ACTH)
iii. Gonadotropic hormone

These hormones exert their effects solely by stimulating their respective target glands. Hypophysectomy leads to atrophy of thyroid, adrenal cortex and gonads.

13.5.1 THYROTROPIC HORMONE

It is also called as thyroid stimulating hormone-TSH. It is a glycoprotein and has high cystine content. It has a molecular weight of about 30,000. Each molecule has 8–9 cystine residues and the disulphide groups are present as intra-chain linkages rather than inter-chain linkages.

Functions TSH plays an important role in regulating the secretion of thyroxine. Secretion of TSH by anterior pituitary on the other hand, is controlled by the level of circulating thyroxine.

13.5.2 ADRENOCORTICOTROPIC HORMONE (ACTH)

It is a protein, containing a straight-chain polypeptide consisting of 39 amino acid residues. Two forms, α-corticotropin and β-corticotropin have been isolated.

Functions ACTH is the main regulator of the activities of adrenal cortex.

13.5.3 GONADOTROPIC HORMONES

The three gonadotropic hormones exercising important effects on the gonads are:

i. Follicle-stimulating hormone (FSH).
ii. Interstitial cell-stimulating hormone (ICSH) or luteinizing hormone (LH).
iii. Lactogenic hormone (prolactin) or luteotropic hormone; (LTH).

In female, they promote growth, maturation and expulsion of the ova and stimulate production of internal secretions of the ovary. In male, they stimulate spermatogenesis and the production of androgen.

13.5.4 FOLLICLE STIMULATING HORMONE (FSH)

It is a water-soluble glycoprotein, containing about 3 to 11% carbohydrate. It is a glycoprotein that contains galactose, mannose, galactosamine, glucosamine, sialic acid, fucose and uronic acid.

Functions In female, FSH stimulates the growth and maturation of the graafian follicles and prepares them for ovulation. In the male, FSH stimulates spermatogenesis. Low level of thyroxine in the blood stimulates the pituitary to release more TSH which in circulation reaches the thyroid and stimulates its secretion. When the concentration of thyroxine in the circulating blood rises above a certain level, secretion of TSH from pituitary is reduced resulting in the reduced production of thyroxine from thyroid.

13.5.5 INTERSTITIAL-CELL-STIMULATING HORMONE (ICSH)

It is also called as luteinizing hormone (LH). It is a peptide hormone with molecular weight of about 26,000 (in man) or 100,000 (in swine).

It lacks tryptophan but has a high content of cystine and proline. Each molecule contains 10 glucosamine and 3 galactosamine residues.

Functions In the female, ICSH (LH) acts in conjunction with FSH to promote maturation of the graafian follicles, ovulation of mature follicles and development of corpus luteum. It stimulates secretion of estrogen and progesterone. In the male, ICSH stimulates the production of testosterone by the testes.

13.5.6 LUTEOTROPIN OR LUTEOTROPIC HORMONE (LTH)

It is also called prolactin. It is the most versatile of all the adenohypophyseal hormones. This is also a peptide hormone with 198 amino acid residues and a molecular weight of about 23,500. It has 3 disulphide bonds between cysteine residues at 4–11, 58–173 and 190–198. It differs from FSH and LH in that it contains no carbohydrate. It is thermolabile and is destroyed by trypsin. In pure form, it has no growth-promoting, thyrotropic, diabetogenic, adrenotropic or gonadotropic activities. However, in association with estrogen, luteotropin promotes the growth of the mammary glands, and induces secretion of milk (lactation) at the time of child birth (parturition). It also stimulates glucose uptake and lipogenesis. Along with androgens, it causes the development of secondary male sexual characters. It also acts as an anabolic agent mimicking the effects of growth hormone, although it is less active in this respect.

Chorionic gonadotropin The hormone chorionic gonadotropin, although of placental origin, resembles hypophyseal hormones in its biological effects. Human chorionic gonadotropin, HCG, is a glycoprotein with molecular weight of about 30,000. The carbohydrate moiety contains some six components, viz., D-galactose, D-mannose, N-acetylglucosamine, N-acetylgalactosamine, L-fucose and N-acetylneuraminic acid.

As to its biological role, HCG factor supplements hypophysis in maintaining growth of the corpus luteum during pregnancy. The hormone appears sharply after pregnancy in the urine.

13.5.7 GROWTH HORMONE (SOMATOTROPIN)

It has 190 amino acids and consists of 2 disulphide bridges between adjacent cysteine residues at 53-164 and 181-188. The N-terminal and C-terminal residues are both phenylalanine.

Functions Administration of growth hormone results in **gigantism** in young and **acromegaly** in adults. It exerts direct action on the metabolism of protein, carbohydrates and lipids.

Protein metabolism It stimulates anabolism of protein in general. It promotes synthesis of protein and induces positive nitrogen balance. It retards amino acid catabolism.

Carbohydrate metabolism

i. It produces hyperglycemia (diabetogenic effect).
ii. It inhibits insulin action resulting in decreased utilisation of glucose.
iii. It inhibits the hexokinase reaction.

Lipid metabolism

i. It stimulates mobilisation of fats from fat depots to blood.
ii. It increases ketogenesis.

Mineral metabolism Growth hormone enhances absorption of calcium from the intestine. It helps in retaining phosphate, sodium and potassium and stimulates the growth of bones.

HORMONES OF NEUROHYPOPHYSIS

The posterior lobe of pituitary, called par nervosa or neurohypophysis secretes vasopressin and oxytocin.

Vasopressin (antidiuretic hormone, ADH), raises the blood pressure and decreases the secretion of urine.

Oxytocin (pitocin), causes contraction of uterine muscles.

Vasopressin and oxytocin are octapeptides and contain 8 amino acids, out of which six are common to both. These are, glycine, proline, aspartic acid, glutamic acid, tyrosine and cystine. These are represented in a cyclic form with two cysteinyl residues forming disulphide bridges. Three of these eight amino acids are present in the form of amides.

Oxytocin contains in addition, leucine and isoleucine, while vasopressin contains in those places, arginine or lysine and phenylalanine (Figures 13.11 & 13.12).

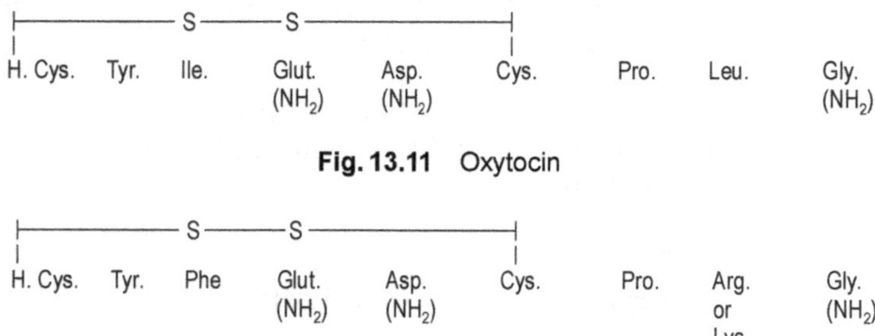

Fig. 13.11 Oxytocin

Fig. 13.12 Vasopressin

Vasopressin and oxytocin are formed in the neurons of the hypothalamic nuclei. They migrate along their axons and are stored in the nerve cell endings in the neurohypophysis. The discharge of ADH, particularly from this structure is controlled by the hypothalamic neurons and is influenced by the degree of concentration of the plasma. Thus, when there is a tendency towards reduction in plasma water content, antidiuresis is stimulated which results in renal tubular reabsorption of increased amounts of water.

In the absence of ADH, urine cannot be concentrated and large volumes of urine are excreted. This condition is *diabetes insipidus*.

Functions

i. Vasopressin stimulates the peripheral blood vessels and causes a rise in the blood pressure. Vasopressin is used to counteract the low blood pressure of shock following surgery, postpartum hemorrhage and uterine inertia. The rise in blood pressure is more gradual than that of epinephrine.

ii. Being an antidiuretic, it helps in regulating water balance. It acts by reabsorption of water from the epithelial cells of renal tubules (distal portion). Deficiency or lack of ADH causes *diabetes insipidus*.

iii. Oxytocin produces contraction of the uterus. It causes ejection of milk from lactating breast.

iv. It stimulates contraction of gall bladder, intestines and urinary bladder.

HORMONES OF PARS INTERMEDIA

The middle lobe of the pituitary secretes a hormone, originally known as intermedin. It is now called melanocyte-stimulating hormone (MSH), because it increases the deposition of pigment melanin by the melanocytes of the human skin. Secretion of MSH is inhibited by epinephrine, norepinephrine, hydrocortisone and cortisone. In adrenocortical deficiency, as in Addison's disease, where the production of corticosteroids is diminished, MSH secretion is increased resulting in increased synthesis of melanin. Brown pigmentation of the skin is a characteristic finding in Addison's disease. The pigmentation increases as the disease progresses.

Two types of melanocyte stimulating hormones, α-MSH and β-MSH have been isolated. They are straight chain peptides and contain 13 and 18 amino acid residues respectively. The amino acid sequences for seven amino acids in both MSH are found to be identical with the sequence of seven amino acids (4 to 10) in ACTH. The similarity in structure between MSH and ACTH explains the mild but definite melanophore-expanding activity of ACTH.

13.6 HORMONES OF CORPUS LUTEUM

13.6.1 RELAXIN

This is a hormone produced during pregnancy and parturition by the tissues of reproductive system, viz. ovaries, placenta and uterus. It is concerned with the phenomenon of pelvic relaxation, which involves the expansion of the size of pelvic cavity to facilitate parturition. It acts by increasing the vascularisation of the connective tissue of the symphysis pubis. Its production is stimulated by progesterone.

13.7 HORMONES OF THE PARATHYROID

The parathyroids were first discovered by Sandstrom in1880. In human beings, there are usually four small parathyroid glands closely associated with the thyroid gland. Of the four parathyroids, two lie embedded in the thyroid and are called as internal parathyroids; the other two lie close and behind the thyroid and are known as external parathyroids.

Histologically, the parathyroid of the adult human being consists mainly of chief cells and oxyphil cells. The oxyphil cells are usually absent in young humans. The chief cells are concerned with the secretion of the parathyroid hormone. The oxyphil cells are rich in mitochondria but lack glycogen. Their function is uncertain (Figure 13.13).

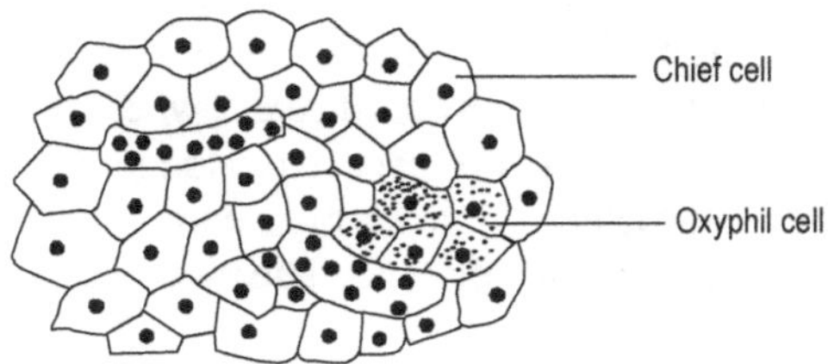

Fig. 13.13 Section of parathyroid

The parathyroid glands secrete two hormones, parathormone and calcitonin, that together with vitamin D are the major regulatory factors in homeostatic control of calcium and phosphate metabolism. Parathormone synthesis and secretion are limited normally to the parathyroid glands, whereas calcitonin is secreted by the thyroid and thymus . In lower vertebrates like fishes, calcitonin is secreted by the ultimobranchial body. This distribution is probably a reflection of the common embryological origin of the above organs, which contain C cells, the production site of calcitonin.

13.7.1 PARATHORMONE

Parathormone, a polypeptide (MW=9,500) is a single polypeptide chain of 84 amino acid residues. Secretion of parathormone by the parathyroid glands is regulated by the concentration of serum Ca^{2+}. If serum Ca^{2+} is lowered by injection of fluoride, oxalate or a calcium-chelating agent, secretion of parathormone promptly rises. Conversely, elevation of serum Ca^{2+} results in diminished hormone secretion. Thus, the blood Ca^{2+} provides a basis for a feedback mechanism for regulation of parathormone secretion.

Functions The principal sites of parathyroid action are bones, kidney and gastrointestinal tract. The following physiological functions are attributed to this hormone:

i. ***Bone resorption*** It exerts a direct influence on the metabolism of bone, leading to an increased release of bone Ca^{2+} into the blood. The exact mechanism behind this phenomenon is not exactly known. It has, however, been suggested that the hormone stimulates the production of citric acid in the bone tissues and an increased concentration of citrate ions leads to the removal of phosphate from calcium phosphate, the bone material. The bone is, thus, made soluble.

ii. ***Renal reabsorption of calcium*** In the kidney, parathormone affects renal tubular reabsorption of calcium. It increases the elimination of calcium and phosphorus in the urine. It is interesting to note that the secretion of this hormone is controlled by Ca^{2+} ion concentration of the blood itself. As the Ca^{2+} ion concentration increases, PTH secretion decreases tending to preserve the original condition. This affords an excellent example of feedback mechanism of metabolic control.

iii. ***Increase in osteoelastic activity*** Parathormone increases osteoelastic activity with augmented growth of the connective tissue.

Hypoparathyroidism Hyposecretion of PTH causes a decrease in Ca^{2+} content of the blood (hypocalcemia), which leads to excessive contraction of the muscle (convulsions). However, during this period, the phosphorus in plasma increases from a normal 5 mg per 100 ml to 9 mg per 100 ml and even higher (hyperphosphotemia). These changes develop into a fatal disease called muscular twitchings or tetanus. It is characterised by locking up of the jaw, rapid breathing, increased heartbeat, rise in temperature and ultimately death due to asphyxia. Tetanus can be relieved either by the administration of a soluble calcium or PTH.

Hyperparathyroidism An increase in PTH production is usually due to a tumour of the gland (parathyroid adenoma). Over-secretion of PTH in man results in a cystic bone disease variously called as osteitis fibrosa cystica or von Recklinghausen's disease or neurofibromatosis.

It is characterised by increased calcium content of the blood (hypercalcemia). Over-production of parathormone causes calcium and phosphorus to move out of the bones and teeth, making them soft and fragile. Such patients, therefore, suffer fractures of the bones very frequently. Cysts in the bones are another characteristic of this disease. Other disorders of the excessive secretion include haemorrhages in the stomach and the intestine as well.

13.7.2 CALCITONIN

Calcitonin is a single chain polypeptide (MW=4,500), containing 32 amino acid residues. The secretion of calcitonin appears to be continuous at physiological concentrations of blood Ca^{2+} and is directly responsive to blood Ca^{2+}. The hormone increases in the blood in response to elevation of Ca^{2+} and decreases with lowering of blood Ca^{2+}. These responses are thus opposite in direction to the control of parathormone secretion by Ca^{2+}.

A second group of regulatory factors may be operative in calcitonin secretion. Glucagon, a hormone of the pancreas also stimulates calcitonin secretion. In addition, gastrin which is a hormone of the gastrointestinal tract has been reported to stimulate calcitonin secretion.

Action of calcitonin The principal effect of calcitonin is to inhibit bone resorption and thus decrease loss of Ca^{2+} from the bone.The overall action of calcitonin on bone opposes the effects of parathormone in that there is reduced bone resorption, with accompanying hypocalcemia and hypophosphatemia and decreased urinary Ca^{2+}. Hydroxyproline excretion in the urine is also reduced perhaps reflecting the inhibitory action of calcitonin on both osteocytes and osteoblasts. An enhanced excretion of phosphates is probably secondary to the alterations in Ca^{2+}.

13.8 HORMONES OF GASTROINTESTINAL TRACT

The mucosa of gastrointestinal tract secretes a number of hormones which help in the process of digestion by regulating release of various digestive secretions.

13.8.1 GASTRIN

Gastric mucosa has been found to secrete two types of gastrins, namely gastrin I and gastrin II. Each contains 17 amino acids. Gastrin I contains unmodified tyrosine whereas, gastrin II contains sulphated tyrosine at position 12. The secretion of gastrin has been found to be stimualted by ingested food, amino acids, acetylcholine and distension of stomach. Glycine has been found most potent amongst amino acids in stimulating gastrin secretion. Gastrin secretion has been found to increase with age. High rate of HCl secretion causes feedback inhibition of gastrin secretion. It is secreted in the blood stream and thence carried to the parietal cells where it causes flow of hydrochloric acid. Gastrin is also involved in the secretion of gastric juice.

13.8.2 SECRETIN

This hormone is produced by intestinal mucosa and is a polypeptide. Its liberation is stimulated by hydrochloric acid secreted from stomach. It is carried by the blood stream to the pancreas where it stimulates secretion of pancreatic juice rich in HCO_3^- ions.

13.8.3 CHOLECYSTOKININ-PANCREOZYMIN (CCK-PZ)

Cholecystokinin-pancreozymin has been found to be a polypeptide containing 33 amino acids. Its hormonal action resides in the octapeptide of the *c*-terminal end. It is secreted by the mucosa of small intestine. HCl secreted from stomach, amino acids, fatty acids and cholinergic influences have been found to stimulate CCK-PZ secretion of pancreatic enzymes and gall bladder contraction thereby causing bile secretion.

13.8.4 ENTEROCRININ

Another hormone, enterocrinin is also released from intestinal mucosa. It is a polypeptide. It stimulates secretion of fluid and enzymes from intestinal mucosa.

13.8.5 GASTRIC INHIBITORY POLYPEPTIDE (GIP)

This hormone has been found to be secreted by duodenal and jejunal mucosa. This is a polypeptide containing 43 amino acids having a molecular weight of 5105. Glucose and fat in the duodenum stimulate the secretion of GIP. This hormone has been found to inhibit gastric acid secretion and gastric motility and stimulates insulin secretion.

13.8.6 HEPATOCRININ

This hormone, released by intestinal mucosa stimulates liver cells to produce more bile juice low in salt content.

13.8.7 ENTEROGASTRONE

Duodenal mucosa secretes a hormone, enterogastrone that has not yet been crystallised. Its secretion is increased by fatty diets. It inhibits gastric secretion and motility. Gastric juice becomes poorer in pepsin and HCl content under the influence of enterogastrone. This inhibition leads to a more complete fat digestion.

13.8.8 PAROTIN

Salivary glands secrete this protein hormone, which stimulates calcification of teeth, decreases serum calcium and increases phosphate level.

13.8.9 VASOACTIVE INTESTINAL POLYPEPTIDE (VIP)

The entire length of small intestinal mucosa and colon has been found to secrete a polypeptide containing 28 amino acids and having a molecular weight of 3100 which has been found to inhibit gastric acid and pepsin secretion and stimulate pancreatic HCO_3^- secretion and secretion of intestinal mucosa. This hormone also inhibits gastric and gall bladder motility. This hormone has been named as vasoactive intestinal polypeptide (VIP).

13.8.10 MOTILIN

The mucosa of duodenum and jejunum has been shown to secrete a polypeptide containing 22 amino acids and having a molecular weight of 2700. It stimulates gastric motility.

13.8.11 ENTEROGLUCAGON

The intestinal mucosa has been found to secrete a polypeptide having a molecular weight between 3500 and 7000. This hormone has been found to cause glycogenolysis. Its secretion is stimulated by glucose or fat in the diet.

13.8.12 CHYMODENIN

This hormone is secreted by mucosa of small intestine. It contains 43 amino acids having a molecular weight of 4900. Its secretion is stimulated by fat in the intestine. It stimulates the chymotrypsin secretion by pancreas.

13.8.13 BULBGASTRON

The duodenal bulb has been found to secrete this hormone. Its secretion is stimulated by acid in the duodenum. This hormone has been found to inhibit gastric acid secretion.

13.8.14 DU CRININ

It controls the secretion of Brunner's gland, which are located in the submucosa of the upper duodenum.

13.8.15 VILLIKININ

It is secreted by the intestinal mucosa. It stimulates the movements of the intestinal villi and hence accelerates intestinal absorption.

AMINO ACID DERIVATIVES

13.9 THYROIDAL HORMONES

The thyroid is the largest endocrine gland in the body. It was first described by Whartonin in 1659 who gave it the descriptive name, thyroid because of its resemblance to a shield (*thyreoides* in Greek means shield-shaped). In man, the gland consists of two lobes and connected across the ventral surface of the trachea by a narrow bridge called isthmus.

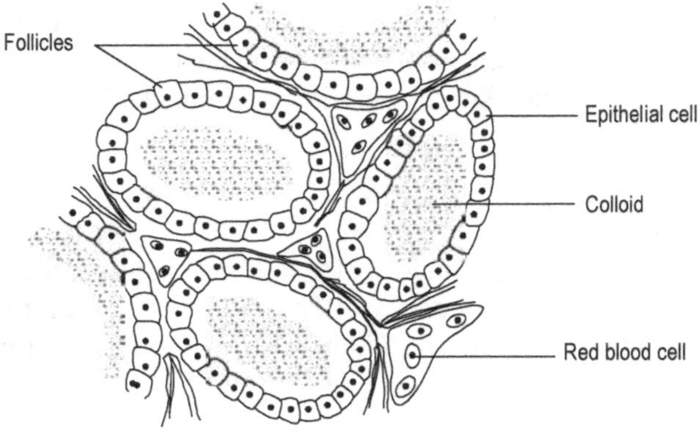

Fig. 13.14 Histology of thyroid gland

Histologically, the thyroid gland is composed of a large number of tiny closed vesicles called follicles, 150 to 300 microns in diameter. The follicles are held together by areolar tissue and are surrounded by a rich network of capillaries. Each follicle is lined with a single peripheral layer of columnar or cuboidal epithelial cells that secrete into the interior of the cells. Its lumen is filled with a secretory substance called colloid (Figure 13.14). The major constituent of colloid is a protein called thyroglobulin, which contains the thyroid hormones. Once the secretion has entered the follicles, it must be absorbed back through the follicular epithelium into the blood before it can perform its function in the body. The thyroid gland is thus unique amongst the endocrine glands in that it stores its hormone as a colloid in small vesicles in the gland. The other endocrine glands, however, store their hormones in the cells themselves.

Thyroxine is the hormone secreted by thyroid gland. Biochemically it is an iodinated protein. Cells of thyroid gland can selectively take up the oral iodide and utilise it for the synthesis of thyroxine. On oxidation by peroxidase, dietary iodide is converted into elemental iodine (I_2), which is stored in the thyroid gland in the form of iodothyroglobulin. It is a glycoprotein with a molecular weight of about 650,000. This protein represents the storage form of the hormone in the gland.

Thyroglobulin is hydrolysed, in the presence of thyrotropin, to release thyroxine (=3, 5, 3′, 5′–tetraiodothyronine) in the blood. The release of thyrotropin is, in turn, controlled by the level of thyroxine in the blood. Thyroxine was first isolated by Kendell, in 1915. Harington and Barger (1925) established its chemical formula. It is an iodine-containing aromatic amino acid and closely resembles tyrosine in structure. Diiodotyrosine is believed to be the precursor of thyroxine.

Besides thyroxine, 3, 5, 3′-triiodothyronine is also produced from enzymatic hydrolysis of thyroglobulin. It is 5 to 10 times more potent in biological activity than thyroxine. The structure of triiodothyronine and tetraiodothyronine is given below:

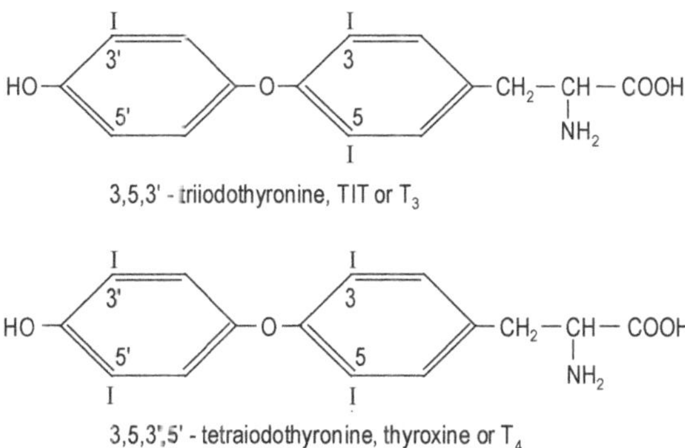

Functions The following functions are attributed to thyroid hormones.

i. They bring about deamination reactions in the liver.

ii. They also carry on deiodination in the extrahepatic tissues.

iii. These influence oxidative phosphorylation by altering the permeability of the mitochondrial membrane.

iv. Their presence accelerates metamorphosis in amphibians. This is a sensitive test and tadpoles have been widely used for the assay of the potency of these hormones.

v. These may increase the level of cytochrome c in the tissues.

Hypothyroidism Underactivity of the thyroid may result from two causes: a degeneration of thyroid cells or a lack of sufficient iodide in the diet. The disease that results from thyroid cell degeneration is **cretinism** in children and **myxoedema** in adults. Cretinism (feeble-mindedness) is characterised by dwarfism and mental suppression. The cretinic children possess a large head and an apathetic face; their teeth erupt late and the speech is retarded. Early treatment with thyroid active compounds partially prevents this disease. Myxoedema is characterised by an abnormally low basal metabolic rate (BMR). Adults with this disease become mentally lethargic and possess thick puffy skin (oedema) and dry hair. The patient shows bagginess under the eyes and swelling of the face. The hair is thin on the eyebrows and scalp. As there is deposition of semi-fluid material under the skin, the name myxoedema is given to this condition. Myxoedema also responds well to administration of thyroid-active compounds. Myxoedema is less severe than cretinism.

Lack of sufficient iodide in the diet results in thyroid gland enlargement, known as **simple goitre**. It is also associated with a low BMR. This type of goitre is also known as **endemic goitre**, since it is prevalent in areas where the soil and drinking water lack iodide. Simple goitre was once common in some mountainous parts of Switzerland and the United States, where soil and water were deficient in iodine compounds.

Hyperthyroidism Abnormally high activity of this gland may occur due to either over-secretion of the gland or an increase in size of the gland. Swelling of the gland results in an **exophthalmic goitre**, characterised by protrusion of the eyeballs. In this, the BMR increases considerably above the normal figures. Consequently, apetite is increased in hyperthyroid individuals. In spite of this, they lose weight and often feel hot because of the increased heat production. Their pulse rate is also increased and excessive sweating occurs. Other symptoms of this disease are dilated pupils, mental excitement, and irritability and cardiac dilatation. Hyperthyroid individuals are, in general, characterised by an above-normal rate in many physiological activities. The clinical syndrome is generally termed **Graves' disease**, after its discoverer Robert James Graves. **Basedow's disease** and **thyrotoxic exophthalmos** are other names of this disease.

Hyperthyroidism can be cured by surgical removal of the thyroid (thyroidectomy), treatment with X-rays, injection of radioactive iodide (^{131}I) or by treating with antithyroid drugs or with agents like thiocyanates or perchlorate, which compete with iodide for the uptake mechanism. Propylthiouracil is being particularly used against the Graves' disease.

13.10 ADRENAL MEDULLARY HORMONES

The adrenal medulla forms the central core of adrenal gland and originates from the neural canal. It is composed of densely packed polyhedral cells containing chromaffin granules. It is highly vascular and receives 6 to 7 ml of blood per gram of tissue per minute. The chromaffin granules store large quantities of adrenal medullary hormones.

Adrenal medulla, whose secretion is under nervous control, produces two hormones: (a) epinephrine or adrenalin ($C_9H_{13}O_3N$) and (b) norepinephrine or noradrenalin ($C_8H_{11}O_3N$) (Figure 13.15 a, b). Epinephrine was the first hormone to be isolated in a crystalline form by Abel. It has been produced synthetically by Stoltz. Chemically, these two hormones are **catecholamines** and are closely related to tyrosine and phenylalanine. Norepinephrine, however, differs from epinephrine structurally in having a hydrogen atom in place of the methyl group.

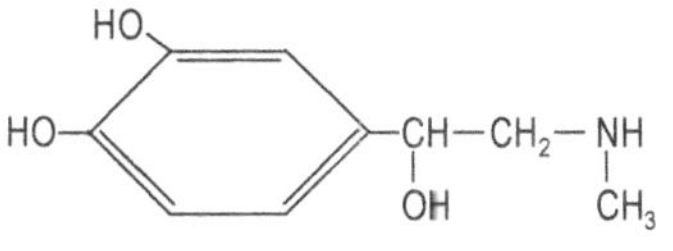

(a) Epinephrine (adrenalin) 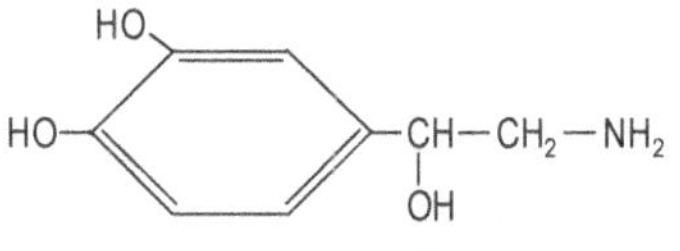 (b) Norepinephrine (noradrenalin)

Fig. 13.15 (a) Epinephrine (Adrenalin) (b) Norepinephrine (Noradrenalin or arterenol)

Since these hormones possess an asymmetric carbon atom, two stereoisomers are possible for each one of them. The naturally occurring epinephrine is the L-isomer and is levorotatory. It is 15 times more active than the D-form. Similarly, the natural D (-) form of norepinephrine is about 20 times more potent than the unnatural isomer.

Functions In general, the adrenal medullary hormones reinforce the functions performed by the sympathetic nervous system. Although both these hormones exert similar effects in regulating carbohydrate metabolism and blood pressure; epinephrine is more closely related to carbohydrate metabolism and norepinephrine to blood pressure.

Epinephrine performs a wide variety of functions, which are as follows:

i. It promotes glycogenolysis in muscles and liver, resulting in an increase of blood glucose level and an increased lactic acid formation in muscles. These changes are then followed by an increase in oxygen consumption.

ii. It causes an increase in blood pressure because of arteriolar vasoconstriction of the skin and splanchnic vessels.

iii. It brings about an increase in the heart rate and in the cardiac output.

iv. It causes dilation of blood vessels (vasodilation) of skeletal muscles, corona and the viscera. This results in an increase of blood flow in these areas.

v. It relaxes the muscles of gastrointestinal tract and bronchials of the lungs, but causes contraction of the pyloric and ileocaecal sphincter muscles.

vi. It also serves in cases of emergency. Under emotional stress, fear or anger, it is secreted in the blood stream and the blood is shifted from the viscera to the brain and the muscles so that the individual becomes aggressive. It is for this reason that the adrenals are frequently referred to as the **'emergency glands'** or the 'glands of flight, fright and fight' and the two adrenal medullary hormones as **'emergency hormones'**.

Norepinephrine, on the other hand, does not relax bronchiolar muscles and has little effect on cardiac output. It augments both systolic and diastolic blood pressure.

13.11 LOCAL HORMONES (TISSUE HORMONES)

Among the local hormones are included those substances which occur in tissues, usually in physiologically inactive forms but which may be activated under certain circumstances to exert profound effects in their immediate neighbourhood. Among local hormones are included the following compounds:

i. Acetylcholine

ii. Histamine

iii. 5-Hydroxytryptamine (5-HT, Serotonin)

iv. Substance-P

13.11.1 ACETYLCHOLINE

Acetylcholine may be formed in the nervous tissues and striated muscles by the action of acetyl coenzyme A on choline. This reaction is carried out in the presence of choline acetylase system. Acetylcholine is a neurohormone, which helps in the transmission of nerve impulses at myoneural junctions. Stimulation of vagus occurs by the release of acetylcholine from its protein-bound form. Stimulation of the preganglionic sympathetic, parasympathetic, post-ganglionic parasympathetic fibres and somatic motor fibres also releases acetylcholine. The nerve fibres, which release acetylcholine, are known as cholinergic.

Acetylcholine helps in the generation of end-plate potential (EPP) at myoneural junctions. After it has completed its action, it is destroyed by acetylcholinesterase.

13.11.2 HISTAMINE

It is a constituent of both animal and plant cells. It occurs in the nettle-sting, wasp and bee venoms in high concentration, and can cause pain and itching. In the mammalian tissues, it is widely distributed particularly in the skin, intestine and lungs.

Histamine is produced because of decarboxylation of the amino acid histidine by an enzyme histidine decarboxylase. Free histamine from the blood is not taken up by the tissues.

The exact form in which histamine occurs in the cells is not known but it is definite that in the mast cells, it occurs in combination with heparin. Histamine is also present in the basophils and blood platelets. In the mast cells histamine occurs in the mitochondrial fraction and when cell membrane is ruptured, histamine is immediately released and becomes physiologically active.

Action of histamine

i. Histamine is a blood pressure-lowering agent because of its ability to dilate capillaries. In man, histamine causes arteriolar as well as capillary dilation. Dilation of meningeal vessels stimulates sensory nerve-endings and causes headache. Histamine increases capillary permeability, which allows escape of some plasma into the tissue spaces thereby reducing blood volume.

ii. Histamine increases the tone of most of the smooth muscles. In the asthmatic patients, histamine can cause an acute asthmatic attack by increasing the tone of bronchiolar muscles.

iii. Histamine causes liberation of gastric juice rich in HCl in the stomach. Cyclic-AMP might be an intermediate in histamine stimulated gastric secretion. There is considerable evidence to show that histamine interacts with the membrane-bound adenylcyclase system of parietal cells resulting in increased intracellular levels of cyclic-AMP which, as second messenger, takes over the role of histamine inside the cells. The rise in acid secretion is paralleled by a dose-dependent increase in the cyclic-AMP content of gastric mucosa.

Histamine liberators Several chemical compounds, when they come in contact with the tissue, can release histamine from the tissues, and hence these are known as histamine liberators. These agents include morphine, pethidine, tubocurarine and certain diamidines used in trypanosomiasis. A substance known as compound 48/80, a condensation product of p-methoxyphenylmethylamine with formaldehyde is a powerful histamine liberator.

Anti-histamines Anti-histamines are those substances, which can antagonise the effect of histamine, such as adrenaline, mepyramine, promethazine, etc. Antagonism of histamine may be effected in the following manner:

By acting as competitive inhibitors The anti-histaminic drugs get attached to the same site where histamine would have normally attached. Naturally, the response of histamine does not appear.

By acting against histamine Anti-histamines act by counteracting histamine by relaxing bronchial muscles, constricting blood vessels of skin and raising blood pressure.

13.11.3 5-HYDROXYTRYPTAMINE (SEROTONIN, ENTERAMINE)

5-hydroxytryptamine (5-HT) is also a naturally occurring biologically active amine and it is of immense pharmacological importance. It occurs in blood platelets, central nervous system, skin and veins. 5-hydroxytryptamine is synthesised from tryptophan.

Functions

i. 5-hydroxytryptamine is a cardiac stimulant and a potent vasoconstricting agent. In man, it can raise the systolic and diastolic blood pressure.

ii. It increases the tone of the smooth muscles, such as muscles of arterioles, bronchioles, bladder and intestine.

iii. 5-HT also exerts an anti-diuretic effect.

iv. It can stimulate chemoreceptors and certain cholinergic ganglia. It is a very potent stimulant of nerve endings percepting pain in the human skin.

13.11.4 SUBSTANCE P

This substance is a polypeptide and has been isolated from intestine and nervous tissue. It stimulates smooth muscles and causes vasodilation.

The complete sequence of the amino acids in substance-P has been found to be as follows.

Arg.—Pro.—Lys.—Pro.—Gln.—Gln.—Phe.—Phe.—Gly.—Leu.—Met.—NH_2

13.12 VASOACTIVE PEPTIDES

In addition to these local hormones which are liberated from the tissues, there exist certain other hormones, which are formed in the blood itself and perform their functions in blood. These hormones are called as vasoactive peptides. There are three groups of well-defined peptides that possess vasoactive properties.

i. Neurohypophysial hormones: (a) oxytocin (b) vasopressin

ii. Angiotensin

iii. Kinin - (a) Bradykinin (b) Kallidin

13.12.1 NEUROHYPOPHYSIAL HORMONES

See section 13.5.9.

13.12.2 ANGIOTENSIN

Angiotensin is synthesised in the blood by the action of renin, a proteolytic enzyme secreted by ischemic renal cortex, an α_2-globulin which is normally present in the blood plasma. This globulin is produced by the liver. This specific α_2-globulin is known as angiotensinogen. The enzyme renin, splits off a decapeptide from this globulin. The decapeptide is known as angiotensin-I. Another enzyme known as converting enzyme present in blood, acts on angiotensin-I, and splits off two amino acids. The residual octapeptide is known as angiotensin-II, (Figure 13.16)

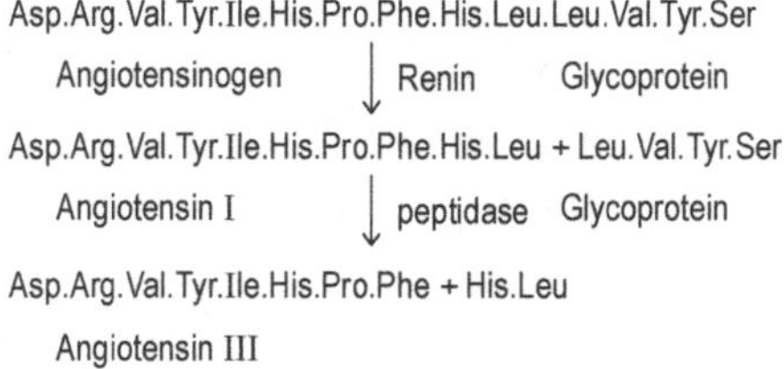

Fig. 13.16 Formation of angiotensin II

which possesses pressure activity. The pressure activity of angiotensin-II is 200 times more as compared to norepinephrine. Angiotensin increases the force of heartbeat and raises blood pressure. Besides this, it also brings about contraction of the smooth muscles. Recently, it has been shown that administration of angiotensin-II increases aldosterone secretion. The normal kidney and other tissues contain a proteolytic enzyme angiotensinase, which destroys angiotensin.

13.12.3 KININS

'Kinins' is the generic name for a group of peptides with potent biological activities in causing smooth muscle contraction, vasodilation, lowering of blood pressure, increasing blood flow and microvascular permeability, and inducing the emigration of granulocytic leukocytes. Kinins are liberated from plasma protein called kininogen exposed to snake venom or to the proteolytic enzyme. Bradykinin and kallidin are typical examples of kinins.

Bradykinin This is a nanopeptide, formed by the action of trypsin or proteolytic enzymes in snake venoms, on plasma α_2-globulin. Bradykinin is destroyed by carboxypeptidase present in RBC, which destroys it within minutes.

Kallidin It is a decapeptide found in saliva and plasma. In plasma, it is formed from a protein α_2-globulin, known as kallidinogen. The decapeptide is split off by a proteolytic enzyme present in snake venoms, and by the enzymes of high substrate specificity known as kallikrines, which occur in saliva, pancreatic secretion, blood and other body fluids. Plasma kallikrines exist in inactive form known as kallikrinogens. Conversion of kallikrinogen into kallikrine is brought about by changes in pH, dilution with saline and contact with foreign substances. It has been shown that Hageman factor of blood clotting system is essential for the conversion of plasma kallikrinogen to kallikrine.

Functions of bradykinin and kallidin The plasma kinins, e.g. bradykinin and kallidin are the most potent vasodilator substances known. These are ten times more active than histamine. The blood vessels of muscles, brain, kidney, viscera and several other glands are dilated by these substances. The pulmonary vascular bed however is dilated at considerably higher dose. The dilator effects result from a direct action of the polypeptide on vascular smooth muscles. A sharp fall in systolic as well as diastolic blood pressure is observed after administration of these kinins.

13.13 OTHER ALLEGED HORMONES

13.13.1 PINEAL HORMONE OR MELATONIN

The pineal gland (epiphysis) is a tiny structure lying deep in the groove between the cerebellum and the cerebral hemispheres. Bruce Fellman (1987) calls this as a **clockwork gland.** It secretes a hormone called melatonin, which inhibits the secretion of an adenohypophyseal hormone called luteinizing hormone. Melatonin is synthesised from its precursor, serotonin, under the influence of an enzyme called **hydroxyindole-o-methyl transferase** (HIOMT), which is present in rich quantity in the pineal gland (Figure 13.17). Serotonin was first isolated and crystallised by Page and others in 1948.

The removal of pineal gland causes the ovaries to undergo hypertrophy whereas, the administration of melatonin decreases the weight of the ovaries.

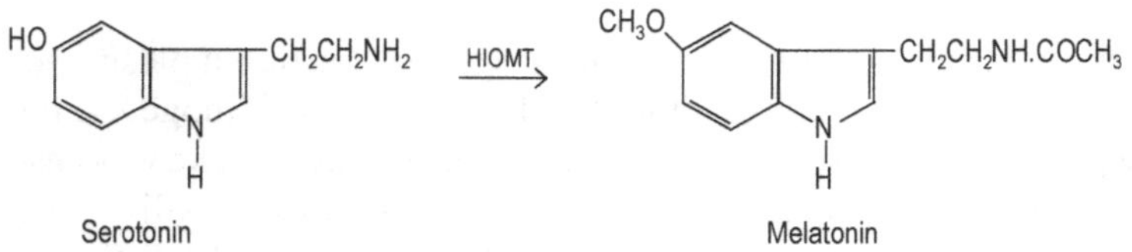

HO–(indole)–$CH_2CH_2NH_2$ $\xrightarrow{\text{HIOMT}}$ CH_3O–(indole)–$CH_2CH_2NH.COCH_3$

Serotonin — Melatonin

Fig. 13.17 Synthesis of melatonin from serotonin

13.13.2 RENAL HORMONES

The kidney secretes two hormones, erythropoietin and renin.

Erythropoietin (or erythrocyte stimulating factor, ESF) It is secreted by the kidney. Its secretion is stimulated by tissue anoxia and also by androgenic hormones and cobalt. In fact, erythropoietin is secreted as an inactive protein called **renal erythropoietin factor** (REF), which is enzyme-like in behaviour. The REF converts a plasma globulin to the active erythropoietin.

$$\text{Plasma globulin} \xrightarrow{\text{REF (Enzyme)}} \text{Active erythropoietin}$$

Erthropoietin has been prepared from the plasma of anaemic sheep and from the urine of anaemic human beings. It is a glycoprotein with

8–12% total hexose and has a molecular weight of about 60,000. Its molecule contains all the common amino acids except methionine. Its activity is much retarded by proteolytic enzymes and also by the antibiotic actinomycin-D.

Erythropoietin stimulates the differentiation of the stem cell (hemohistoblast) of the bone marrow into the erythroid series, increasing the number of proerythroblasts in the bone marrow. This is followed by increase in other nucleated erythrocytes and erythrocyte precursor cells and mature erythrocytes in the peripheral circulation. Recent studies indicate that the earliest effect of erythropoietin is stimulation of the synthesis of a very large RNA by bone marrow cells.

Renin It catalyses the synthesis of angiotensins, which causes vasoconstriction in the kidneys, thereby causing electrolyte and water retention in the body. This system has been referred to as the *reninangiotensin system*. An increase in the renal pressure and the levels of plasma Na^+ and angiotensins cause a decrease in renin production.

13.13.3 PROSTAGLANDINS (PG)

Prostaglandins are the most important group of physiologically active substances. The prostaglandins comprise a group of fatty acids which exhibit highly potent and manifold effects in biological systems. The name "prostaglandins" was given by von Euler in 1935 to a principal compound found in human seminal plasma and vesicular glands of sheep, which induced contractions in isolated smooth muscle preparations. The prostaglandins are known to occur widely in almost all kind of tissues, in very small quantities (nanogram to micrograms) and they have been identified from a great number of human tissues, such as seminal plasma, menstrual fluid, lungs, etc.

Prostaglandins are derivatives of a hypothetical C_{20} saturated fatty acid called *prostanoic acid* (Figure 13.18). It consists of 5-carbon rings with two hydrocarbon chains attached to two neighbouring carbon atoms. Naturally occurring prostaglandins are of 4 types: PGE, PGF, PGA and PGB types.

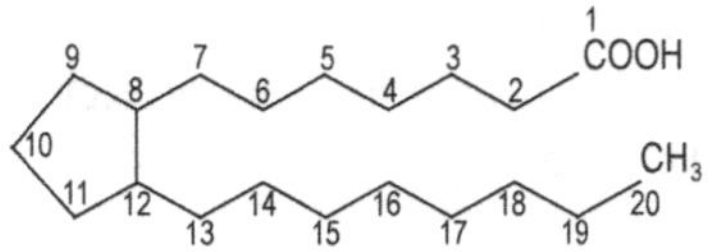

Fig. 13.18 Prostanoic acid

The prostaglandins as a class are represented as 'PG', followed by the appropriate letter as A, B, E, F with numerical subscripts to denote the specific number of each class. Further subdivisions utilises 'α' or 'β'. The numerical subscripts in each class are given depending upon the number of double bonds in the extra-ring structure (i.e. in the side chains). The double bond in the ring is not counted for this purpose. Alpha after the numerical subscript denotes the placing of hydroxyl group.

The different prostaglandins differ from one another in the number and position of the double bonds and hydroxyl group substituents.

Prostaglandins E_1, E_2, E_3, $F_1\alpha$ and $F_3\alpha$ are considered as primary prostaglandins. PGAs are produced by loss of the elements of water from the five-membered ring of the PGEs, whereas PGBs are produced from PGAs by isomerisation of the double bond present in the ring.

The structures of some important prostaglandins are given in Figure 13.19.

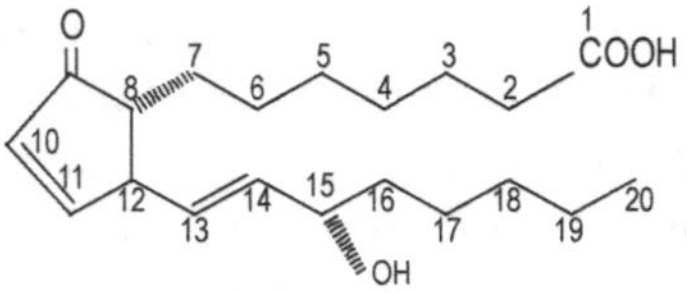

PGA_1
(15-Hydroxy-9 keto (10-ene), prosta-13t, enoic acid)

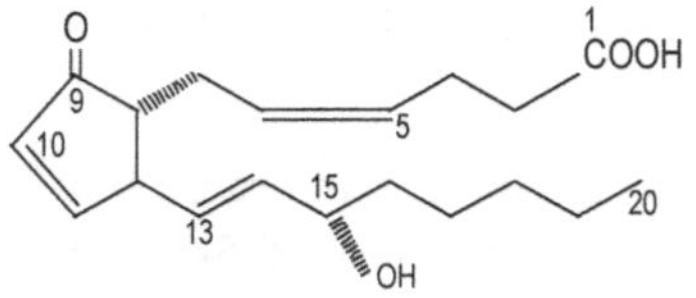

PGA_2
(15-Hydroxy-9 keto (10-ene), prosta-5c, 13t, dienoic acid)

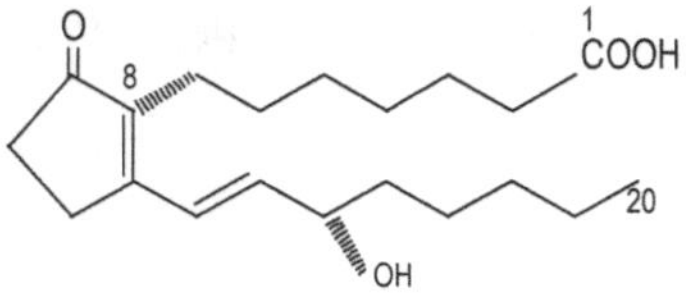

PGB_1
(15-Hydroxy-9 keto (8-ene), prosta-13t, enoic acid)

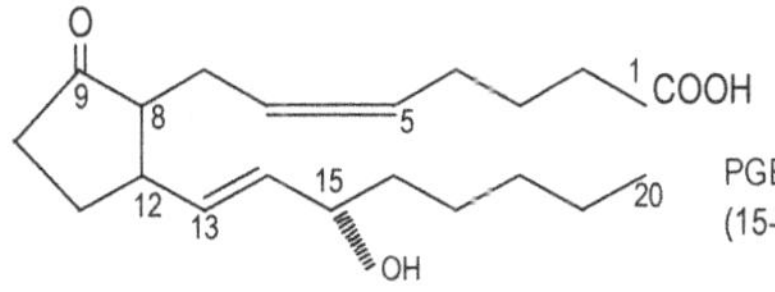

PGB_2
(15-Hydroxy-9 keto (8-ene), prosta-5c, 13t, dienoic acid)

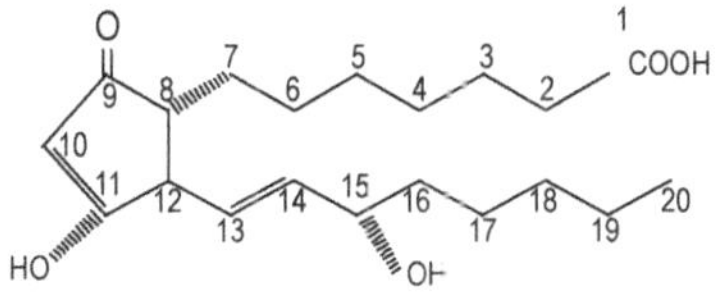

PGE_1
(11-α, 15-Dihydroxy-9 keto-prosta 13t-enoic acid)

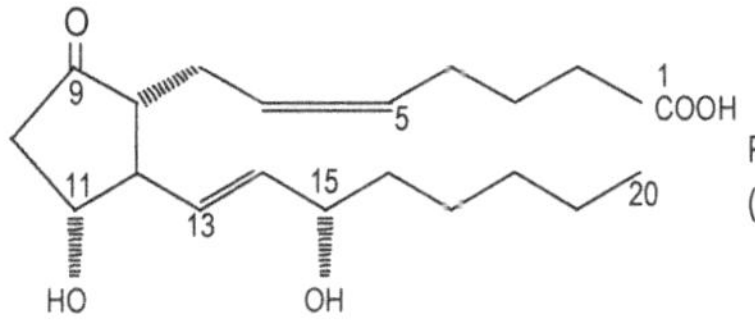

PGE_2
(11-α,15-Dihydroxy-9 keto-prosta-5c, 13t-enoic acid)

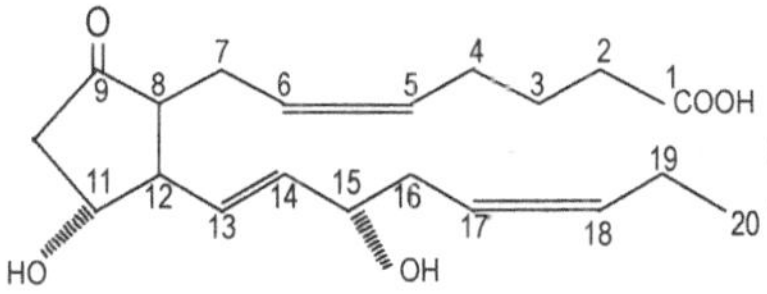

PGE_3
(11-α,15-Dihydroxy-9 keto-prosta-5c, 13t, 17c-trienoic acid)

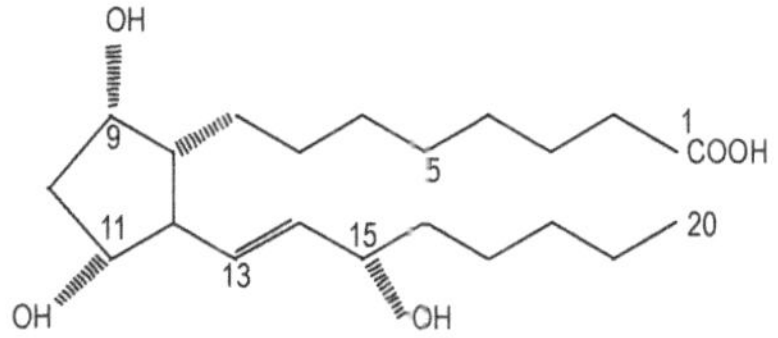

$PGF_1\alpha$
(9-α,11-α, 15-Trihydroxy prosta 13t-enoic acid)

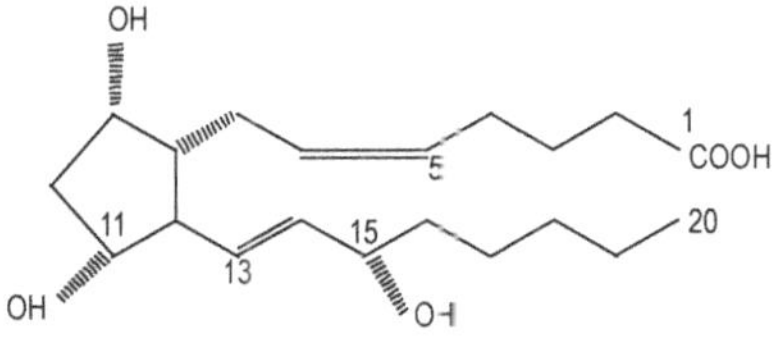

$PGF_2\alpha$
(9-α, 11-α, 15-Trihydroxy prosta 5c, 13t-dienoic acid)

Fig. 13.19 Structures of some important prostaglandins

Various members of the E-series differ from the F-series only in possessing 9α -hydroxyl groups in place of 9-keto group.

13.13.4 PHYSIOLOGICAL ROLE OF PROSTAGLANDINS

Prostaglandins influence numerous aspects of cell metabolism.

Effect on muscle contraction They induce powerful contractions in smooth muscle like rat uterus muscle, gastrointestinal smooth muscle, respiratory tract smooth muscle, rabbit duodenum muscle, etc. However, in several species, PGE has been shown to relax the circular muscles of intestine but on the other hand induce contractions in longitudinal muscles.

Effect on reproductive system Prostaglandins stimulate the activity of the uterus during pregnancy and this has found practical utility in inducing expulsion of products of conception i.e. foetus and placenta. Very small quantities of prostaglandins are required either intravenously or for introducing directly into the uterine cavity to induce abortion.

A functional role for prostaglandins in parturition is suggested by increase of prostaglandin concentration in human blood and amniotic fluid during labour and by the observation that exogenously administered prostaglandins can induce labour or abortion. Only a few prostaglandins i.e. PGE_3 and $PGF_2\alpha$ and a number of their analogs, in particular 15-methyl $PGF_2\alpha$ and 15-methyl PGE_2 have been found to be effective in inducing abortion. Mechanical stimuli such as uterine distention can promote prostaglandin production so that the increase in prostaglandin release during labour could be a consequence of increased mechanical activity of uterus.

Prostaglandins $PGF_1\alpha$, $PGF_2\alpha$, PGE_1 and PGE_2 have shown to exert potent oxytocic properties and all have been successfully used in induction of labour and abortion.

Prostaglandins $F_2\alpha$ induce luteolysis at early stages of pregnancy in a number of species including rat. Prostaglandins have also been shown to contract the proximal (uterine) end of the fallopian tube but relax the distal three quarters. Such prostaglandins occur in human semen; and their action could result in the retention of the ovum within the tube and thus increase the possibility of fertilisation.

Effect on nervous system Prostaglandins are natural constituents of nervous tissue. They are released from the brain upon stimulation of afferent pathways. The possibility of prostaglandins acting as mediators of synaptic transmission has also been considered. An important metabolic effect is that the PGE_1 antagonises the stimulatory effects of a number of hormonal compounds on the release of free fatty acids and glycerol from epididymal fat pads of rats.

Effect on inflammatory response Prostaglandins are known to have a unique role in the inflammatory response. Gastric acid is an important factor in causing gastrointestinal ulcers. Prostaglandins have some effect on the inhibition of gastric secretion to check ulcer formation.

Effect on renal function There are reports suggesting the role of prostaglandins in renal functions. Intravenous infusion of norepinephrine has been shown to release prostaglandin-like substances into the artery. It has been reported that the diuretic function of norepinephrine can be mediated by prostaglandin release. The presence of prostaglandins in the kidney and their release under various circumstances suggests a relationship with renal hypertension.

The numerous reports on the action of prostaglandins in the water flow and sodium transport across toad bladder, frog skin, guinea pig ileum, have suggested that prostaglandins function in the regulation of ion fluxes across epithelial membranes.

Effect on osmotic and ionic regulation All prostaglandins enhance the ionic flux across epithelial membranes. Infusion of PGE_1 and $PGF_2\alpha$ into the renal artery of dogs increases urinary volume and excretion of Na^+, K^+ and Cl^-. Prostaglandins can stimulate intestinal adenylate cyclase and thus can inhibit Na^+ entry into the mucosal cells with an accompanying increase in Cl^- secretion. These effects coupled with increased H_2O secretion and increased intestinal motility cause diarrhoea. $PGF_2\alpha$ may be secreted by some thyroid tumours, which is accompanied by diarrhoea. Similarly, prostaglandins when given to terminate pregnancy produce diarrhoea as an undesirable effect. It is also suggested that cholera toxin causes synthesis of E type prostaglandin in the human intestine and thus causes diarrhoea.

Effect on cardiovascular function Prostaglandins can augment blood flow by inducing a generalised vasodilation with decreased peripheral resistance. These changes cause increased cardiac output and decreased arterial blood pressure.

Effect on cyclic-AMP level In most of the tissues, prostaglandins stimulate the formation of cyclic AMP, e.g. endocrine glands, by stimulating adenyl cyclase activity. PGE_1 and PGE_2 stimulate this activity. However, these two PGs inhibit this enzyme in the adipose tissue and lower the cyclic AMP level.

Effect on plasma free fatty acids Low doses of PGE_1 have been reported to elevate plasma free fatty acid concentration, whereas high doses decrease FFA concentration.

13.14 HORMONE FROM THYMUS

Thymus is present in all jawed vertebrates. It is a flat, pinkish, bilobed structure, located in the chest behind the sternum. The gland arises as a proliferation of the gill pouch epithelium, which becomes infiltrated with lymphocytes. It grows rapidly to acquire a large size in the young animals and atrophies in the adults. In man, the gland reaches its greatest development at the age of 14 to 16, after which it atrophies because of the activity of the sex cells. Thymus is, however, absent in the hagfishes and in the lamprey, it is represented by a group of cells beneath the gill epithelium. This is called as a prothymus. An active glycopeptide is present in thymus.

Functions Certain biochemical functions are attributed to this gland. These are as follows:

It is the primary source of lymphocytes in the mammals. The lymphocytes are responsible for the immunological functions of the body by producing antibodies.

13.15 MECHANISM OF HORMONAL ACTION

Hormones can influence diverse cellular activities including increase in permeability of the cell membranes to different solutes and ions, altering the rate of synthetic and catabolic processes, stimulatory or inhibitory effects on enzyme systems, effects on nuclear material and expression of genes. The anterior pituitary affects target glands through its tropic hormones which in turn stimulate the activity and differentiation of other cells and tissues assisting in their growth and development and the synthesis of characteristic molecules. Thyroxine brings about the metamorphosis of the tadpole into the adult frog which is accompanied by an alteration from excretion of ammonia as a product of nitrogen metabolism to the production of urea, such a change being coupled with induction of urea-synthesising enzymes. Estrogens have a stimulating effect on the synthesis of all types of RNA—mRNA, tRNA and rRNA.

Adrenaline influences catabolic processes involving the breakdown of glycogen and lipolysis of triglycerides, whereas, insulin promotes glycogenesis and lipogenesis. The action of adrenaline has been traced to its effect on stimulation of adenylate cyclase, a membrane-bound enzyme which acts on ATP to liberate intracellular cyclic AMP which mediates the action of several hormones. Receptors are present on the target cells and bind tightly to the specific hormone. Binding of the hormone to the active site of the cell leads to the activation of adenylate cyclase with formation of cyclic AMP, a second messenger, which activates a cascade of metabolic enzymes and amplifies the activity of the hormone inside the cell. Inactive phosphorylase consists of a regulatory unit, a cyclic AMP binding site, and a catalytic unit. The catalytic unit is activated when cyclic AMP binds to the regulatory unit (Figure 13.20).

Calcium ions are also concerned with regulation of these reaction mechanisms, probably through calmodulin interaction. Calmodulin is a small protein consisting of a single polypeptide chain containing 149 amino acids and is found in nearly all eukaryotic cells. Each molecule of calmodulin binds with four calcium ions and the complex influences cellular activities in various ways. For example, the calcium–calmodulin complex may serve to directly activate several enzymes. Such effect has been demonstrated for phosphorylase kinase, an enzyme controlling glycogen breakdown. However, the effect of this complex on cell metabolism may be slow and indirect. The Ca–calmodulin complex acts to stimulate adenylate cyclase activity in the plasma membrane of the cells, resulting in the production of cAMP which influences cell metabolism as a secondary messenger.

Cyclic AMP mediates the action of a variety of hormones including the hypothalamic releasing hormones, the tropic hormones of the anterior pituitary vasopressin, MSH, parathyroid hormone, adrenaline and glucagon as well as some neurotransmitters.

Four general theories of hormone action have been proposed to explain the mechanism of hormonal regulation: (a) Hormone action on enzyme system of a cell. (b) Hormone action on membrane system of a cell. (c) Hormone action on genes and (d) Hormone action on controlled release of ions or other micromolecules.

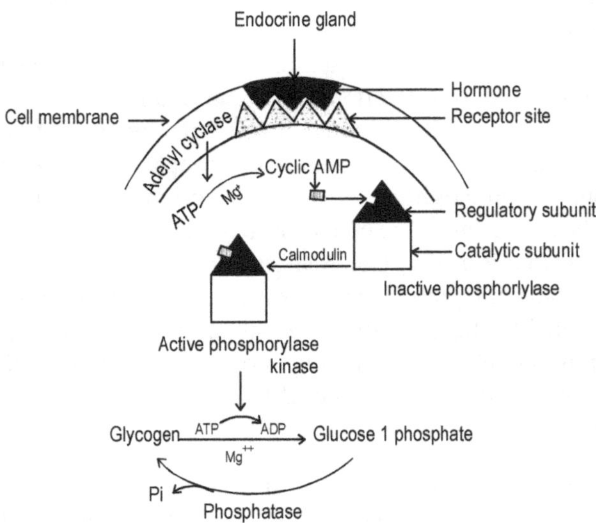

Fig. 13.20 Hormonal action: the release of cyclic AMP as a second messenger and activation of phosphorylase

Hormonal action on enzyme system of a cell Hormones exert a direct quantitative effect upon intracellular enzyme systems. There is also a considerable interaction between hormones and enzyme systems as are evident from the fact that many vitamins are cofactors both for enzymes and hormones. From the experiments on cell free extracts, it has been proved that even the minute amounts of hormones like epinephrine and glucagon bring about the conversion of inactive phosphorylase *b* into the active form. Some hormones have been shown to exert influence upon complex metabolic systems by selectively affecting a particular enzyme step. For example, insulin selectively influences the hexokinase system of complex glycogenolytic pathway.

Hormonal action on membrane system of cell Hormones are found to control permeability relationships at the cell surface or elsewhere and hence, indirectly condition the enzymatic reactions. It has been proved that many enzyme proteins are the structural proteins of the membrane system.

Hormonal action on genes Hormones can regulate the formation of chromosomal puff by activating or suppressing specific gene loci. This hypothesis is an outcome of studies on insect development, particularly the effect of ecdysone on the giant chromosomes of the salivary glands. The chromosomes are actually bundles of chromosomal threads arranged in a regular manner. Studies on the midge chironomus indicate that the transverse bands tend to loosen or puff up sequentially during the course of larval development. The puffs have been identified as the sites of protein synthesis and are produced because of hormonal action on genes. Thus, a hormone is supposed to bring about the synthesis of required enzymes through DNA-directed protein synthesis.

Hormonal action by controlled release of ions Some hormones control the release and uptake of ions. TSH (Thyrotrophic hormone) which regulates iodine is important in the synthetic processes and metabolism. Parathormone and calcitonin regulate blood calcium level.

13.16 SYNTHETIC HORMONES

The normal course of ovarian cycle and all events of female sexual cycle, right from ovulation to parturition are controlled by hormones, primarily from pituitary and secondarily from ovary and placenta. Thus, a specific hormonal milieu coordinates and controls all the phases right from development of an ovum to the birth of a child.

Synthetic chemical substances can be used to alter the hormonal milieu in such a way that neither ovulation nor implantation will occur; fertility will be effectively checked. Synthetic hormones are nothing but hormones like estrogen and progesterone or analogous synthetic steroids, which are physiologically analogous to natural steroids. When orally administered, they disturb the hormonal milieu so that pregnancy fails even if fertilisation or implantation occurs.

Examples: Analogs of progesterone: norethynodrel, norethydrone, norethydrone acetate, medroxyprogesterone acetate, ethynodiol diacetate, dimethisterone, chlormadinone acetate, norethindrone, etc.

Analogs of estrogen: mestranol, ethynyl estradiol, etc.

Analogs of protein hormones: synthetic insulin and synthetic thyroxine.

Synthetic hormones produce many side effects. Some of the common side effects are irregular bleeding, weight gain, gastrointestinal symptoms, i.e. headache, dizziness, oedema and chloasma, yellow patches on back and neck.

REVIEW QUESTIONS

1. Give an account of steroid hormones.
2. Explain the chemical nature and functions of corticosteroids.
3. What are the different hormones secreted by pituitary gland? Discuss their functions.
4. Enumerate the hormones secreted by gastrointestinal tract. Add a note on their functions.
5. Discuss the synthesis and functions of thyroxine.
6. What are local or tissue hormones? Discuss their functions.
7. Give an account of vasoactive peptides.
8. Discuss the chemical nature and physiological role of prostoglandins.
9. Explain the mechanism of hormonal action.
10. Write short notes on:
    i. C18 steroids
    ii. C19 steroids
    iii. C21 steroids
    iv. Addison's disease

v. Cushing's syndrome
vi. Mineralocorticoids
vii. Glucocorticoids
viii. Insulin
ix. Parathormone
x. Calcitonin
xi. Catecholamines
xii. Melatonin
xiii. Graves' disease
xiv. Conn's syndrome
xv. Histamine
xvi. Angiotensin
xvii. Kinin
xviii. Renal hormones
xix. Secondary messenger
xxi. Calmodulin
xxii. Synthetic hormones

14

BIOLOGICAL DETOXIFICATION

The term **biological detoxification** can be defined as the mechanism by which the foreign substances and the toxic substances produced *in vitro* are converted into non-toxic or less toxic substances, under the influence of enzymes. Foreign substances include all those substances, which are not ordinarily ingested or utilised by the body. Most of the toxic substances whether endogenous or exogenous are converted after biological detoxification into products which are non-toxic or less toxic to the body than the parent toxic substances. Such detoxified products can be readily excreted through various excretory channels. The toxic substances may be produced inside the body as a result of metabolism or these may find entry into the body along with the food material, or may be administered orally or parenterally in the form of medicine. If such compounds are not detoxified immediately, these might produce certain harmful or undesired effects in the body and sometimes may prove to be fatal.

14.1 PRINCIPLES OF DETOXIFICATION

Detoxification may be brought about by **oxidation, reduction** and **hydrolysis**. The products in most cases are then subjected to a final process, **synthesis** or **conjugation** or coupling with some substances in the body which will render it to a form suitable for excretion. Such reactions are mainly carried out in the liver with the help of enzymes.

14.2 SITES OF DETOXIFICATION

In most cases, liver is the site of detoxification but kidney and some other organs may also participate in this process. In some cases, intactness of the tissue structure has been found to be essential in order to bring about detoxification.

14.3 MECHANISM OF DETOXIFICATION

Detoxification may be brought about by any one or more of the following reactions:

a. Oxidation b. Reduction c. Hydrolysis d. Conjugation

14.3.1 DETOXIFICATION BY OXIDATION REACTIONS

Several foreign compounds are detoxified in the body by the process of oxidation. Such compounds include methyl alcohol, ethyl alcohol, benzyl alcohol, some secondary alcohols, and some amines such as benzyl amine; aromatic hydrocarbons such as benzene; drugs such as meprobamate, chloral hydrate; and aromatic aldehydes such as, vanilline, benzaldehyde, etc. Some of the oxidised compounds later on get conjugated with glucuronic acid and may be excreted in the form of glucuronides. The oxidation products formed during detoxification of the above mentioned compounds are given below:

1. Alcohols

a) $CH_3OH \longrightarrow H.COOH$

Methyl alcohol → Formic acid

b) $CH_3CH_2OH \longrightarrow CH_3.C(=O)H$

Ethyl alcohol → Acetaldehyde

c) $C_6H_5CH_2OH$ ⟶ C_6H_5COOH

Benzyl alcohol ⟶ Benzoic acid

d) Phenylethyl alcohol ⟶ Phenylacetic acid

2. Amines

a) Benzylamine ⟶ Benzoic acid
b) Aniline ⟶ p-aminophenol

3. Aldehydes

a) Benzaldehyde ⟶ Benzoic acid
b) Vanilline ⟶ Vanillic acid

4. Aromatic hydrocarbons

a) Toluene ⟶ Benzoic acid
b) Benzene ⟶ Phenol (OH)

5. Anilides

Acetanilide ⟶ p-acetylaminophenol

6. Drugs

Meprobamate ⟶ Hydoxymeprobamate

14.3.2 DETOXIFICATION BY REDUCTION REACTIONS

Detoxification by this process is less common in human beings. The different foreign compounds detoxified by this mechanism are picric acid, p-nitrobenzaldehyde, p-nitrophenol, 2,4-dinitrophenol, 2,4,6-trinitrotoluene and chloral. The detoxified products formed are listed below:

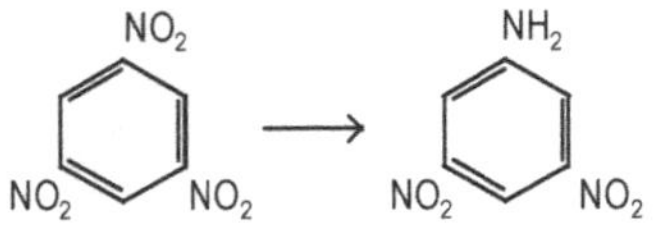

NO$_2$ (picric acid ring, NO$_2$, NO$_2$) ⟶ NH$_2$ (ring, NO$_2$, NO$_2$)

Picric acid ⟶ Picramic acid

p-Nitrobenzaldehyde ⟶ p-Aminobenzoic acid

p-Nitrobenzene ⟶ p-Nitrophenol

2,4,6-Trinitrotoluene ⟶ 2,6-Dinitro-4-aminotoluene

Such compounds are also partially excreted in the form of glucuronides besides being excreted in the free form.

14.3.3 DETOXIFICATION BY HYDROLYSIS

Drugs like procaine, acetyl salicylic acid and cardiac glycosides like digitalin undergo hydrolysis.

$$NH_2C_6H_4COO\,CH_2\,CH_2N(C_2H_5)_2 + H_2O \longrightarrow NH_2C_6H_4COOH + HO\,CH_2CH_2N(C_2H_5)_2$$

Procaine → p-Aminobenzoic acid + Diethylaminoethanol

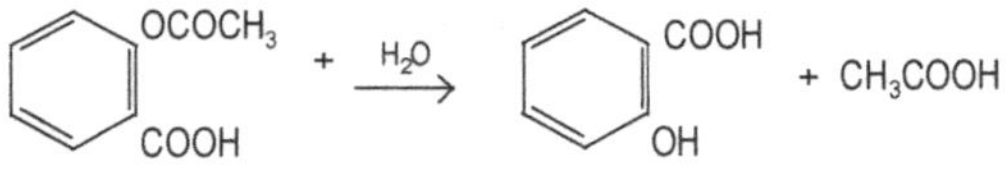

Acetyl salicylic acid (C_6H_4(OCOCH$_3$)(COOH)) + H_2O ⟶ Salicylic acid (C_6H_4(COOH)(OH)) + CH_3COOH

14.3.4 DETOXIFICATION BY CONJUGATE REACTIONS

In this process, the foreign substances get coupled with compounds known as conjugating agents which can be readily excreted. In man such reactions predominantly occur in the liver tissue, and to some extent in the kidney. In cases where oxidation fails, conjugation becomes an alternative method of detoxification. The conjugated products are eliminated from the body. The different conjugating agents normally occurring in the body are glucuronic acid, sulphuric acid, glycine, acetic acid, glutamine, ornithine, cysteine and methyl group.

i. **Conjugation with glucuronic acid** Many compounds which contain free hydroxy or carboxyl groups, or those compounds which may be converted into compounds bearing such groups, form conjugates with glucuronic acid. Two types of glucuronides are usually formed, namely ether type and ester type depending upon whether glucuronic acid is conjugating with an alcoholic or carboxylic group. Some of the substances detoxified by glucuronide formation are given below:

Glucuronic acid + Phenol ⟶ Phenyl glucuronide (Ether linkage)

Glucuronic acid + Benzoic acid ⟶ Benzoyl glucuronide (Ester linkage)

Stilbestrol ⟶ Stilbestrol glucuronide

Phenylacetic acid ⟶ Phenylacetyl glucuronide

Female sex hormones (steroids) ⟶ Steroid glucuronides

Sulphapyridine ⟶ Hydroxysulphapyridine glucuronide

An enzyme β-glucuronidase which hydrolyses glucuronides, can also catalyse the reverse reaction bringing about coupling of varoius substances with glucuronic acid. This enzyme is mostly present in the liver tissue.

ii. **Conjugation with sulphuric acid** In general, phenolic compounds have a tendency to conjugate with sulphuric acid forming sulphates which can be readily excreted. The examples of different compounds forming such conjugates are given below:

Phenol + H_2SO_4 ⟶ Phenyl sulphuric acid

Indole ⟶ Indican (*indoxyl sulphate*)
Androsterone ⟶ Androsterone sulphate
p-cresol ⟶ p-cresol sulphate
α-naphthol ⟶ α-naphthol sulphate
Acetanilide ⟶ p-acetaminophenol sulphate

It has been observed that administration of epinephrine increases conjugation of the above substances with sulphates.

iii. **Conjugation with glycine** The amino acid glycine mostly conjugates with the acidic compounds. Glycine conjugates with benzoic acid and forms hippuric acid, and with niacin forms nicotinuric acid.

Benzoic acid (COOH) + H_2NCH_2COOH (Glycine) ⟶ Benzoyl glycine (hippuric acid) (CO.NH.CH_2COOH) + H_2O

Niacin (COOH) + H_2NCH_2COOH (Glycine) ⟶ Nicotinuric acid (CO.NH.CH_2COOH) + H_2O

iv. **Conjugation with acetic acid** Acetic acid usually conjugates with the amino compounds. Insulin has been reported to increase such acetylation reactions:

Sulphanilamide (SO_2NH_2, NH_2) + CH_3COOH (Acetic acid) ⟶ Acelytated sulphanilamide (SO_2NH_2, $NHCOCH_3$)

p-Aminobenzoic acid (PABA) ⟶ p-acetylaminobenzoic acid

v. **Conjugation with ornithine** Animals utilise ornithine for detoxification of phenylacetic acid.

Phenylacetic acid + Ornithine ⟶ Diphenylacetylornithine

vi. **Conjugation with cysteine** Cysteine is involved in the detoxification of mainly halogen compounds and forms their mercapturic acids:

Bromobenzene ⟶ p-bromophenylmercapturic acid
Chlorobenzene ⟶ p-chlorophenylmercapturic acid
Iodobenzene ⟶ p-iodophenylmercapturic acid
Flurobenzene ⟶ p-fluorophenylmercapturic acid
Naphthalene ⟶ Naphthalenemercapturic acid

vii. **Conjugation by methylation reactions** This type of detoxification is not very common. The methyl group required for such reactions is derived from active methionine:

Nicotinamide ($CONH_2$) + CH_3 (Methyl group) ⟶ N-methyl nicotinamide ($CONH_2$, N–CH_3)

Pyridine ⟶ Methylhydroxypyridine
Norepinephrine ⟶ Normetanephrine

14.3.5 DETOXIFICATION OF CYANIDES

Cyanides are received by the body from fruits, by breakdown of proteins and from tobacco smoke. These are toxic to the body and are detoxified to thiocyanates. This conversion is brought about by an enzyme rhodanase, which is formed in the liver in the presence of suitable sulphur compound.

$$\underset{\text{Cyanide}}{HCN} + \underset{\text{Sodium thiosulphate}}{Na_2S_2O_3} \longrightarrow \underset{\text{Thiocyanate}}{HCNS} + Na_2SO_3$$

14.3.6 TOXIC COMPOUNDS OF LARGE INTESTINE

The toxic products of large intestine, such as indole, skatole, histamine and putrescine are mostly eliminated in the stool. A small amount of these substances may be absorbed. These substances after absorption are conjugated in the liver and kidney with sulphuric acid, and glucuronic acid and are eliminated in the urine in the form of glucuronides.

REVIEW QUESTIONS

1. Describe the mechanism of detoxification.
2. Write short notes on:
   i. Biological detoxification
   ii. Detoxification by oxidation reaction
   iii. Detoxification by reduction
   iv. Detoxification by hydrolysis
   v. Detoxification by conjugate reaction

15

ANTIBIOTICS

An antibiotic is a chemical substance produced by a living organism that demonstrates inhibitory or germicidal activity towards microorganisms. The term antibiosis was first defined by Vuillemin in 1889 as a condition in which one creature destroys the life of another in order to sustain its own, the first being entirely active and the second entirely passive. Later, in 1945, Waksman coined the term antibiotics for the chemical substances of microbial origin, which in small amounts exert antimicrobial activity.

Antibiotic substances are produced by certain members of the plant kingdom, chiefly by microorganisms and green plants. Most of the antibiotics isolated and studied up to the present time have been produced by fungi and bacteria. The most important sources of antibiotics among the fungi are the penicillia, actinomycetes, aspergilli, and higher fungi. The two most important antibiotics isolated to date, the penicillins and streptomycin, are produced by penicillia and an actinomycete, respectively. Antibiotics of considerable promise, such as bacitracin, have been isolated from bacteria. Tomato plants, horse chestnuts, radish seeds, and other plants are also sources of antibacterial substances.

Several thousand antibiotic substances have been isolated and identified since 1940. The significance of antibiotics lies in their ability to destroy many kinds of pathogens and to their being relatively nontoxic to the host. The ideal antibiotic should prevent the ready development of resistant forms. They should not produce undesirable side effects in the host, such as sensitivity or allergic reactions, nerve damage, or irritation of the kidneys and gastrointestinal tract. They should not eliminate the normal microbial flora of the host.

15.1 MODE OF ACTION OF ANTIBIOTICS

Antibiotics interfere with the pathogens at a number of vulnerable sites in the cell. They may interfere with (i) cell wall synthesis, (ii) cytoplasmic membrane function, (iii) nucleic acid function, (iv) protein synthesis and (v) enzyme system.

15.1.1 ANTIBIOTICS AFFECTING CELL WALL SYNTHESIS

Any substance that destroys the cell wall or that prevents the synthesis of cell wall polymers in growing cells leads to the development of osmotically sensitive cells and ultimately result in death. The component of the cell wall that confers rigidity is murein or peptidoglycan layer. This substance consists of polysaccharide chains composed of alternating units of N-acetyl-glucosamine and N-acetyl muramic acid. Short peptides linked to the carboxyl group of muramic acid are covalently cross-linked with peptides of neighbouring polysaccharide chains.

Among the antibiotics that affect the bacterial cell wall are (a) penicillin (b) cephalosporins (c) cycloserine (d) vancomycin and (e) bacitracin.

Penicillin It is the first and still the most widely used antibiotic for general therapeutic use. Penicillins are a class of β-lactam antibiotics of related structure with slightly different properties and activities. All penicillins have a common basic nucleus, a fused β-lactam thiazolidine ring with different side chains, which give each its unique properties. Penicillin is produced by *Penicillium notatum*, *Penicillium chrysogenum*, and by other species of molds. Penicillin is selective for gram-positive bacteria, some spirochetes, and the gram-negative diplococci (*Neisseria*). Although it is rarely toxic to humans, it may give rise to sensitivity reactions, which vary from a mild skin reaction to severe anaphylaxis. Benzyl penicillin (penicillin G) is the most useful natural penicillin (Figure 15.1).

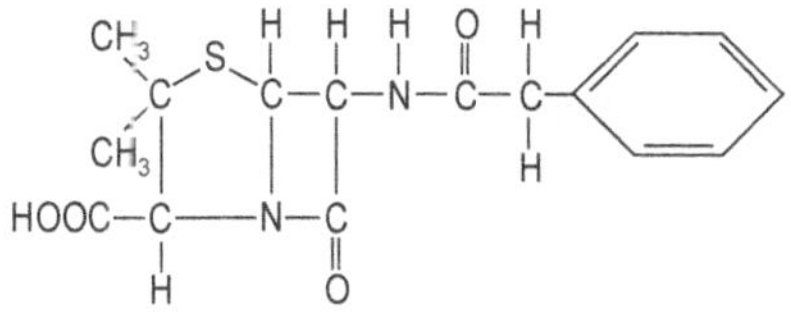

Fig. 15.1 Benzylpenicillin

Semisynthetic penicillins One of the first semisynthetic penicillins to be produced for clinical use was phenethicillin. It is just as effective as penicillin G. Another of the semisynthetic penicillins, methicillin, is more resistant to pencillinase and therefore is less likely to be inactivated. Ampicillin, yet another semisynthetic penicillin, acts against

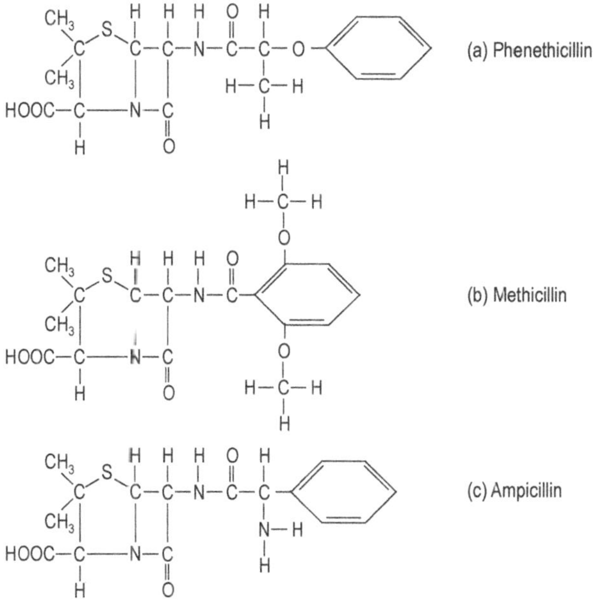

(a) Phenethicillin

(b) Methicillin

(c) Ampicillin

Fig. 15.2 Semisynthetic penicillins

a broad spectrum of bacteria. It is strongly bactericidal and lacks toxicity, but it is not resistant to penicillinases. It is relatively stable to gastric acid and hence can be administered orally. The chemical structures of these three penicillins are shown in Figure 15.2.

Penicillin is used to prevent and control infections caused by staphylococci, streptococci, pneumococci, gonococci and certain other gram-positive organisms. Penicillin is indicated for use in preventing secondary infections, which may follow tooth extraction, tonsillectomy, or other operations. Penicillin has proved to be helpful in the treatment of actinomycosis, diphtheria, and syphilis. However, penicillin is ineffective against gram-negative bacillary infections such as typhoid fever, dysentery, undulant fever and tubaremia.

Cephalosporins These are a group of antibiotics produced by a species of marine fungus, *Cephalosporium acremonium*, which bears considerable resemblance to *Penicillium* sp. They are effective against gram-positive and gram-negative bacteria. The cephalosporins have antibacterial properties similar to those of the semisynthetic penicillins. They are effective therapeutically and have a low toxicity. The nucleus of the cephalosporins resembles that of penicillin (Figure 15.3). Several semisynthetic cephalosporins have been manufactured commercially for therapeutic use. The mode of action of the cephalosporins is to inhibit transpeptidase.

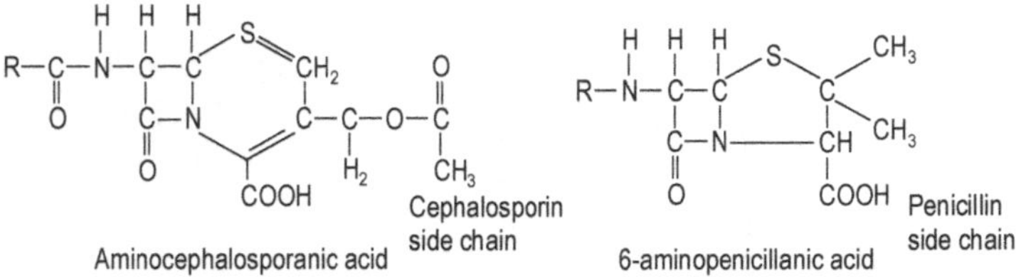

Aminocephalosporanic acid — Cephalosporin side chain

6-aminopenicillanic acid — Penicillin side chain

Fig. 15.3 A comparison of the nucleus of cephalosporin (amino cephalosporanic acid) with the nucleus of penicillin (6-aminopenicillanic acid)

Cycloserine It is related in structure to alanine (Figure 15.4). It was originally discovered as an antibiotic produced by streptomyces and is now manufactured through chemical synthesis. This is used in tuberculosis therapy. Cycloserine manifests its inhibitory effect on peptidoglycan synthesis by interference with the synthesis of the peptide moiety of the peptidoglycan. Specifically, it inhibits both **alanine**

racemase and **D-alanyl-D-alanine synthetase**, the enzymes involved in the synthesis of the pentapeptide side chains.

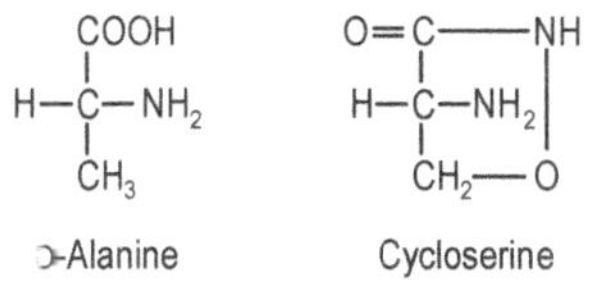

$$\begin{array}{c} COOH \\ | \\ H-C-NH_2 \\ | \\ CH_3 \end{array} \qquad \begin{array}{l} O{=}C\text{——}NH \\ \quad\; | \qquad\quad | \\ H-C-NH_2 \\ \quad\; | \qquad\quad | \\ \quad CH_2\text{——}O \end{array}$$

D-Alanine Cycloserine

Fig. 15.4 The chemical structures of D-alanine and the antibiotic cycloserine

Vancomycin It is an antibiotic produced by *Streptomyces orientalis*. It is a complex chemical entity consisting of amino acids and sugars. Vancomycin inhibits peptidoglycan synthesis by binding with the D-alanyl-D-alanine group on the peptide side chain of one of the membrane-bound intermediates.

Bacitracin It is a polypeptide antibiotic that is bactericidal for many gram-positive species. With the exception of *Neisseria*, it has little effect on gram-negative organisms. Resistance to the drug does not develop readily, and cross-resistance between it and other antibiotics does not occur either *in vitro* or *in vivo*. Bacitracin resembles penicillin in its mode of action, causing spheroplast formation and the accumulation of nucleotide precursors of the cell wall. Its specific site of action however differs from that of penicillin in that it interferes with cell wall synthesis at an earlier stage. It interferes with the second stage of cell wall synthesis.

15.1.2 ANTIBIOTICS AFFECTING THE CYTOPLASMIC MEMBRANE

The cell membrane plays a vital role in the cell. Several polypeptide antibiotics produced by *Bacillus* sp. have the ability to damage cell-membrane structure. Polymyxins, gramicidins, and polyene tyrocidines are antibiotics which damage the cytoplasmic membrane.

The polymyxins are particularly effective against gram-negative organisms while the tyrocidines and gramicidins are more effective against gram-positive organisms. The source of polymyxins is *Bacillus polymyxa*. Polymyxins bind specifically to the cell membrane, destroy its osmotic properties and cause a leakage of metabolites from the cytoplasm of the cell. Nystatin produced by *Streptomyces noursei* and amphotericin produced by *Streptomyces nodosus* are polyene antibiotics acting upon cells, which have sterols in their cytoplasmic membrane. They act upon fungi and animal cells but do not affect bacteria. Their antimicrobial action is attributed to their ability to increase cell permeability.

15.1.3 ANTIBIOTICS INTERFERING WITH NUCLEIC ACID FUNCTIONS

Mitomycin, nalidixic acid and novobiocin are the antibiotics affecting nucleic acid function. When mitomycin is added to growing bacterial cells, it leads to inhibition of cell division and death. It has no clinical value and is useful only as a biochemical tool. Nalidixic acid is a synthetic product of 1,8- napthyridine. It has been used in the clinical management of urinary tract infections caused by various gram-negative organisms. Novobiocin is bactericidal for gram-positive organisms and its range of antibacterial activity is very similar to that of penicillin and erythromycin. Its primary effect probably is on DNA synthesis, although numerous secondary effects are also induced. DNA polymerase activity is inhibited immediately.

15.1.4 ANTIBIOTICS INHIBITING PROTEIN SYNTHESIS

A number of antibiotics interfere with the metabolism of proteins. Major categories of antibiotics affecting protein synthesis are described below.

Rifamycins Some of the members of this group occur in nature, but most are semisynthetic. These agents at very low concentrations inhibit the growth of *Mycobacterium tuberculosis*. Its selectivity inhibits protein synthesis by inactivating the DNA-dependent RNA polymerase.

Streptomycin It is produced by *Streptomyces griseus*. It is an aminoglycoside antibiotic composed of streptidine and streptobiosamine, a disaccharide containing a methyamino group (Figure 15.5). Streptomycin is particularly important because it inhibits many organisms resistant to sulfonamides and penicillin. Its antibacterial spectrum includes many gram-negative bacteria, including *Francisella tularensis* and some organisms in the salmonella group. It is inhibitory for several species of *Mycobacterium,* including *Mycobacterium tuberculosis.* Highly purified streptomycin is nontoxic to humans and other animals when given in small doses, but it appears to have a cumulative detrimental effect on a specific region of the nervous system when given as a medication over long periods. Streptomycin inhibits protein synthesis by combining irreversibly with the 30S subunit mRNA. Thus, the normal synthetic sequence is disrupted.

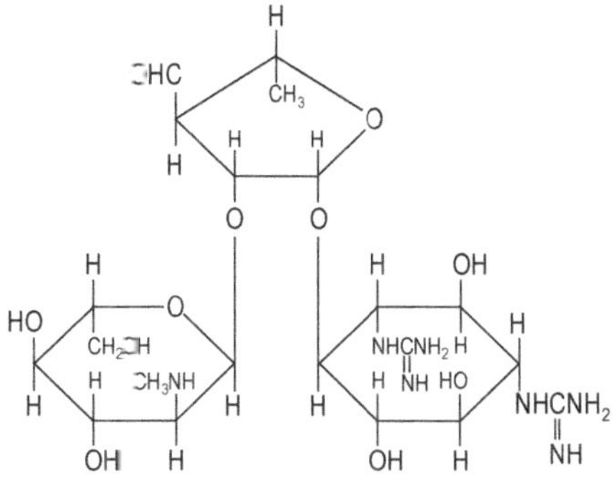

Fig. 15.5 Chemical structure of streptomycin

Tetracyclines The tetracyclines are a family of closely related antibiotics, consisting of tetracycline, the parent compound, chlortetracycline (Aureomycin) and oxytetracycline (Tetramycin) (Figure 15.6). The antimicrobial, pharmacologic and therapeutic properties of all the tetracyclines are similar. The tetracyclines inhibit protein synthesis through interference by binding of aminoacyl-t RNA to the 30S subunit ribosomes.

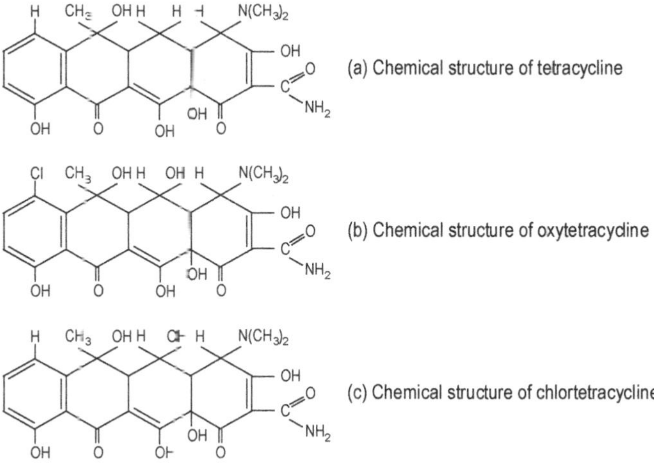

(a) Chemical structure of tetracycline

(b) Chemical structure of oxytetracydine

(c) Chemical structure of chlortetracycline

Fig. 15.6 Tetracyclines, broad-spectrum antibiotics produced from streptomyces

Chloramphenicol It is a broad-spectrum antibiotic, active against many gram-positive and gram-negative bacteria. Its antimicrobial spectrum is similar to that of tetracycline. It is also bacteriostatic. Chemically, it is a nitrobenzene ring with non-ionic chlorine (Figure 15.7). The possibility of serious side effects such as blood dyscrasias have limited the use of this antibiotic as a general antibacterial agent. Chloramphenicol inhibits protein synthesis by combining with the 50S subunit ribosome. The transpeptidation and translocation functions associated with this site are blocked.

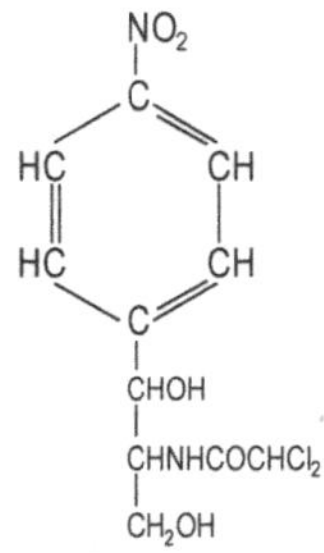

NO_2
C
HC CH
HC CH
C
CHOH
$CHNHCOCHCl_2$
CH_2OH

Fig. 15.7 Structure of chloramphenicol, a broad-spectrum antibiotic from *Streptomyces venezuelae*

Erythromycin It is produced by a strain of *Streptomyces erythraeus*. Erythromycin is active against the gram-positive bacteria, some gram-negative bacteria, and pathogenic spirochetes. With regard to antimicrobial spectrum and clinical usefulness, it resembles penicillin, but it is also active against organisms that become resistant to penicillin and streptomycin. Erythromycin belongs to the chemical class of antibiotics known as macrolides. Structurally it contains a large lactone ring linked with amino sugars through glycosidic bonds. Erythromycin inhibits protein synthesis because of binding on the 50S subunit ribosome; the steps of transpeptidation and translocation in protein synthesis are blocked.

Neomycin, gentamycin All these antibiotics are bactericidal. They contain both an aminoglycoside and an aminocyclitol moiety joined via a glycosidic linkage. With the exception of gentamycin, which is produced by a species of *Micromonospora*, they are produced by species of *Streptomyces*. Their antimicrobial spectrum includes many gram-positive and gram-negative bacteria but their major clinical application is in *E.coli* gastroenteritis, suppression of bowel flora, local application in burns and wounds.

15.1.5 ANTIBIOTICS AFFECTING ENZYME SYSTEMS

The structure of these antibiotics is related to the compound *p*-aminobenzoic acid. Many bacteria require *p*-aminobenzoic acid (PABA) as a precursor to their synthesis of the essential coenzyme tetrahydrofolic acid (THFA). PABA is a structural part of the THFA acid molecule. Sulphonamide acts as a competitive inhibitor and blocks the synthesis of cellular constituent THFA. The basic structure for the sulphonamide is shown in Figure 15.8. The cellular functions of the THFA coenzyme include amino acid synthesis, thymidine synthesis, etc. Lack of this coenzyme will disrupt normal cellular activity. Sulphonamides will inhibit growth of those cells which synthesise their THFA from PABA and will not interfere with the growth of those cells (including mammalian host cells) which require the vitamin folic acid and reduce it directly to THFA. This accounts for the selective antibacterial action of sulphonamides and makes them useful in the treatment of many infectious diseases.

Fig. 15.8 The basic structure of sulphonamides

15.2 DRUG RESISTANCE

Drug resistance is the major problem in controlling the microorganism. Drug resistance may be due to a pre-existing factor in the microorganism, or it may be due to some acquired factor(s). Penicillin resistance, for example, may result from the production of penicillinase by resistant organisms, which converts penicillin into inactive penicilloic acid. On the other hand, some normally susceptible strains of bacteria may acquire resistance to penicillin. Acquired resistance is also due to penicillinase production in genetically adapted varieties of microorganisms. Many organisms, which do not produce penicillinase, are also resistant to penicillin. This suggests an alternative metabolic pathway or enzyme reaction not susceptible to inhibition by penicillin.

Other mechanisms of drug or antibiotic resistance may be due to (1) competitive inhibition between an essential metabolite and a metabolic analog (drug), (2) development of an alternative metabolic pathway which bypasses some reaction that would normally be inhibited by the drug, (3) production of an enzyme altered in such a way that it functions on behalf of the cell but is not affected by the drug, (4) synthesis of excess enzyme over the amount that can be inactivated by the antibiotic or drug, (5) inability of the drug to penetrate the cell due to some alteration of the cell membrane, and (6) alteration of ribosomal protein structure.

Antibiotic resistance represents a serious problem for clinicians, and great effort is being made to understand the mechanisms involved and to prevent its occurrence. The development of resistance can be minimised by (1) avoiding the indiscriminate use of antibiotics where they are of no real clinical value, (2) refraining from the use of antibiotics commonly employed for generalised infections for topical applications, (3) using correct dosages of the proper antibiotic to overcome an infection quickly, (4) using combinations of antibiotics of proved effectiveness, and (5) using a different antibiotic when an organism gives evidence of becoming resistant to the one used initially.

REVIEW QUESTIONS

1. Explain the mode of action of antibiotics
2. Write short notes on:
    i. Penicillin
    ii. Streptomycin
    iii. Antibiotics affecting cell wall synthesis
    iv. Antibiotics affecting the cell membrane
    v. Antibiotics inhibiting protein synthesis

16

BIOCHEMICAL TECHNIQUES

A knowledge of the basic principles and applications of biochemical techniques are prerequisite for students of biochemistry before they start doing any biochemical analysis. Hence, the principles and applications of a few instrumental methods of biochemical analysis like centrifuge, pH meter, colorimetry, chromatography and electrophoresis have been explained in this chapter.

16.1 CENTRIFUGE

The centrifuge is an instrument which is used to spin substances at high speed. It is often used to separate particles present in a liquid.

Principle If a solution of large particles is allowed to stand for sometime, the particles will tend to sediment under the influence of gravity. If a force which is greater than the gravitational force of the earth is applied, the particles sediment more rapidly. This is the basis of centrifugal separation technique. Particles that differ in density, shape or size can be separated easily because they sediment at different rates in the centrifugal field. The sedimentation rate of each particle is directly proportional to the applied centrifugal force.

16.1.1 TYPES OF CENTRIFUGATION TECHNIQUES

Centrifugation techniques are broadly classified into two major types, namely, **preparative centrifugation** and **analytical centrifugation**.

Preparative centrifugation technique is concerned with the separation, isolation and purification of sub-cellular organelles, plasma membranes, ribosomes, chromatin, nucleic acids and viruses. This technique needs very large amount of biological samples and helps in the study of their morphology, composition and biological activity.

In contrast, analytical centrifugation technique is devoted mainly to the study of purified macromolecules. This technique requires only small amount of material. It yields information regarding the purity, relative molecular weight and shape of the material.

16.1.2 TYPES OF CENTRIFUGE

Different types of centrifuges are available. All centrifuges can be divided into four major types on the basis of their operating speed.

Hand centrifuge Hand centrifuge is manually operated consisting of two centrifuge tube holders (Figure 16.1).

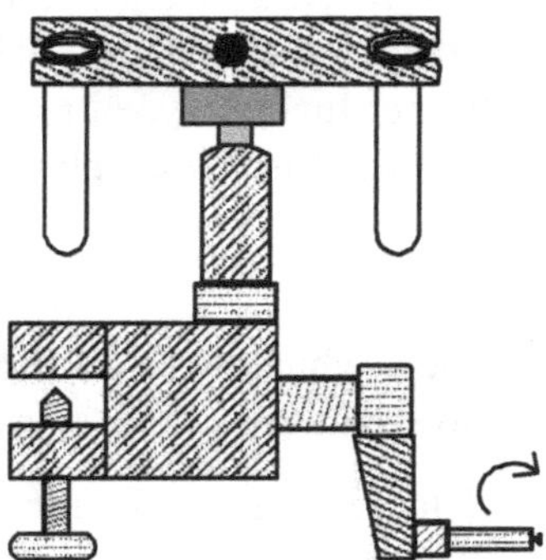

Fig. 16.1 Hand centrifuge

Desktop centrifuge or small bench centrifuge These are very simple and small and hence can be placed atop a desk. They are also known as **clinical centrifuges** since most of the clinical work is done by these models (Figure 16.2). They are normally used to isolate red blood cells, yeast cells or bulky precipitate of chemical reactions. Their maximum speed is usually 3000 rpm with maximum relative centrifugal force of 7000g. They do not usually have any temperature regulatory system. They are useful for the separation of large volumes of crude preparation. The rotors are mounted vertically on a rigid shaft. Therefore, centrifuge tubes must be placed diametrically opposite to each other after balancing their weights accurately.

Fig. 16.2 Clinical centrifuge

High speed centrifuge These centrifuges operate with a maximum speed of 25,000 rpm providing about 90,000g centrifugal force. They are usually equipped with refrigeration equipment to remove the heat generated due to friction between the air and the spinning rotor. The temperature can easily be maintained in the range 0°C to 4°C (Figure 16.3). A maximum volume of about 150 ml can be loaded. This is used to:

i. Isolate subcellular organelles such as the nuclei, mitochondria and lysosomes and

ii. Collect microorganisms, cell debris, precipitates of chemical reactions and immunoprecipitates.

Fig. 16.3 High speed centrifuge

Ultracentrifuge The ultracentrifuge can operate at speeds up to 75,000 rpm providing centrifugal force in excess of 500,000g. At such speed the friction between air and the spinning rotor generates significant amount of heat. To eliminate this source of heating, the rotor chamber is sealed and a vacuum is created inside. It has a refrigeration system which can maintain the temperature of the rotor between 0°C and 4°C. The shaft is made up of aluminium or titanium alloy of high tensile strength to withstand the great forces generated during centrifugation. Ultracentrifuge is used to isolate subcellular organelles such as endoplasmic reticulum and cytoplasmic components and to harvest viruses in pure form.

16.2 pH METER

pH meter is a potentiometer which measures the voltage between two electrodes placed in a solution (Figure 16.4).

Principle An electric potential is generated when a thin glass membrane separates two solutions of different H^+ ion concentrations. When a solution of unknown pH is used, the difference in potential between the glass electrode and the reference electrode is measured, amplified, and converted into a direct pH reading on the meter.

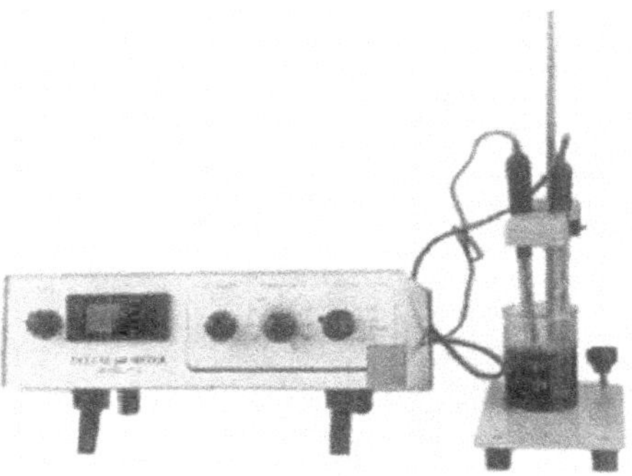

Fig. 16.4 pH meter

The two electrodes used in pH meter are a **calomel electrode** and a **glass electrode**. The calomel electrode is the external reference electrode whose electrical potential is always constant, whereas the glass electrode is the standard test electrode whose electrical potential depends on the pH of the test solution.

16.2.1 CALOMEL ELECTRODE

The calomel electrode contains mercury, mercury chloride and a saturated solution of potassium chloride (Figure 16.5). Each of these compounds exists in ionised state although the extent of ionisation may vary widely. Their dissociation constants are:

a) $Hg \Leftrightarrow Hg^{+} + e^{-}$

$$K_a = \frac{[Hg^{+}]\ [e^{-}]}{[Hg]}$$

b) $Hg_2Cl_2 \Leftrightarrow 2Hg^{+} + 2Cl^{-}$

$$K_b = \frac{[2Hg^{+}]\ [2Cl^{-}]}{[Hg_2\ Cl_2]}$$

The calomel electrode is dipped in saturated solution of KCl. The electrical contact between the calomel electrode and the test solution is achieved by the KCl salt bridge through a fine capillary in the glass casing known as porous plug.

16.2.2 GLASS ELECTRODE

The glass electrode contains silver, silver chloride and 0.1M HCl solution (Figure 16.5). Their dissociation constants are as follows.

c) $Ag \Leftrightarrow Ag^{+} + e^{-}$

$$K_c = \frac{[Ag^{+}]\ [e^{-}]}{[Ag]}$$

d) $AgCl \Leftrightarrow Ag^{+} + Cl^{-}$

$$K_d = \frac{[Ag^{+}]\ [Cl^{-}]}{[Ag\ Cl]}$$

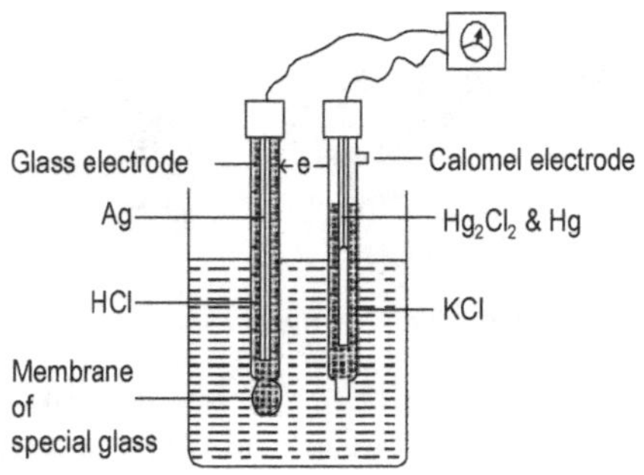

Fig. 16.5 Glass and calomel electrode assembly for pH measurements

This electrode is dipped in 0.1M HCl solution.

The calomel electrode is made of a thick glass that is impermeable to H^{+} ions. Therefore, its potential is independent of pH. In contrast, in the glass electrode, the tip of electrode is made of special thin (0.05 to 0.1 mm) borosilicate glass bulb, which is permeable to H^{+} ions only but not to other cations or anions.

All these equilibrium reactions from (a) to (d) are delicately balanced and when the electrodes are connected, the electrons will move from more positive electrode to the other.

If the electrodes are placed in a solution containing a high concentration of H^{+} ions (low pH, acidity), the calomel electrode will not respond as it is not permeable to H^{+}, the H^{+} ions pass through the glass membrane and neutralise the electrons of the electrodal reaction and hence electrons flow from calomel to glass. When the test solution has high concentrations of OH^{-} ions (high pH, alkalinity), the H^{+} move out of the glass bulb rendering a momentary negative charge and hence the electrons flow from glass to calomel. Because of this passage of ions, an electrical potential develops across the glass electrode and calomel electrode, which results in a flow of current between the electrodes. The magnitude of this current depends on the concentration of H^+/OH^- ions in the test solution. In the pH meter the current is fed into a calibrated dial in such a way that the dial reading directly gives the pH of the solution.

16.2.3 COMBINED ELECTRODE

Combination electrodes consist of a glass and a reference electrode in a single unit (Figure 16.6). The advantage of this type of electrode is that smaller volumes of solution can be measured. The disadvantage is its high cost and the fact that it must be discarded if one of the elements fail.

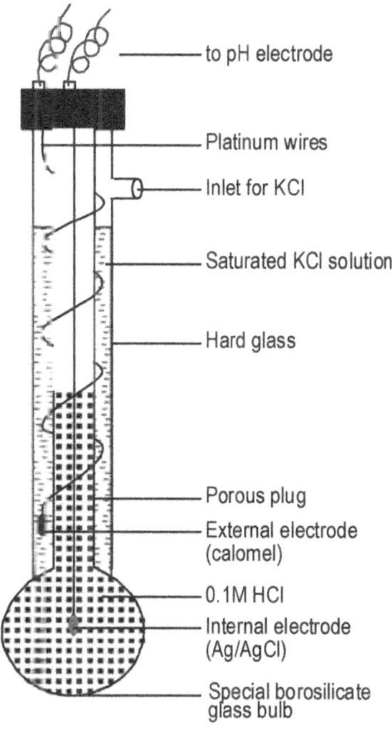

Fig. 16.6 Combined pH electrode

Application pH meter is used to measure pH of a given solution and to prepare buffer for biological research. Buffer solution resists a change in hydrogen ion concentration on addition of an acid or alkali. Buffer systems provide protection to cells against sudden changes in pH. They have a role in the maintenance of osmotic pressure between the cell and extracellular fluid. Blood contains a number of buffer systems. They maintain a constant blood pH of about 7.4. Enzymes exhibit a maximum catalytic action at specific pH. Buffers are responsible for providing the desired pH. Phosphate buffers and tris buffers are widely used in the clinical laboratory and in biochemical studies.

16.2.4 PHOSPHATE BUFFERS

Phosphate buffers are widely used because of their very high buffering capacity. Sodium and potassium salts are highly soluble in water. Since these ions are strongly charged, high ionic strength is obtained without the need for excessive molarity. But it is not possible to prepare a phosphate buffer that has a low ionic strength and a high buffering capacity. By choosing appropriate mixtures of $H_3PO_4/H_2PO_4^-$, $H_2PO_4^-/HPO_4^{2-}$ or HPO_4^{2-}/PO_4^{3-}, buffer solutions having a pH range from 2 to 12 can be prepared. Phosphate buffers have the following major disadvantages.

i. They are somewhat toxic to mammalian cells *in vitro*.

ii. Ca^{2+} and Mg^{2+} ions are bound by phosphate ions.

16.2.5 TRIS BUFFERS

A buffer that is widely used in the clinical laboratory and in biochemical studies is that prepared from tris (hydroxy methyl) aminomethane [$(OHCH_2)_3CNH_2$ - Tris or THAM] and its conjugate acid [$(OHCH_2)_3CNH_3^+$]. It has a high solubility in physiological fluids. It is non-hygroscopic and does not absorb CO_2 appreciably. It does not appear to inhibit many enzyme systems. It has a pK close to physiological pH (pKa for the conjugate acid = 8.08). But its buffering capacity starts diminishing below pH 7.5, which is a disadvantage. These buffers are usually prepared by adding hydrochloric acid or glycine solution to a solution of tris and by adjusting the pH to the desired value. Tris-Glycine buffer is commonly used in disc-gel electrophoresis. Tris also has the following drawbacks.

i. It reacts with some metal ions such as Cu^{2+}, Ni^{2+}, Ag^+ and Ca^{2+}.

ii. Its pH varies slightly with temperature.

iii. It reacts with certain electrodes.

16.3 COLORIMETRY

Colorimetry is a form of photometry, which deals with measurement of light absorption by coloured substances in solutions. The instrument, which measures the intensity of the colour, is known as colorimeter.

Principle The colorimeter is one of the most widely used instruments in biological research. It is based on the principle that when a beam of incident light passes through a coloured solution, the coloured substances in the solution absorb a part of the light and hence, the intensity of the transmitted light is always less than that of the incident light. As the number of light-absorbing molecules increases, the intensity of light coming out of the medium decreases exponentially and vice-versa. The difference in intensities between the incident and transmitted light, in turn reflects the number of absorbing molecules or in other words, the concentration of the absorbing molecules in that solution.

Beer–Lambert's Law Absorption of light by a substance in solution is governed by two basic laws discovered by Beer and Lambert.

Beer's law states that when a parallel beam of monochromatic light passes through an isotopic, light absorbing medium, the amount of light that is absorbed is directly proportional to the number of light-absorbing molecules in that medium or in other words, the concentration of the substance in that medium.

That is, $A \propto C$, where A is the absorbance or optical density and C is the concentration of the light-absorbing substance in the medium.

Lambert's law states that when a parallel beam of monochromatic light passes through an isotopic, light-absorbing medium, the amount of light that is absorbed is directly proportional to the length of the medium through which the light passes.

That is $A \propto L$, where A is the absorbance and L is the length of the medium. Since the measurement of light absorption depends on both the laws, it is popularly known as *Beer–lambert's Law*.

Thus, $$A \propto C \times L$$

or $$A = eCL$$

where 'e' is known as the extinction coefficient.

Also

$$A = \log \frac{I_i}{I_t}$$

$$= \log \frac{100}{T}$$

$$= \log 100 - \log T$$

$$= 2 - \log T$$

where I_i is the intensity of the incident light, I_t is the intensity of the transmitted light and T is transmittance.

Molar extinction coefficient In the above equation, when *concentration* is 1 M i.e. one mole per litre and *path length of light* is 1 cm, then 'e' is known as *molar extinction coefficient* (ΣM). ΣM is constant for a particular compound at particular wavelength and has a maximal value when the compound is in its purest state.

Specific extinction coefficient When the molecular weight of compounds such as proteins, nucleic acids, etc. are not known, then the term specific extinction coefficient can be conveniently used. This is the extinction of a 1% (W/V) solution of the compound when the light path is 1 cm. For example, the specific extinction coefficient of BSA = 0.667, IgG = 1.35 and IgM = 1.2 at 280 nm.

Wavelength Light is supposed to have dual characteristics, namely **corpuscular** and **waveform.** Thus, a beam of light is considered as an electromagnetic waveform. Electromagnetic radiation is composed of an electric vector and a magnetic vector, which oscillate in planes at right angles to each other and both at right angles to the direction of propagation. The distance along the direction of propagation for one complete cycle is known as wavelength (λ).

Absorption spectrum The pattern of energy absorption by a solution of any substance when light of different wavelengths passes through it is a characteristic of that substance. This pattern is known as **absorption spectrum**. The absorption spectrum of a substance is established by measuring either optical density or transmittance of a particular concentration of the substance at different wavelengths.

Emission spectrum When a solid is heated to a high temperature, light is emitted. On examining this light by a spectroscope, a spectrum is found to be produced which is known as **emission spectrum**. The emission spectra of sodium and potassium are simple, consisting of only a few wavelengths. For other elements such as iron and uranium, thousands of distinct reproducible wavelengths are present.

Visible spectrum Light from the sun contains the entire visible spectrum. Daylight or white light is a combination of seven colours (i.e. it is polychromatic). VIBGYOR is a code to remember the component colours in the order. The wavelength of light, which the human eye can perceive, ranges from 400 to 700 nanometres (nm). This is called **visible spectrum** (Table 16.1). Radiation of shorter wavelength, less than 400 is called *ultraviolet* (UV) light and radiation of longer wavelength, more than 700 is called *infrared* (IR) light.

Table 16.1 Visible spectrum

Colour of the solution	Range of Wavelength	Complementary colour
Violet	400–465	Greenish yellow
Blue	465–482	Yellow
Green	498–530	Red purple (magenta)
Yellow	576–580	Blue
Orange	587–610	Greenish blue
Red	617–660	Bluish green
Purple red	670–720	Green

Photoelectric colorimeters may be classified into two main categories. The first category comprises filter photometers (colorimeters) and the second category comprises spectrophotometers which contain a diffraction grating or a prism instead of filters for the same purpose.

16.3.1 COLORIMETER

Colorimeter is a device that measures the intensity of the light before and after it has passed through a coloured solution (Figure 16.7). A colorimeter consists mainly of six parts.

i. a light source
ii. a condensing lens to render the light rays parallel
iii. a filter to generate monochromatic light
iv. a sample holder
v. a photocell to convert light energy (photons) into electrical energy
vi. a galvanometer to measure the electrical energy (current) thus generated

Fig. 16.7 Colorimeter

The light from a tungsten lamp (composed of wavelengths between 400 and 900 nm) passes through a slit, a condensing lens, and a filter and finally emerges as a parallel beam of monochromatic light. Filters consist of selected glass (or sometimes dyed gelatin), which is capable of transmitting light over a limited portion of the spectrum only (Figure 16.8). The instrument is provided with a set of replaceable filters marked "V, B, G, Y or R" violet, blue, green, yellow or red filters. Instead of the above mark a number may be written on each which indicates the wavelength of light that the filter transmits. For example, a filter of number '54' absorbs all light except that of wavelengths around 540 nm, which pass through. Other filters bear the following numbers: 42, 49, 59 and 65. In some type of instruments, the filters are not removable and they are fixed onto a disc, which can only be rotated to bring the appropriate filter in the light path. Filters are of limited specificity, and one that is designed to transmit 540 nm may actually transmit light between 520 and 560 nm with a peak transmittance at 540 nm.

The monochromatic light passes through the sample solution and the transmitted light falls on the photocell. The photocell converts the transmitted light energy into electrical energy, which is amplified and measured by the galvanometer. The galvanometer is calibrated to read the absorbancy/transmittance directly.

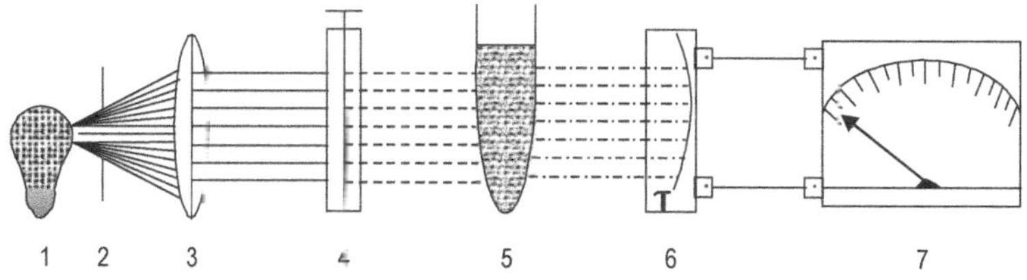

Fig. 16.8 Schematic diagram of colorimeter

1. Light source 2. Slit 3. Condensing lens 4. Filter
5. Sample well 6. Photocell 7. Galvanometer

Application Colorimetry has perhaps the widest application in biological sciences. The procedure described above helps in measuring the concentration of any unknown substance. If the substance by itself is colourless, it can be chemically converted to a coloured substance stoichiometrically by adding a chromophoric group and then the concentration measured.

It is also possible to measure the turbidity of solutions using a colorimeter. Largely these solutions follow Beer's law. But in the case of turbid solutions, the scattering can be determined with a nephelometer, which is a better measure of turbidity.

16.3.2 SPECTROPHOTOMETER

Spectrophotometer is an instrument which measures light absorption as a function of wavelength in the UV as well as visible regions (Figure 16.9). It also follows essentially the laws of light absorption, viz. the Beer–Lambert's Law. Unlike colorimeters, in spectrophotometers the compounds can be measured at precise wavelengths.

A spectrophotometer consists essentially of six parts.

i. light sources
ii. condensing lens

iii. a monochromator
iv. sample holder(s)
v. detector(s) and
vi. recorder

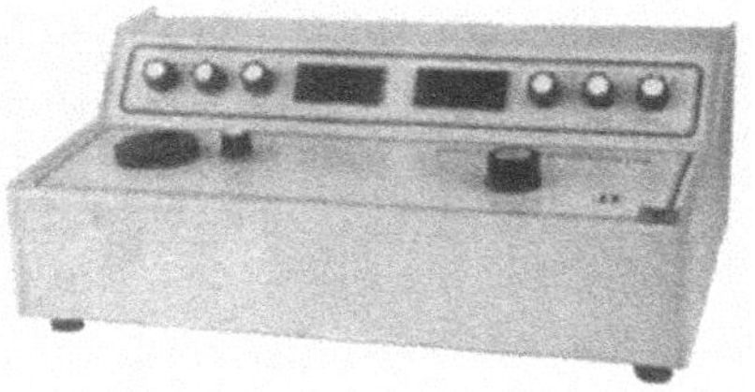

Fig. 16.9 Spectrophotometer

A spectrophotometer has two light sources, a UV light (for measuring light absorption from ~200 to ~400 nm) and white light (for measuring light absorption from ~400 to 900 nm). With the help of a shutter, only one of the light is allowed to fall on a silvered mirror (SM). The reflected light from the mirror passes through an entrance slit and a condensing lens.

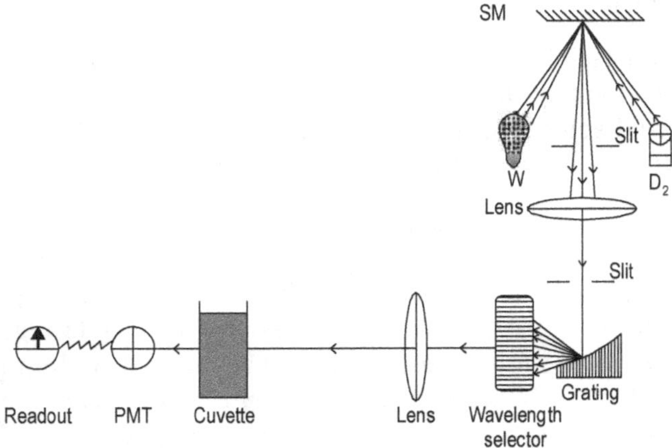

Fig. 16.10 Schematic diagram of a single-beam spectrophotometer

The lens renders the light rays into parallel beams and the parallel beams of light now fall on a monochromator (grating). The monochromator disperses the light into its component wavelengths. Using the wavelength selector, the desired wavelength is selected. Now the selected beam of monochromatic light passes again through a lens to a light-tight compartment where the sample is kept in a cuvette. After passing through the sample, the transmitted light falls on a photomultiplier (PMT). The PMT converts the light energy into electrical energy, which is amplified, measured and recorded on the analog/digital Readout (Figure 16.10).

In double-beam spectrophotometers, the monochromatic light coming out from the lens is split into two halves by placing a half-silvered mirror (HSM) on its path.

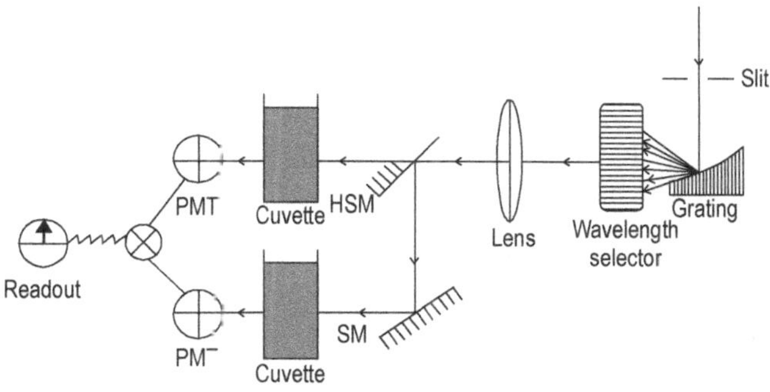

Fig 16.11 Schematic diagram of a double-beam spectrophotometer

Now 50% of the light passes directly through the mirror and falls on the reference cuvette and 50% of the light is reflected onto a second silvered mirror and then allowed to fall on the sample cuvette. At any given time, the intensities of the transmitted lights from the reference sample cuvettes are measured, amplified, the difference in intensities computed and sent to the Readout (Figure 16.11).

PARTS OF A SPECTROPHOTOMETER

Light sources A spectrophotometer has two light sources, a tungsten lamp for visible light and a deuterium or a hydrogen lamp, for UV light, respectively (deuterium lamp gives wider and more intense light in UV region than a hydrogen lamp). The light from the light source is composed of a wide range of wavelengths. This light is called polychromatic or heterochromatic light. The polychromatic light reflected back using a plane mirror, passes through an entrance slit, a condensing lens and falls onto a monochromator. The monochromator disperses the light and the desired wavelength is focussed on the exit slit using the wavelength selector.

Monochromators The monochromators, which produce radiations of single wavelength, are based either upon refraction by a prism or diffraction by a grating. Prisms are made of glass for visible region and of quartz or silica for UV region. A grating consists of ruled lines (as many as 2000 lines per millimetre) on a transparent or reflecting base. Resolving power of a grating is directly proportional to the closeness of these lines. Gratings are superior to prisms as they yield linear resolution of the spectrum for the entire range of wavelengths. The efficiency of monochromation is enhanced by using double monochromators in which a selected part of the spectrum from the first grating is further resolved by a second grating, resulting in a bandwidth of as low as 0.1 mm.

Cuvettes The optically transparent cells (cuvettes) are made up of glass, plastic, silica or quartz. Glass and plastic absorb UV light below 310 nm. Hence, they cannot be used for light measurements in UV region. Silica and quartz do not absorb UV light and hence they are used for both UV and visible light measurements. Since quartz absorbs light below 190 nm, cuvettes of lithium fluoride can be used which transmits radiation down to 110 nm. Oxygen also absorbs light at wavelengths less than 200 nm. Therefore, if spectra are required in this region the apparatus must be evacuated. The standard cuvettes are made up of quartz, have an optical path of 1 cm, and hold a volume of 1–3 ml. Microcuvettes (0.3–0.5 ml) are used for measurement of expensive chemicals.

Photocell and photo-multiplier tubes A *photocell* is a photoelectric device, which converts light energy into electrical energy, which is then amplified, detected and recorded. In photocells, the photons strike on a semi-cylindrical photo-emissive cathode in vacuum. This causes emission of electrons, which is proportional to the intensity of the radiation. When a potential difference is applied across the electrodes, the emitted electrons flow to the anode wire generating a photocurrent. This current is amplified electronically and measured. The theoretical accuracy of this photocell is 1±0.003 OD, i.e. 0.3%.

Applications Spectrophotometer is a more refined instrument and it gives a far better precision and resolution than a colorimeter. It finds a variety of applications in biological research.

i. It is used to estimate the concentration of both coloured as well as colourless solutions, which could absorb light.
ii. Because of its higher sensitivity, it is used to estimate extremely small quantities of substances in a matter of a few minutes.
iii. It usually does not degrade or modify the materials studied (unless a photochemical reaction occurs) and hence the materials can be recovered and reused.
iv. It is also used to find out absorption maxima of compounds in a wide range of wavelengths.
v. It offers selectivity in that each component in a solution or reaction mixture can be singled out and estimated.
vi. It also enables one to follow details of fast reactions and fast enzyme kinetics.
vii. It is also used to measure the growth of bacteria and yeasts and to determine the number of cells in a culture.
viii. Small volumes (as small as 0.3 ml) can be used for estimation of precious samples.

16.3.3 ULTRAVIOLET AND INFRARED SPECTRA

Absorption of light for ultraviolet light is measured at the wavelength region of 240 nm to 380 nm. Special devices have to be used to generate and measure the ultraviolet light. Infrared light (800 nm to 1500 nm) is known to interact with the molecules at the level of atomic vibrations. An infrared spectrum gives valuable information about the molecular structure and not concentration.

16.4 CHROMATOGRAPHY

The term chromatography was originally applied by Michael Tswett, a Russian botanist in 1906 to a procedure where a mixture of different coloured pigments (chlorophylls and xanthophylls) is separated from each other. He used a column of $CaCO_3$ to separate the various components of petroleum ether chlorophyll extract into green and yellow zones of pigments. He termed such a preparation as **chromatogram** and the technique as **chromatography**.

Chromatography may be defined as a technique in which the components of a mixture are separated based upon the rates at which they are carried through a stationary phase by a liquid or gaseous mobile phase. Separation starts to occur when one component is held more firmly by the stationary phase than the other which tends to move on faster in the mobile phase. Thus, the underlying principle of chromatography is first to adsorb the components of a mixture on an insoluble material and then suitable liquid solvents. The adsorbent can be in the form of sheet (paper and thin layer chromatography) or it can be packed into a column (column chromatography). A third form of chromatography is obtained with columns containing ion exchange resins (ion exchange chromatography).

In all the chromatographic techniques, difference in affinity involves the process of either adsorption or partition. In adsorption, the binding of a compound to the surface of the solid phase takes place; whereas, in partition the relative solubility of a compound in two phases results in the partition of the compound in two phases. Thus, all types of chromatography known so far have been grouped in either of the two mentioned forms:

Partition Chromatography	1. Paper chromatography 2. Thin layer chromatography 3. Gel filtration chromatography 4. Partition column chromatography	Liquid as mobile and solid as stationary phase
	5. Gas liquid chromatography or vapour phase chromatography	Gas as mobile and liquid as stationary phase

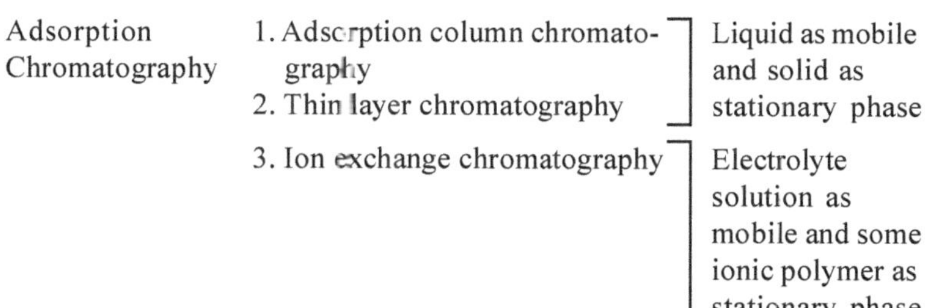

Adsorption Chromatography	1. Adsorption column chromatography 2. Thin layer chromatography	Liquid as mobile and solid as stationary phase
	3. Ion exchange chromatography	Electrolyte solution as mobile and some ionic polymer as stationary phase

A few chromatographic techniques are described below.

16.4.1 PAPER CHROMATOGRAPHY

Paper chromatography was first used by Martin Consden and Gordon to separate a mixture of amino acids. The stationary phase is the filter paper on which the mixture of amino acids is made to migrate. The filter paper is made of cellulose molecules that are hydrated. The mobile phase consists of a mixture of immiscible solvent system of which one is usually water which has a stronger affinity to the supporting medium. The other solvent has lower affinity to the stationary phase and carries the solute across the stationary phase.

As the solvent passes through an area of the paper containing a solute, the solute will begin to partition itself between the aqueous and the organic phases in proportion to its relative solubility in the two phases. The distance the solute moves in relation to the distance the solvent moves serves as a convenient means for identifying the solute. This relative rate of flow is the R_f value for the compound under the specified conditions of the experiment.

$$R_f = \frac{\text{Distance travelled by solute}}{\text{Distance travelled by solvent}}$$

Ascending paper chromatography This is the simplest chromatographic technique. It consists of a reservoir and the paper is held in position by clamps. The lower end of the paper is dipped into the solvent. The sample is spotted in a position just above the surface of the solvent so that as the solvent moves vertically up the paper by capillary action, separation of sample is achieved (Figure 16.12). When the solvent front has moved about 1cm from the

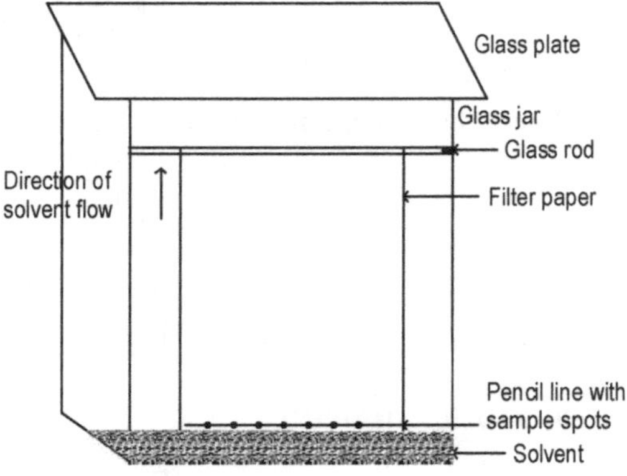

Fig. 16.12 Ascending chromatography

upper edge of the paper, the paper is removed and the solvent front is marked with a pencil and air-dried. The paper is sprayed with the ninhydrin and placed in an oven at 90°C until the development of purple colour. From this chromatogram R_f values are determined and the amino acids are identified. This method is preferable for quick analysis of a large number of substances.

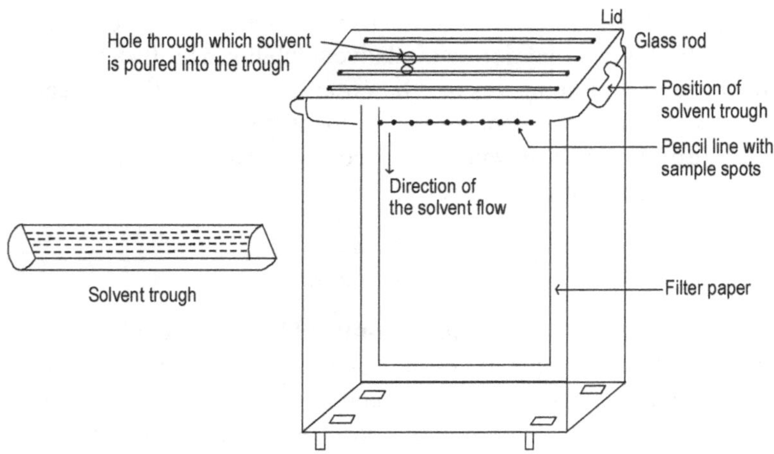

Fig. 16.13 Descending chromatography

Descending paper chromatography In this method, the solvent is kept in a trough at the top of the chamber and is allowed to flow down the paper (Figure 16.13). The liquid moves down by capillary action as well as by the gravitational force. In this case the flow is more rapid as compared to ascending method. Because of this rapid speed the chromatogram is completed in a comparatively shorter time. The solvent is placed in a trough at the top, which is usually made up of an inert material. The paper is then suspended in the solvent and the lid is placed at the top.

The method of paper chromatography may be one-dimensional or two-dimensional depending on the complexity involved in the analysis. The method given above is one-dimensional chromatography. Several compounds may have the same R_f value in a particular solvent system; running more than one chromatogram each with a different solvent

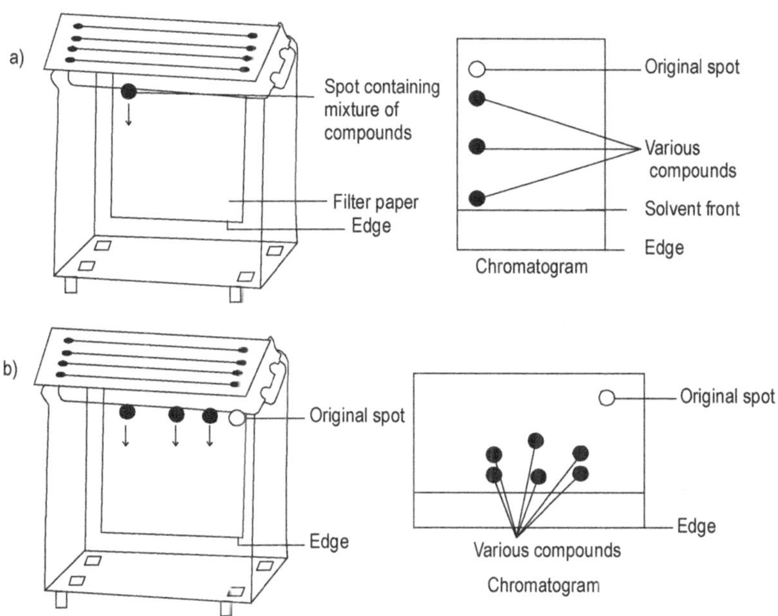

Fig. 16.14 Two dimensional paper chromatography

system can separate these compounds. For obtaining better separation of components a two-dimensional paper chromatography is applied. In this technique a square sheet of filter paper is taken. The test sample is applied to the upper left corner and chromatographed for few hours with one solvent mixture (e.g. butanol: acetic acid: water) (Figure 16.14a). The paper is dried and is turned 90° and again chromatographed in a second solvent mixture (e.g. collidine: water) (Figure 16.14b). This technique thus allows both vertical and horizontal separation of the amino acids.

Forces in operation The movement of the solute molecules on the chromatogram depends on the net result of a number of forces operating in the system. These forces are of two types: propelling and retarding.

i. Propelling forces These include the capillary force and the solubility force of the solvent. The Whatman paper is made up of numerous fibrils which are placed very close to each other, thus forming a network of capillaries. The solute rises through these capillaries as a result of capillary force. The smaller the pore of the capillaries, the greater is the height to which the solute rises.

   The solubility force of the solvent refers to the capacity of the solvent to dissolve the solute. The rise of solute also depends on its solubility in the solvent being used. The greater the solubility of the solute in the solvent, the greater is the height to which it rises in the chromatogram.

ii. Retarding forces Concurrent with the forces of propulsion, certain retarding forces also operate in the system, which drag the solute molecules from moving in either direction. These retarding forces include the gravitational force and the partition force. In ascending chromatography, the solute molecules have to move against gravitational force, which acts from below and tends to retard the movement of solute molecules in the upward direction. In descending chromatography, the gravitational force does not act as a retarding force, rather, it assists in propulsion and henceforth, acts as a propelling force.

   The partition force refers to the force between liquid and liquid molecules. The solvent molecules occupy the interstices between the fibrils on the Whatman paper and the solute molecules, for their movement, have to displace them. If the space were not filled with the solvent molecules, the movement of the solute molecules through them would be much more facilitated. Thus the movement of the solute molecules on the Whatman paper in either direction is the net result of the interaction between various forces of propulsion and retardation. Movement of the solute molecules is exhibited only when the propelling forces exceed the retarding forces in magnitude.

Application Paper chromatography is used in the separation of a variety of compounds of biochemical interest including fatty acids, carbohydrates and amino acids. In clinical chemistry, chromatography is applied to separate and identify amino acids, sugar and various other compounds of clinical interest present in urine and other biological fluids. In general, paper chromatography is the most widely used procedure for the separation and identification of amino acids. This method is so common that almost every type of compound is separated in this manner. Fields like biology, medicine and other intelligence departments use this method.

16.4.2 THIN LAYER CHROMATOGRAPHY (TLC)

Thin layer chromatography is used mainly for the separation of low molecular weight compounds. This technique is more or less similar to paper chromatography.

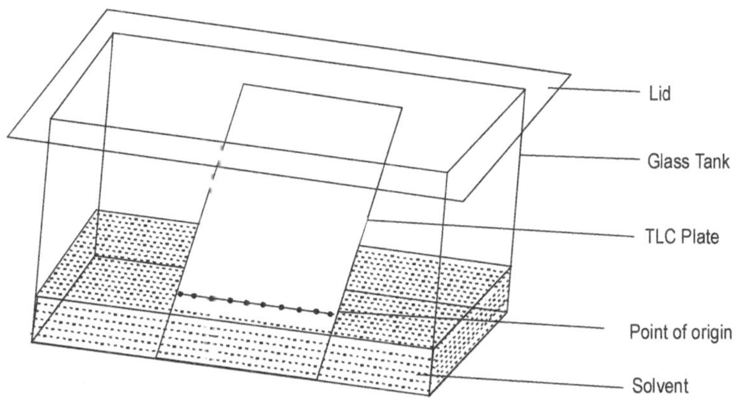

Fig. 16.15 Thin layer chromatography

Thin layer chromatography may be either carried out by the adsorption principle (if the thin layer is prepared by an adsorbent such as Kieselgur or alumina) or by the partition principle (if the layer is prepared by a substance such as silica gel which holds water like the paper).

A slurry of the stationary phase is made in a solvent. The slurry is poured onto the glass plate and spread evenly. The plates are then dried by keeping in an oven at 40°C. The sample is applied using a micropipette and dried. The plate is placed in a chamber containing the solvent and developed by ascending chromatography (Figure 16.15). After the solvent front has almost reached the top, the plate is dried. It can be rechromatographed at right angles with a second solvent for two-dimensional work. Spots are located by natural colour, by fluorescence or by spraying various reagents. Materials can be eluted from the chromatogram by scraping off the stationary phase and by eluting the powder with a suitable solvent.

Application Thin layer chromatography is used exclusively for the separation of small molecules. Amino acids, nucleic acids, lipids, steroids, terpenoids, hydrocarbons and carbohydrates can be separated using thin layer chromatography. It is often used to identify drugs, contaminants and adulterants. It is also widely used to resolve plant extracts and many other biochemical preparations. TLC is commonly used in studying the incorporation of ^{14}C-labelled compounds into metabolites and for the analysis of urine and blood in pathological laboratories.

16.2.3 COLUMN CHROMATOGRAPHY

Column chromatography is defined as a separation process involving uniform percolation of a liquid through a column packed with finely divided material. The separation in the column is effected either by direct interaction between the solute components and the surface of the stationary phase or by adsorption of solute by the stationary phase. Column chromatography involves ion exchange, molecular sieve, adsorption or partition phenomenon. In adsorption column chromatography the substances are preferentially adsorbed by the adsorbent (stationary phase) packed in the column, while in partition column chromatography, the components of a mixture distribute themselves and then get separated.

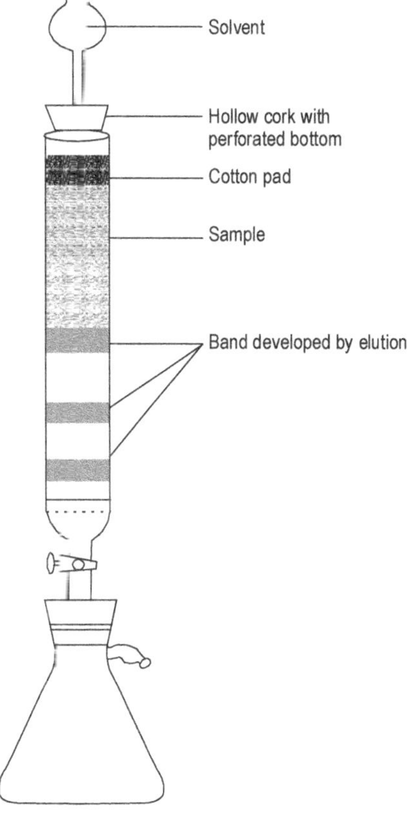

Fig. 16.16 Column chromatography

Adsorption Chromatography It was first developed by the American petroleum chemist D.T.Day in 1900. Later, M.S.Tswett, the Polish botanist, in 1906 used adsorption columns in his investigations of plant pigments. It was not until about 1930 that the method was used extensively by chemists. Rate of adsorption varies with a given adsorbent for different materials. This principle of selective adsorption is used in column chromatography. In this method, the mixture to be separated is dissolved in a suitable solvent and allowed to pass through a tube containing the adsorbent. The component, which has greater adsorbing power, is adsorbed in the upper part of the column. The next component is adsorbed in the lower portion of the column, which has lesser adsorbing power than the first component. This process continues. As a result, the materials are partially separated and adsorbed in the various parts of the column. The initial separation of the various components can be improved by passing either the original or some other suitable solvent slowly through the column. The various bands present in the column become more defined. The banded column of adsorbent is termed a chromatogram (Figure 16.16).

Adsorption chromatography is used for the separation of

i. polycyclic aromatic compounds, phenols, amines etc.
ii. urinary 17 ketosteroids and their glucuronides,
iii. plasma cortisol
iv. aliphatic hydrocarbons form aromatic hydrocarbons,
v. geometrical isomers and
vi. technical products in a highly purified state.

Ion Exchange Chromatography The principle of ion exchange chromatography is based upon the simple fact that different cations (or anions) have different capacity to undergo exchange reaction on the surface of a given exchanger. The capacity of an ion to undergo exchange reaction has been found to depend upon the charge and the size of the hydrated ion in solution. Under similar conditions, the capacity has been found to increase in the charge on the ion (i.e. the valency of the ion) but has been found to decrease with the increase in the size of the hydrated ion.

Applications Ion exchange chromatography is used for

i. Separation of ions from one another, because the different ions undergo exchange reactions to different extents.
ii. Removal of interfering radicals.
iii. Softening of hard water.
iv. Complete demineralisation of water.

v. Purification of organic compounds extracted in water.
vi. Separation of sugars.
vii. Separation of amino acids.

Gel filtration chromatography Gel filtration chromatography is a separation method which depends only upon molecular size. This method is also known as molecular sieve chromatography, gel permeation chromatography or molecular exclusion chromatography. Samples diffuse by about 25 per cent in molecular dimensions and may be totally separated by this technique.

In gel filtration chromatography, a column is prepared by filling tiny particles of an inert substance that contains small pores. If a solution containing molecules of various dimensions is passed through the column, molecules larger than the pores move only in the space between the particles and hence are not retarded by the column material. However, molecules smaller than the pores diffuse in and out of particles. Molecules are eluted from the column in the order of decreasing size.

The gels used as molecular sieves consist of cross-linked polymers that are generally inert. They do not bind or react with the material being analysed and are even uncharged. The gels currently in use are of three types namely, dextran, agarose and polyacrylamide.

Gel filtration chromatography has widespread use in purifying enzymes and other proteins and in fractioning nucleic acids. Various species of RNA and viruses are successfully fractionated and purified using agarose gels.

Gas liquid Chromatography (GLC) Gas liquid chromatography (GLC) is used to separate and quantify volatile compounds in a heated column. The columns consist of an inert support material coated with the stationary phase which must be a liquid at the elevated temperatures (up to 400°C) of the column, nonvolatile and non-reactive with the samples or solvents moving through the column. The movable phase is an inert carrier gas (e.g. argon or nitrogen). It carries the volatile compounds through the column to a detector. Volatile molecules with the lowest boiling points move very rapidly along by the carrier gas. The compounds that are soluble in the liquid phase are retained in the liquid and their forward movement is slowed down. Thus, both volatile and soluble differences form the basis of gas liquid chromatography.

Application

i. Volatile organic compounds are separated by GLC.

ii. Alcohols, esters, fatty acids and amines present in the biological samples are often separated by GLC.

iii. Concentration of individual elements such as carbon and hydrogen can be determined very accurately.

High-performance liquid chromatography (HPLC) Liquid chromatography is a slow separating technique that is performed in vertical columns and under gravitational flow. Recent developments have improved its speed and versatility. The increase in speed has been achieved by pumping the solution through a column at pressures up to more than 10,000 pSe.

The characteristic features of HPLC are sensitivity, ready adaptability to accurate quantitative determination, suitability for separating nonvolatile species or thermally fragile species. HPLC allows separation and measurements to be made in a matter of minutes. This technique has also been given a few other names, such as High Pressure Liquid Chromatography and High Speed Liquid Chromatography. HPLC system consists of four major components: (i) Pump, (ii) Injector, (iii) Column, and (iv) Detector (Figure 16.17).

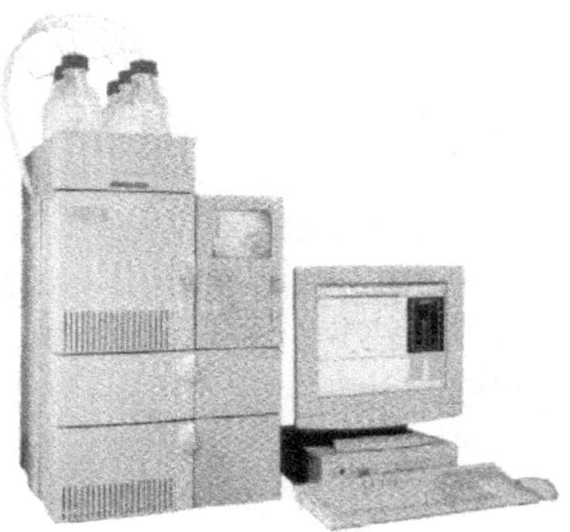

Fig. 16.17 High performance liquid chromatography (HPLC)

i. Delivery pump The pump delivers a steady stream of solvent from the reservoir to the detector through the column.

ii. Sample Injection Unit An unknown sample is introduced into the flowing stream of solvent with an injector.

iii. Columns The column is made of stainless steel, the length and inside diameter of which differs depending on their application. Several kinds of column packing are used. Although there are great variety of packing materials, all are based on the principle of separation by size, absorption, solubility and ion exchange.

iv. Solvent A successful separation could be achieved by establishing proper balance between the attraction of the solvent and solid support for the sample molecules.

v. Detectors Each component is retained by the column for different lengths of time, the individual components leave the column one after another. The relative concentration of each component is recorded with respect to time with the help of detectors.

Advantages of HPLC

HPLC has a number of advantages over gas chromatographic method, which are given below:

i. Gas chromatographic method is not applicable to thermally unstable substances. On the other hand, there is no decomposition of substances in HPLC method which is operated at room temperature.

ii. The separation of each component of substances is easily done by HPLC method whereas this is not true with gas chromatographic method.

iii. The nonvolatile substances are converted into volatile substances by chemical means in gas chromatographic method such as methyl esterification of fatty acids. Such a kind of chemical treatment is not needed in HPLC.

iv. HPLC performs more difficult separations than gas chromatography. This is because HPLC controls a greater number of separating variables. Besides, a greater variety of packing materials contributes to wider separating possibilities by HPLC.

16.5 ELECTROPHORESIS

Electrophoresis was first developed by Tisselius in 1930. The movement of charged particles under the influence of an electric current to oppositely charged electrodes is called electrophoresis. The movement of the charged particles in an electric field depends upon time, electric current, conductivity of the solvent and charge of the molecule to be separated. Thus, the electrophoretic mobility is defined as the distance travelled by the particles in one second under the potential gradient of one volt per centimetre. The different compounds in a mixture will have different electrophoretic mobilities and hence they can be separated. Mixture of amino acids, proteins, and nucleic acids can be separated by electrophoresis.

There are two main types of electrophoretic methods:

i. Moving boundary electrophoresis where, separation is carried out in the absence of a supporting medium.

ii. Zone electrophoresis where paper or gel is used as supporting medium for separation.

i. **Moving boundary electrophoresis** Moving boundary electrophoresis was first developed by Tiselius of Sweden in 1930. This method of separation is carried out in the absence of

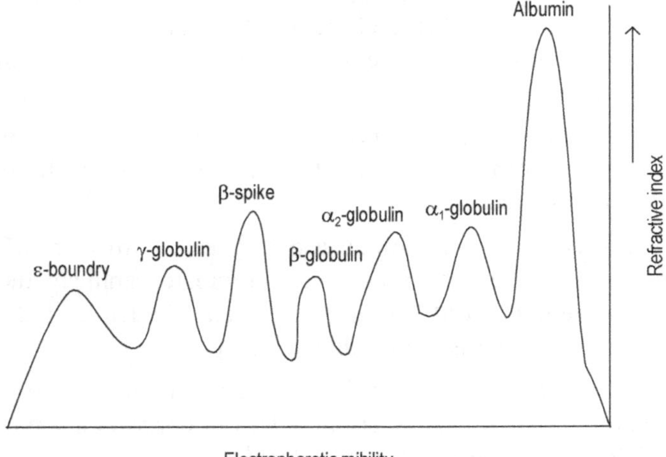

Fig. 16.18 Separation of serum proteins by moving boundary electrophoresis

supporting medium and requires large volumes of sample. In this method, buffered, negatively charged macromolecules are placed in a U-shaped cell. When an electrical current is applied, proteins migrate towards the anode. The migration of negatively charged proteins from the macromolecule solution to the pure buffer forms a boundary. As a result of this, there is sharp change in the refractive index of the solution at this boundary. The changes in the refractive index are measured by Schleirn optical device. The electrophoretic pattern, shows the direction and relative rate of migration of the molecules in the sample (Figure 16.18).

ii. **Zone electrophoresis** This is the most commonly used type of electrophoresis. This method requires very small quantity of the sample. Paper or gel is used as the supporting medium for the separation of complex mixtures.

16.5.1 PAPER ELECTROPHORESIS

Paper electrophoresis is a type of zone electrophoresis (Figure 16.19). A great advantage of this technique is the necessity of small volume of the sample. Upon separation, the molecules are immobilised by fixation at different zones. The molecules are then detected by the following methods:

i. staining them on the supporting medium,
ii. visualising by ultraviolet light,
iii. by virtue of enzymic reaction and
iv. by radioactivity.

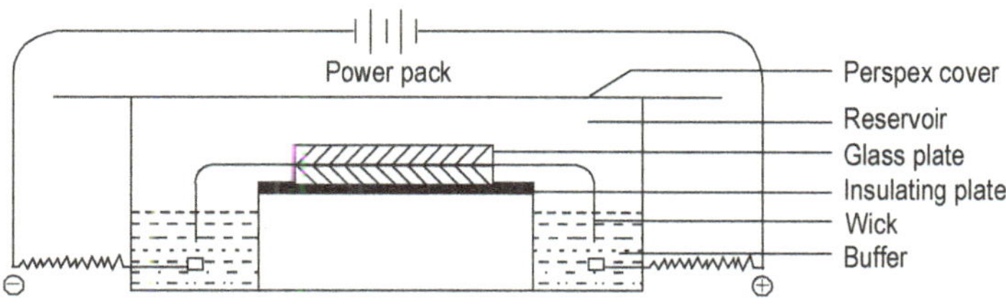

Fig. 16.19 Paper electrophoresis

Paper electrophoresis is very popular for the study of normal and abnormal plasma proteins. The equipment required for electrophoresis consists of two units, a power pack and an electrophoretic cell. The serum under investigation is mixed with bromophenol blue, a blue coloured stain, and spotted at the centre of a strip of a special filter paper, saturated with barbitone buffer of pH 8.6.

When an electric current of proper amperage and voltage is passed through the paper, charged protein fractions bearing different charges migrate at different rates. If the pH of the serum is adjusted by the addition of a proper buffer to a value alkaline to the isoelectric points of all the fractions of plasma protein, they all will carry negative charges, but of different magnitudes. The different fractions of plasma will migrate toward the anode at characteristically different rates. After a run of about 5 to 6 hours, the paper is dried and stained with a solution containing bromophenol blue. In human serum, five different bands can be identified on paper electrophoresis. They are designated in the order of decreasing mobility as albumin, alpha$_1$-globulin, alpha$_2$-globulin, beta-globulin and gamma-globulin. Albumin, being the fastest moving fraction of the proteins of plasma, forms the last band of the paper. Gamma globulin, which is the slowest moving protein, forms a band at the other end. The rest of the fractions take their positions in between these two bands (Figure 16.20).

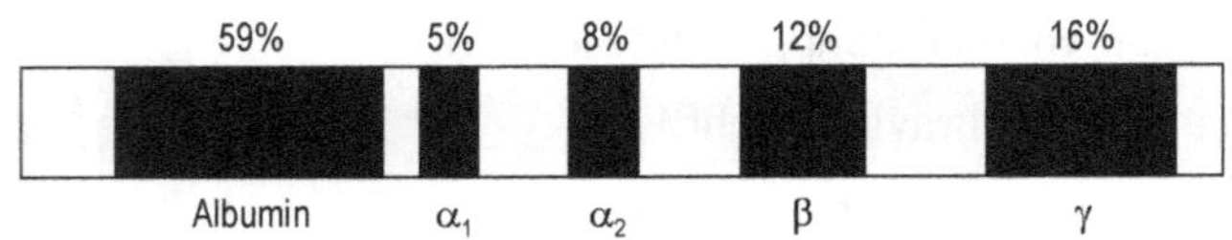

Fig. 16.20 Electrophoretogram of human serum protein

16.5.2 GEL ELECTROPHORESIS

Various types of gels are used as the supporting medium to separate complex mixtures. They are: (i) Starch gel, (ii) Agar gel, (iii) Polyacrylamide gel and (iv) Agarose acrylamide gel. The use of gels as supporting media has enhanced resolution, particularly for proteins and amino acids.

Gel electrophoresis is usually carried out through any one of the following methods.

i. Column electrophoresis
ii. Slab gel electrophoresis

i. **Column electrophoresis** In this type, the gel is polymerised in a thin glass column of known diameter. Both ends of the column are not closed. After polymerisation this column is fitted between the upper and lower buffer reservoirs (Figure 16.21).

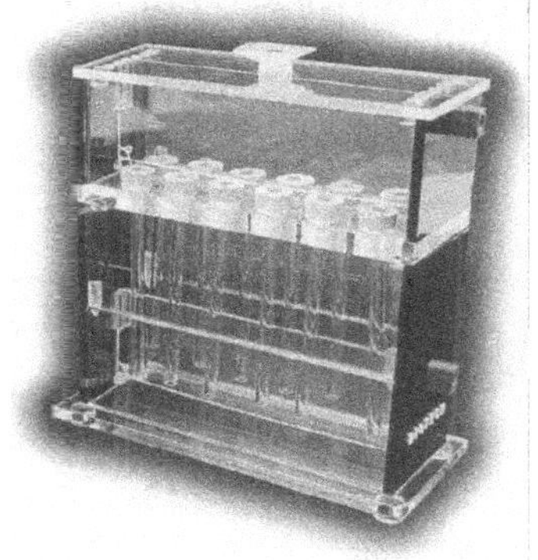

Fig. 16.21 Column electrophoresis

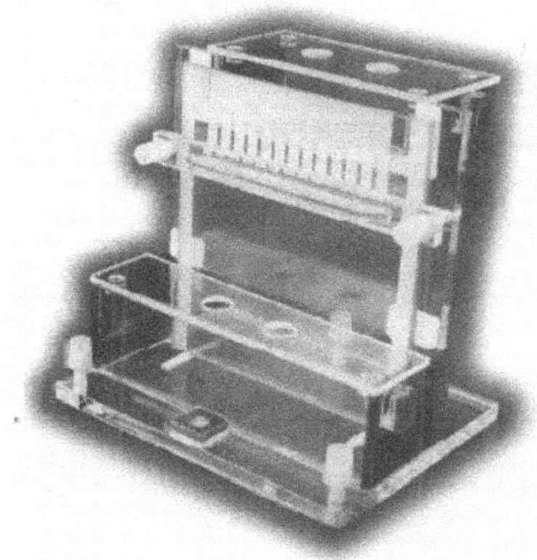

Fig. 16.22 Slab gel electrophoresis

ii. **Slab gel electrophoresis** In this type, the gel is set or polymerised into a thin slab between two glass plates (Figure 16.22). A Sample well is made by placing a comb-shaped jig into the glass before it polymerised. After the gel has set, the comb is removed. Since a number of wells can be arranged side by side, a number of samples can be loaded simultaneously and compared under identical conditions. This technique is highly useful in the field of molecular biology.

Identification of separated components Detection of separated components is possible by spot test techniques. Proteins are usually located by staining, and enzymes by their specific activities. Amino acids can be detected by fluorescence under UV light. Radioactive substances can be located by autoradiography or by staining. Lipoproteins can be detected by staining with the fat-soluble dye such as Sudan dye. Glycoprotein is detected by using modified Schiff's reagent.

When serum proteins are separated, five bands are obtained at the end of electrophoretic run. These bands are known as albumin, α_1-globulin, α_2-globulin, β-globulin and γ-globulin respectively, starting with greatest migration towards the anode. The gamma band is usually displaced slightly from the point of application towards the cathode because of electro-endosmotic flow.

The electrophoretogram of serum proteins is helpful in detecting charges in the individual protein fractions and in detecting abnormal bands in certain diseased conditions (Figure 16.23).

The pattern in nephrosis reveals a low level of albumin and a marked rise of α_2-globulins. A small increase in β-globulin can also be seen in this condition.

In chronic infection (hepatitis) a relative decrease in albumin with a notable elevation in γ-globulin is observed.

In multiple myeloma the characteristic feature is the presence of an abnormal band (M protein) usually between β- and γ- globulin bands, closer to the γ band.

In hypogammaglobulinemia a considerable drop in γ-globulin is readily observed. A slight increase in α_2-globulin is also seen.

In chronic liver diseases, a decrease in albumin band is observed. Occasionally an increase in γ- and β-globulin is also seen.

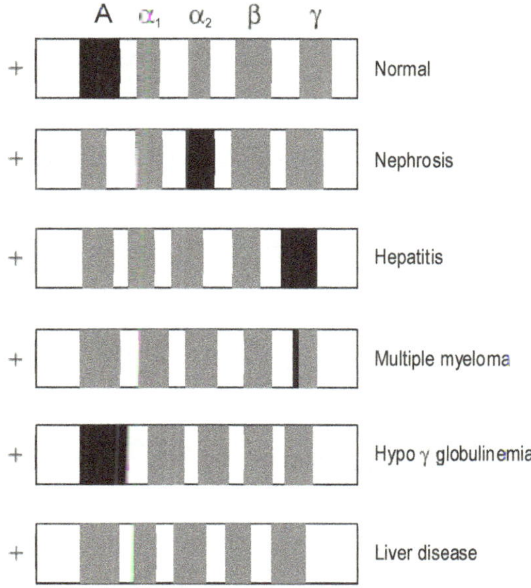

Fig. 16.23 Electrophoretogram of serum proteins of normal and diseased conditions

Applications Starch gel is mainly used for the analysis of isozyme patterns. Agarose gel electrophoresis is used to:

i. Isolate a large number of proteins
ii. Identify the purity of the isolated proteins
iii. Determine the sequences of DNA
iv. Find out the point of mutation in DNA or RNA (Southern and Northern blotting)
v. Detect the precursor molecules of tRNA, rRNA and mRNA
vi. Study the kinetics of interconversions of conformation of many tRNAs
vii. Find out the number of subunits present in a protein
viii. Determine the molecular weight of proteins and DNA

Polacrylamide gel electrophoresis combined with sodium dodecyl sulphate, $CH_3{-}(CH_{2-})_{10}{-}CH_2OSO_3^-Na^+$ is known as SDS-PAGE.

It is the most widely used method for analysing protein mixtures qualitatively. It is particularly useful for monitoring protein purification. Since this method is based on the separation of proteins according to size, it is also used to determine the relative molecular weight of proteins. PAGE is the most versatile electrophoretic system for the analysis and separation of proteins, small RNA molecules and very small fragments of DNA.

REVIEW QUESTIONS

1. Explain the principle, applications and types of centrifuges.
2. Explain the principle and applications of colorimeter and spectrophotometer.
3. Explain the principle and application of paper chromatography.
4. Explain the principle, types and application of column chromatography.
5. Explain the principle, types and applications of electrophoresis.
6. Write short notes on:
   - i. Calomel electrode
   - ii. Glass electrode
   - iii. Beer-Lambert's law
   - iv. Electrophoretogram
   - v. Moving boundary electrophoresis
   - vi. Paper electrophoresis

APPENDICES

APPENDIX-I

PREPARATION OF BUFFERS

1. ACETATE BUFFER

Stock solution

A:0.2M solution of acetic acid (11.55ml in 1000ml)

B:0.2M solution of sodium acetate (16.4g of C_2H_2Na or 27.2 g of $C_2H_3O_2Na.3H_2O$ in 1000ml).

x ml of A, y ml of B diluted to a total of 100ml.

x	y	pH
46.3	3.7	3.6
44.0	6.0	3.8
41.0	9.0	4.0
36.8	13.2	4.2
30.5	19.5	4.4
25.5	24.5	4.6
20.0	30.0	4.8
14.8	35.2	5.0
10.5	39.5	5.2
8.8	41.2	5.4
4.8	54.2	5.6

2. BORIC ACID-BORAX BUFFER

Stock solution

A:0.2M solution of boric acid (12.4g in 1000ml)

B:0.05 solution of borax (19.05g in 1000ml; 0.2M in terms of sodium borate).

50ml of A, xml of B, diluted to a total of 200ml.

x	pH
2.0	7.6
3.1	7.8
4.9	8.0
7.3	8.2
11.5	8.4
17.5	8.6
22.5	8.7
30.0	8.8
42.5	8.9
59.0	9.0
83.0	9.1
115.0	9.2

3. CARBONATE-BICARBONATE BUFFER

Stock solutions

A:0.2M solution of anhudrous sodium carbonate (21.2g in 1000ml)

B:0.2M solution of sodium bicarbonate (16.8g in 1000ml) x ml of A, y ml of B, diluted to a total of 200ml).

x	y	pH
4.0	46.0	9.2
7.5	42.5	9.3
9.5	40.5	9.4
13.0	37.0	9.5
16.0	34.0	9.6
19.5	30.5	9.7
22.0	28.0	9.8
25.0	25.0	9.9
27.5	22.5	10.0
30.0	20.0	10.1
33.0	17.0	10.2
35.5	14.5	10.3
38.5	11.5	10.4
40.5	9.5	10.5
42.5	7.5	10.6
45.0	5.0	10.7

4. CITRATRE BUFFER

Stock solution

A:0.1M solution of citric acid (21.01g in 1000ml)

B:0.1M solution of sodium citrate (29.41g of $C_6H_5O_7$ $Na_3.2H_2O$ in 1000ml)

x ml of A, yml of B diluted to a total of 100 ml.

x	y	pH	x	y	pH	x	y	pH
46.5	3.5	3.0	33.0	17.0	4.0	18.0	32.0	5.2
43.7	6.3	3.2	31.5	18.5	4.2	16.0	34.0	5.4
40.0	10.0	3.4	28.0	22.0	4.4	13.7	36.3	5.6
37.0	13.0	3.6	25.5	24.5	4.6	11.8	38.2	5.8
35.0	15.0	3.8	23.0	27.0	4.8	9.5	41.5	6.0
			20.5	29.5	5.0	7.2	42.8	6.2

5. GLYCINE–HCL BUFFER

Stock solution

A:0.2M Glycine (15.01g in 1000ml)

B:0.2N HCL

25ml of A, x ml of B diluted to a total of 100ml.

x	pH
22.0	2.2
16.2	2.4
12.1	2.6
8.4	2.8
5.7	3.0
4.1	3.2
3.2	3.4
2.5	3.6

6. PHOSPHATE BUFFER

Stock solution

A:0.2M solution of monobasic sodium phosphate (27.8g in 1000ml)

B:0.2M solution of di basic sodium phosphate (53.66g of $Na_2HPO_4.7H_2O$ or 71.7g if $Na_2HPO_4.12H_2O$ in 1000ml).

x ml of A, y ml of B diluted to a total of 200 ml.

x	y	pH	x	y	pH
93.5	6.5	5.7	45.0	55.0	6.9
92.0	8.0	5.8	39.0	61.0	7.0
90.0	10.0	5.9	33.0	67.0	7.1
87.7	12.3	6.0	28.0	72.0	7.2
85.0	15.0	6.1	23.0	77.0	7.3
81.5	18.5	6.2	19.0	81.0	7.4
77.5	22.5	6.3	16.0	84.0	7.5
73.5	26.5	6.4	13.0	87.0	7.6
68.5	31.5	6.5	10.5	89.5	7.7
62.5	37.5	6.6	8.5	91.5	7.8
56.5	43.5	6.7	7.0	93.0	7.9
51.0	49.0	6.8	5.3	94.7	8.0

7. TRIS (HYDROXYMETHYL) AMINOMETHANE (TRIS.HCL) BUFFER

Stock solutions

A:0.2M solution fo Tris (hydroxymethyl) aminomethane (24.2g in 1000ml)

B:0.2N HCl.

50ml of A, x ml of B diluted to a total of 200ml.

x	pH
5.0	9.0
8.1	8.8
12.2	8.6
16.5	8.4
21.9	8.2
26.8	8.0
32.5	7.8
38.4	7.6
41.4	7.4
44.2	7.2

8. GIYCINE (NAOH) BUFFER

Stock solutions

A:0.2M solution of Glycine (15.01g in 1000ml)

B:0.2M NaOH

50ml of A+ xml of B diluted to a total of 200ml.

x	pH
4.0	8.6
6.0	8.8
8.8	9.0
12.0	9.2
16.8	9.4
22.4	9.6
27.2	9.8
32.0	10.0
38.6	10.4
45.5	10.6

APPENDIX-II

AMINO ACID REPRESENTATION

Amino acid	Three letter representation	Single letter representation
Alanine	Ala	A
Valine	Val	V
Leucine	Leu	L
Isoleucine	Ile	I
Proline	Pro	P
Phenylalanine	Phe	F
Tryptophan	Trp	W
Methionine	Met	M
Glycine	Gly	G
Serine	Ser	S
Threonine	Thr	T
Cysteine	Cys	C
Tyrosine	Tyr	Y
Asparagine	Asn	N
Glutamine	Gln	Q
Aspartic acid	Asp	D
Glutamic acid	Glu	E
Lysine	Lys	K
Arginine	Arg	R
Histidine	His	H

APPENDIX–III

KEY TO NUMBERING AND CLASSIFICATION OF ENZYMES

1. OXIDOREDUCTASES

1.1 Acting on the CH–OH group of donors
- 1.1.1 With NAD or NADP as acceptor
- 1.1.2 With a cytochrome as acceptor
- 1.1.3 With O_2 as acceptor
- 1.1.99 With other acceptors

1.2 Acting on the aldehydes or keto group of donors
- 1.2.1 With NAD or NADP as acceptor
- 1.2.2 With a cytochrome as acceptor
- 1.2.3 With O_2 as acceptor
- 1.2.4 With lipoate as acceptor
- 1.2.99 With other acceptors

1.3 Acting on the CH–CH group of donors
- 1.3.1 With NAD or NADP as acceptor
- 1.3.2 With a cytochrome as acceptor
- 1.3.3 With O_2 as acceptor
- 1.3.99 With other acceptors

1.4 Acting on the CH–NH_2 group of donors
- 1.4.1 With NAD or NADP as acceptor
- 1.4.3 With O_2 as acceptor

1.5 Acting on the C-NH group of donors
- 1.5.1 With NAD or NADP as acceptor
- 1.5.3 With O_2 as acceptor

1.6 Acting on reduced NAD or NADP as donor
- 1.6.1 With NAD or NADP as acceptor

1.6.2 With a cytochrome as acceptor
1.6.4 With a disulphide compound as acceptor
1.6.5 With a quinone or related compounds as acceptor
1.6.6 With a nitrogenous group as acceptor
1.6.99 With other acceptors

1.7 Acting on other nitrogenous compounds as donors
1.7.1 With O_2 as acceptor
1.7.99 With other acceptors

1.8 Acting on sulphur groups of donors
1.8.1 With NAD or NADP as acceptor
1.8.3 With a O_2 as acceptor
1.8.4 With a disulphide compound as acceptor
1.8.5 With a quinone or related compounds as acceptor
1.8.6 With a nitrogenous group as acceptor

1.9 Acting on heme groups of donors
1.9.3 With O_2 as acceptor
1.9.6 With a nitrogenous group as acceptor

1.10 Acting on diphenols and related substances as donors
1.10.3 With O_2 as acceptor

1.11 Acting on H_2O_2 as acceptor

1.12 Acting on hydrogen as donor

1.13 Acting on single donors with incorporation of oxygen (oxygenases)

1.14 Acting on paired donors with incorporation of oxygen into one donor (hydroxylases)
1.14.1 Using reduced NAD or NADP as one donor
1.14.2 Using ascorbate as one donor
1.14.3 Using reduced pteridine as one donor

2. TRANSFERASES

2.1 Transferring one-carbon groups
2.1.1 Methyl transferases
2.1.2 Hydroxymethyl-formyl- and related transferases
2.1.3 Carboxyl-and carbamyl transferases
2.1.4 Amidino transferases

2.2 Transferring aldehydic or ketonic residues

2.3 Acyl transferases
- 2.3.1 Acyl transferases
- 2.3.2 Aminoacyl transferases

2.4 Glycosyl transferases
- 2.4.1 Hexosyl transferases
- 2.4.2 Pentosyl transferases

2.5 Transferring alkyl or related groups

2.6 Transferring nitrogenous groups
- 2.6.1 Amino transferases
- 2.6.3 Oximino transferases

2.7 Transferring phosphorus-containing groups
- 2.7.1 Phospho transferases with an alcohol group as acceptor
- 2.7.2 Phospho transferases with a carboxyl group as acceptor
- 2.7.3 Phospho transferases with a nitrogenous group as acceptor
- 2.7.4 Phospho transferases with a phospho- group as acceptor
- 2.7.5 Phospho transferases , apparently intramolecular
- 2.7.6 Pyrophospho transferases
- 2.7.7 Nucleotidyl transferases
- 2.7.8 Transferases for other substituted phospho-groups

2.8 Transferring sulphur- containing groups
- 2.8.1 Sulphur transferases
- 2.8.2 Sulpho transferases
- 2.8.3 CoA-transferases

3. HYDROLASES

3.1 Acting on ester bonds
- 3.1.1 Carboxylic ester hydrolases
- 3.1.2 Thiolester hydrolases
- 3.1.3 Phosphoric monoester hydrolases
- 3.1.4 Phosphoric diester hydrolases
- 3.1.5 Triphosphoric monoester hydrolases
- 3.1.6 Sulphuric ester hydrolases

3.2 Acting on glycosyl compounds
- 3.2.1 Glycoside hydrolases
- 3.2.2 Hydrolysing N-glycosyl compounds

3.2.3 Hydrolysing S-glycosyl compounds

3.3 Acting on ether bonds
- 3.3.1 Thioether hydrolases

3.4 Acting on peptide bonds (Peptide hydrolases)
- 3.4.1 α-Amino-acyl-peptide hydrolases
- 3.4.2 Peptidyl-amino acid hydrolases
- 3.4.3 Dipeptide hydrolases
- 3.4.4 Peptidyl-peptide hydrolases

3.5 Acting on C-N bonds other than peptide bonds
- 3.5.1 In linear amides
- 3.5.2 In cyclic amides
- 3.5.3 In linear amidines
- 3.5.4 In cyclic amidines
- 3.5.5 In cyanides
- 3.5.99 In other compounds

3.6 Acting on acid–anhydride bonds
- 3.6.1 In phosphoryl-containing anhydrides

3.7 Acting on C–C bonds
- 3.7.1 In ketonic substances

3.8 Acting on halide bonds
- 3.8.1 In C-halide compounds
- 3.8.2 In P-halide compounds

3.9 Acting on P-N bonds

4. LYASES

4.1 Carbon-carbon lyases
- 4.1.1 Carboxy lyases
- 4.1.2 Aldehyde lyases
- 4.1.3 Ketoacid lyases

4.2 Carbon-oxygen lyases
- 4.2.1 Hydrolyases
- 4.2.99 Other carbon-oxygen lyases

4.3 Carbon-nitrogen lyases
- 4.3.1 Ammonia lyases
- 4.3.2 Amidine lyases

4.4 Carbon-sulphur lyases

4.5 Carbon-halide lyases

4.99 Other lyases

5. ISOMERASES

5.1 Racemases and epimerases
- 5.1.1 Acting on amino acids and derivatives
- 5.1.2 Acting on hydroxyacids and derivatives
- 5.1.3 Acting on carbohydrates and derivatives
- 5.1.99 Acting on other compounds

5.2 Cis-trans isomerases

5.3 Intramolecular oxidoreductases
- 5.3.1 Interconverting aldoses and ketoses
- 5.3.2 Interconverting keto- and enol groups
- 5.3.3 Transposing C=C bonds

5.4 Intramolecular transferases
- 5.4.1 Transferring acyl groups
- 5.4.2 Transferring phosphoryl groups
- 5.4.99 Transferring other groups

5.5 Intramolecular lyases

5.99 Other isomerases

6. LIGASES

6.1 Forming C–O bonds
- 6.1.1 Amino acid-RNA ligases

6.2 Forming C–S bonds
- 6.2.1 Acid-thiol ligases

6.3 Forming C–N bonds
- 6.3.1 Acid-ammonia ligases (amide synthetases)
- 6.3.2 Acid-amino acid ligases (peptide synthetases)
- 6.3.3 Cyclo-ligases
- 6.3.4 Other C–N ligases
- 6.3.5 C–N ligases with glutamine as N-donor

6.4 Forming C–C bonds

APPENDIX-IV

PREFIXES USED IN THE INTERNATIONAL SYSTEM OF UNITS

Prefix	Multiple	Abbreviation
Exa	10^{18}	E
Peta	10^{15}	P
Tera	10^{12}	T
Giga	10^{9}	G
Mega	10^{6}	M
Kilo	10^{3}	k
deci	10^{-1}	d
centi	10^{-2}	c
milli	10^{-3}	m
micro	10^{-6}	μ
nano	10^{-9}	n
pico	10^{-12}	p
femto	10^{-15}	f
atto	10^{-18}	a

APPENDIX-V

GREEK ALPHABETS

Α	α	alpha	Ο	ο	omicron
Β	β	beta	Π	π	pi
Γ	γ	gamma	Ρ	ρ	rho
Δ	δ	delta	Σ	σ	sigma
Ε	ε	epsilon	Τ	τ	tau
Ζ	ζ	zeta	Υ	υ	upsilon
Η	η	eta	Φ	φ	phi
Θ	θ	theta	Χ	χ	chi
Ι	ι	iota	Ψ	ψ	psi
Κ	κ	kappa	Ω	ω	omega
Λ	λ	lambda	Μ	μ	mu
Ν	ν	nu	Ξ	ξ	xi

APPENDIX-VI

UNITS OF MEASUREMENTS

UNITS OF LENGTH

1 METRE = 100 CM

1 cm = 10 mm

1 mm = 10^3 μm

1 μm = 10^3 nm

1 nm = 10^2 A° or 10^{-9} metre

10^7A°

1 km = 0.062 miles

1 mile = 1.61 kilometers

5280 feet = 1 mile

1760 yards (8 furlongs) = 1 mile

1 light year = 9.461 x 10^{12} kilometers

1 cm = 0.3937 inch

1 inch = 2.5396 cm

1 meter = 3.281 feet

1 feet = 0.3048 metre

UNITS OF VOLUME

10^{-3} moles = 1 millimole

10^{-6} moles = 1 micromole

10^{-9} moles = 1 nanomole

10^{-12} moles = 1 picomoles

UNITS OF MASS

1 kg = 1000 gm 1 kg = 2.205 lbs. (USA)

1 g = 0.001 kg   1 lb=0.4536 kg

1 litre = 1000 ml

1 ml = 1000 μl

1 μl = 1000 nl

1 litre = 0.264 gallon (USA)

1 gallon = 3.785 litres

3.5 gallons = 1 barrel

1 mg = 1000 μg

1 g = 10^3 mg

1 g = 10^6 μg

1 g = 10^9 ng

1 g = 10^{12} pg

1 carat = 0.2 g (200mg)

1 ounce = 28.35 g

1 g = 0.0353 ounce

1 pound = 16 ounces

100 kg = 1 quintal

1000 kg = 1 tonne

GLOSSARY

acidosis A metabolic condition in which the capacity of the body to buffer H^+ is diminished; usually accompanied by decreased blood pH.

activation energy (ΔG^{++}) The amount of energy (in joules) required to convert all the molecules in 1 mole of a reacting substance from the ground state to the transition state.

activator (1) A DNA-binding protein that positively regulates the expression of one or more genes; that is, transcription rates increase when activators are bound to DNA. (2) A positive modulator of an allosteric enzyme.

active site The region of an enzyme surface that binds the substrate molecule and catalytically transforms it; also known as catalytic site.

ADP (adenosine diphosphate) A ribonucleoside 5′-diphosphate serving as phosphate group acceptor in the cell energy cycle.

alcohol fermentation The anaerobic conversion of glucose to ethanol via glycolysis. (See also fermentation)

aldose A simple sugar in which the carbonyl carbon atom is an aldehyde; that is, the carbonyl carbon is at one end of the carbon chain.

alkalosis A metabolic condition in which the capacity of the body to buffer OH^- is diminished; usually accompanied by an increase in blood pH.

allosteric enzyme A regulatory enzyme with catalytic activity modulated by the noncovalent binding of a specific metabolite at a site other than the active site.

allosteric site The specific site on the surface of an allosteric enzyme molecule to which the modulator or effector molecule is bound.

α helix A helical conformation of a polypeptide chain, usually right-handed, with maximal intrachain hydrogen bonding; one of the most common secondary structure in proteins.

aminoacid amino-substituted carboxylic acids, the building block of proteins.

aminotransferases Enzymes that catalyse the transfer of amino groups from one metabolite to another, also called transaminases.

amphoteric A compound capable of donating and accepting protons, and thus is able to serve both as an acid and as a base.

amphipathic compound A compound whose molecules contain both polar and nonpolar domains.

anabolism The phase of intermediary metabolism concerned with the energy-requiring biosynthesis of cell components from smaller precursor molecules.

antibiotics An antibiotic is a chemical substance produced by a living organism that demonstrates inhibitory or germicidal activity towards microorganisms.

apoenzyme The protein portion of an enzyme, exclusive of any organic or inorganic cofactors of prosthetic groups that might be required for activity.

asymmetric carbon atom A carbon atom that is covalently bonded to four different groups and thus may exist in two different tetrahedral configurations.

ATP (adenosine triphosphate) A ribonucleoside 5′–triphosphate functioning as a phosphate group donor in the cell energy cycle; carries chemical energy between metabolic pathways by serving as a shared intermediate, coupling endergonic and exergonic reactions.

β conformation An extended, zigzag arrangement of a polypeptide chain; a common secondary structure in proteins.

β oxidation Oxidation of fatty acids at the carbon atom in the β position to the carboxyl group, resulting in the splitting of the two terminal carbon atoms, leaving a fatty acid chain shorter than the original acid by two carbon atoms.

β turn A type of secondary structure in polypeptides consisting of four amino acid residues arranged in a tight turn so that the polypeptide turns back on itself.

base pair Two nucleotides in nucleic acid chains that are paired by hydrogen bonding of their bases; for example, A with T or U, and G with C.

bile salts Amphipathic steroid derivatives with detergent properties, participating in digestion and absorption of lipids.

binding energy The energy derived from noncovalent interactions between enzyme and substrate or receptor and ligand.

biomolecule An organic compound normally present as an essential component of living organisms.

biotin A vitamin; an enzymatic cofactor involved in carboxylation reactions.

bond energy The energy required to break a bond.

buffer A system capable of resisting

changes in pH, consisting of a conjugate acid–base pair in which the ratio of proton acceptor to proton donor is near unity.

calmodulin It is a calcium binding protein, that activate certain enzymes like phosphorylase kinase, adenylate cyclase etc.

carbohydrate A polyhydroxylic aldehyde or ketone

catabolism The phase of metabolism involved in the energy-yielding degradation of nutrient molecules.

catabolite activator protein (CAP) A specific regulatory protein that controls initiation of transcription of the genes of enzymes required for a cell to use some other nutrient when glucose is lacking.

c AMP see cyclic AMP.

catecholamines Hormones such as adrenaline, that are amino derivatives of catechol.

chromatography A process in which complex mixtures of molecules are separated by many repeated partitionings between a flowing (mobile) phase and a stationary phase.

cistron A unit of DNA or RNA corresponding to one gene.

citric acid cycle A cyclic enzymatic reaction for the oxidation of acetyl residues to carbon dioxide, in which formation of citrate is the first step; also known as the Krebs cycle or tricarboxylic acid cycle.

coenzyme An organic cofactor required for the action of certain enzymes; often contains a vitamin as a component.

coenzyme A A pantothenic acid-containing coenzyme serving as an acyl group carrier in certain enzymatic reactions.

column chromatography It is a separation process involving the uniform percolation of a liquid through a column packed with finely divided material. The separation in the column is effected either by direct interaction between the solute components and the surface of the stationary phase or by adsorption of solute by the stationary phase.

competitive inhibition A type of enzyme inhibition reversed by increasing the substrate concentration; a competitive inhibitor generally competes with the normal substrate or ligand for a protein's binding site.

configuration The arrangement in space of substituent groups around an asymmetric carbon atom.

conformation The three-dimensional shape or form of a macromolecule.

conjugate acid–base pair A proton donor and its corresponding deprotonated species; e.g. acetic acid (donor) and acetate (acceptor).

conjugated protein A protein containing a metal or an organic prosthetic group or both.

corticosteroids Steroid hormones formed by the adrenal cortex.

covalent bond A chemical bond that involves sharing of electron pairs.

cyclic AMP (cAMP) A second messenger within cells; its formation by adenyl cyclase is stimulated by certain hormones or other molecular signals.

cytochromes Heme proteins serving as electron carriers in respiration, photosynthesis, and other oxidation–reduction reactions.

deamination The enzymatic removal of amino groups from biomolecules such as amino acids or nucleotides.

dehydrogenases Enzymes catalysing the removal of pairs of hydrogen atoms from their substrates.

denaturation Partial or complete unfolding of the specific native conformation of a polypeptide chain, protein, or nucleic acid.

denatured protein A protein that has lost its native conformation by exposure to a destabilising agent such as heat or detergent.

deoxyribonucleic acid See DNA.

diabetes mellitus A metabolic disease resulting from insulin deficiency, characterised by a failure in glucose transport from the blood into cells at normal glucose concentrations.

dipole A molecule having both positive and negative charges.

disaccharide A carbohydrate consisting of two covalently joined monosaccharide units.

dissociation constant (1) An equilibrium constant (K_d) for the dissociation of a complex of two or more biomolecules into its components; e.g. dissociation of a substrate from an enzyme. (2) The dissociation constant (K_a) of an acid, describing its dissociation into its conjugate base and a proton.

disulphide bridge A covalent cross-link between two polypeptide chains formed by cystine residues (S-S).

DNA (deoxyribonucleic acid) A polynucleotide having a specific sequence of deoxyribonucleotide units covalently joined through 3′-5′-phopshodiester bonds; serves as the carrier of genetic information.

double helix The natural coiled conformation of two complementary antiparallel DNA chains.

electron transport Movement of electrons from substrates to oxygen, promoted by the respiratory chain.

electrophoresis Transport of charged solutes in response to an electrical field; often used to separate mixtures of ions.

elongation factors Specific proteins required in the elongation of polypeptide chains by ribosomes.

end-product inhibition See feedback inhibition.

enzyme A protein that catalyses a specific chemical reaction. It does not affect the equilibrium of the catalysed reaction; it enhances the rate of a reaction by providing a reaction path with lower activation energy.

epimerases Enzymes that catalyse the reversible interconversions of two epimers.

epimers Two stereoisomers differing in configuration at one asymmetric centre, in a compound having two or more asymmetric centres.

equilibrium The state of a system in which no further net change is occurring; the free energy is at a minimum.

equilibrium constant (K_{eq}) A constant, characteristic for each chemical reaction; relates the specific concentrations of all reactants and products at equilibrium at a given temperature and pressure.

essential amino acids Amino acids that cannot be synthesised by humans (and other vertebrates) and must be obtained from the diet.

essential fatty acids The group of polyunsaturated fatty acids produced by plants, but not by humans; required in the human diet.

FAD (flavin adenine dinucleotide) The coenzyme of some oxidation–reduction enzymes; it contains riboflavin.

fatty acid A long-chain aliphatic carboxylic acid found in natural fats and oils; also a component of membrane phospholipids and glycolipids.

feedback inhibition Inhibition of an allosteric enzyme at the beginning of a metabolic sequence by the end product of the sequence; also known as end-product inhibition.

fermentation Energy-yielding anaerobic breakdown of a nutrient molecule, such as glucose without net oxidation; yields lactate, ethanol, or some other simple products.

fibrous proteins Insoluble proteins that serve in a protective or structural role; contain polypeptide chains that generally share a common secondary structure.

Fischer projection formula See projection formula.

flavin nucleotides Nucleotide coenzymes (FMN and FAD) containing riboflavin.

flavo protein An enzyme containing flavin nucleotide as a prosthetic group.

gel filtration A chromatographic technique for separation of a mixture of molecules on the basis of molecular size.

globular proteins soluble proteins where the polypeptide chain is tightly folded in three dimension to yield globular shape.

glucogenic amino acids Amino acid with carbon chains that can be metabolically converted into glucose or glycogen via gluconeogenesis.

gluconeogenesis The biosynthesis of a carbohydrate from simpler, noncarbohydrate precursor such as oxaloacetate or pyruvate.

glycan Another term for polysaccharide; a polymer of monosaccharide units joined by glycosidic bonds.

glycerophospholipid An amphipathic lipid with a glycerol backbone; fatty acids are ester-linked to C-1 and C-2 of glycerol and a polar alcohol is attached through a phosphodiester linkage to C-3.

glycolipid A lipid containing a carbohydrate group.

glycolysis The catabolic pathway by which a molecule of glucose is broken down into two molecules of pyruvate.

glycoprotein A protein containing a carbohydrate group.

glycosidic bonds Bonds between a sugar and another molecule (typically an alcohol, purine, pyrimidine or sugar) through intervening oxygen.

Haworth perspective formula A method for representing cyclic chemical structures so as to define the configuration of each substituent group; the method commonly used for representing sugars.

heat of vapourisation The number of calories required to convert 1 g of a liquid to the vapour state at the same temperature.

helix See also α helix.

heme The iron porphyrin prosthetic group of heme protein.

heme protein A protein containing a heme as a prosthetic group.

heteropolysaccharide A polysaccharide containing more than one type of sugar.

hexose A simple sugar with a backbone containing six carbon atoms.

high-energy compound A compound that on hydrolysis undergoes a large decrease in free energy under standard conditions.

histones The family of five basic proteins that associate tightly with DNA in the chromosomes of all eukaryotic cells.

holoenzyme A catalytically active enzyme including all necessary subunits, prosthetic groups and cofactors.

homeostasis The maintenance of a dynamic steady state by regulatory mechanisms that compensate for changes in external circumstances.

homopolysaccharide A polysaccharide made up of only one type of monosaccharide unit.

hormone A chemical substance synthesised in small amounts by an endocrine tissue and carried in the blood to another tissue, where it acts as a messenger or regulates the function of the target tissue or organ.

hormone receptor A protein in or on the surface of target cells that binds a specific hormone and initiates the cellular response.

hyaluronic acid Hyaluronic acid is acidic polysaccharide and forms larger complexes (proteoglycans) with proteins and other acidic polysaccharides.

hydrogen bond A weak electrostatic attraction between one electronegative atom (such as oxygen or nitrogen) and a hydrogen atom covalently linked to a second electronegative atom.

hydrolases Enzymes (proteases, lipases, phosphatases, nucleases) that catalyse hydrolysis reactions.

hydrolysis Cleavage of a bond, such as an anhydride or peptide bond, by the addition of the elements of water, yielding two or more products.

hydronium ion The hydrated hydrogen ion (H_3O^+).

hydrophilic Polar molecules that are soluble in water.

hydrophobic Nonpolar molecules or groups that are insoluble in water.

induced fit A change in the conformation of an enzyme in response to substrate binding that renders the enzyme catalytically active; also used to denote changes in the conformation of any macromolecule in response to ligand binding such that the binding site of the macromolecule better conforms to the shape of the ligand.

induction An increase in the expression of a gene in response to a change in the activity of a regulatory protein.

initiation code AUG (sometimes GUG in prokaryotes) codes for the first amino acid in a polypeptide sequence N-formylmethionine in prokaryotes, and methionine in eukaryotes.

ion-exchange chromatography It is based upon the simple fact that different cations (or anions) have different capacity to undergo exchange reaction on the surface of a given exchanger. The capacity of an ion to undergo exchange reaction has been found to depend upon the charge and the size of the hydrated ion in solution.

ion-exchange resin A polymeric resin that contains fixed charge groups and is used in chromatographic columns to separate ionic compounds.

ion product of water (K_w) The product of the concentrations of H^+ and OH^- in pure water $K_w = [H^+][OH^-] = 1 \times 10^{-14}$ at 25°C.

isoelectric pH (isoelectric point) The pH at which a solute has no net electric charge and thus does not move in an electric field.

isoenzymes See isozymes.

isomerases Enzymes that catalyse the transformation of compounds into their positional isomers.

isomers Any two molecules with the same molecular formula but different arrangement of molecular groups.

isozymes Multiple forms of an enzyme that catalyse the same reaction but differ from each other in their amino acid sequence, substrate affinity, V_{max} and / or regulatory properties; also called isoenzymes.

keratins Insoluble protective or structural proteins consisting of parallel polypeptide chains in α-helical or β-conformations.

ketogenic amino acids Amino acids with carbon skeletons that can serve as precursors of the ketone bodies.

ketone bodies Acetoacetate, D-β-hydroxybutyrate, and acetone are produced during partial oxidation of fatty acids.

ketose A simple monosaccharide having its carboxyl group at a position other than the terminal position.

ketosis A condition in which the ketone body concentration of the blood, tissues and urine is abnormally high.

kinases Enzymes that catalyse the phosphorylation of certain molecules by ATP.

Krebs cycle See citric acid cycle.

ligand A small molecule that binds specifically to a larger one; for example, a hormone is the ligand for its specific protein receptor.

ligases Enzymes that catalyse condensation reactions in which two atoms are joined using the energy of ATP or another energy-rich compound.

Lineweaver-Burk equation An algebraic transform of the Michaelis-Menten equation, allowing determination of V_{max} and K_m by extrapolation of [S] to infinity.

lipases Enzymes that catalyse the hydrolysis of lipids.

lipid A small water-insoluble biomolecule generally containing fatty acids, and glycerol.

lipoate (lipoic acid) A vitamin for some microorganisms; an intermediate carrier of hydrogen atoms and acyl groups in a-keto acid dehydrogenases.

liposome A small, spherical vesicle composed of a phospholipid bilayer, which forms spontaneously when phospholipids are suspended in an aqueous buffer.

low-energy phosphate compound A

phopshorylated compound with a relatively small standard-free-energy of hydrolysis.

lyases Enzymes that catalyse the removal of a group from a molecule to form a double bond, or the addition of a group to a double bond.

lysis Destruction of a cell's plasma membrane or of a bacterial cell wall, releasing the cellular contents and killing the cell.

messenger RNA (mRNA) A class of RNA molecules, each of which is complementary to one strand of DNA; carries the genetic messages from the chromosome to the ribosomes.

metabolism The entire set of enzyme-catalysed transformations of organic molecules in living cells; the sum of anabolism and catabolism.

metalloprotein A protein having a metal ion as its prosthetic group.

Michaelis constant (K_m) The substrate concentration at which an enzyme-catalysed reaction proceeds at one-half its maximum velocity.

Michaelis–Menten equation The equation describing the hyperbolic dependent of the initial reaction velocity, V_0 on substrate concentration[S], in many enzyme-catalysed reactions.

$$V = \frac{V_{max}(S)}{K_m + (S)}$$

Michaelis–Menten kinetics A kinetic pattern in which the initial rate of an enzyme-catalysed reaction exhibits a hyperbolic dependence on substrate concentration.

mixed inhibition The reversible inhibition pattern resulting when an inhibitor molecule can bind to either the free enzyme or to the enzyme–substrate complex (not necessarily with the same affinity).

modulator A metabolite that, when bound to the allosteric site of an enzyme, alters its kinetic characteristics.

molar solution One mole of solute dissolved in water to give a total volume of 1,000 ml.

mole One gram molecular weight of a compound .

mucopolysaccharide An older name for glycosaminoglycan.

multienzyme system A group of related enzymes participating in a given metabolic pathway.

NAD, NADP (nicotinamide adenine dinucleotide, nicotinamide adenine dinucleotide phosphate) Nicotinamide-containing coenzymes functioning as carriers of hydrogen atoms and electrons in some oxidation–reduction reactions.

negative feedback Regulation of a biochemical pathway achieved when a reaction product inhibits an earlier step in the pathway.

nicotinamide adenine dinucleotide, nicotinamide adenine dinucleotide phosphate See NAD,NADP.

ninhydrin reaction A colour reaction given by amino acids and peptides on heating with ninhydrin; widely used for their detection and estimation.

noncompetitive inhibition Noncompetitive inhibition is defined as that one in which substrate and inhibitor do not exhibit mutual competition. The sites of attachment of substrate and noncompetitive inhibitor are different on the enzyme molecules.

nonessential amino acids Amino acids that can be synthesised by humans and other vertebrates from simpler precursors, and are thus not required in the diet.

nonheme iron protein Proteins, usually acting in oxidation–reduction reactions, containing iron but no porphyrin groups.

nonpolar Hydrophobic; molecules or groups that are poorly soluble in water.

nucleic acids Biologically occurring polynucleotides in which the nucleotide residues are linked in a specific sequence by phosphodiester bonds; DNA and RNA.

nucleoside A compound consisting of a purine or pyrimidine base covalently linked to a pentose.

nucleotide A nucleoside phopshorylated at one of its pentose hydroxyl groups.

oligopeptide A few amino acids joined by peptide bonds.

oligosaccharide Several monosaccharide groups joined by glycosidic bonds.

operator A region of DNA that interacts with a repressor protein to control the expression of a gene or group of genes.

operon a unit of genetic expression consisting of one or more related genes and the operator and promotor sequences that regulate their transcription.

optimun pH The characteristic pH at which an enzyme has maximal catalytic activity.

oxidases Enzymes that catalyse oxidation reactions in which molecular oxygen serves as the electron acceptor, but neither of the oxygen atoms is incorporated into the product.

oxidation The loss of electrons from a compound.

oxidation–reduction reaction A reaction in which electrons are transferred from a donor to an acceptor molecule; also called a redox reaction.

oxidative phosphorylation The enzymatic phosphorylation of ADP to ATP coupled to electron transfer from a substrate to molecular oxygen.

pentose a simple sugar with a backbone containing five carbon atoms.

pentose phosphate pathway A pathway that serves to interconvert hexoses and pentoses and is a source of reducing equivalents and pentoses for biosynthetic processes; present in most organisms. Also called the phosphogluconate pathway.

peptidases Enzymes that hydrolyse peptide bonds.

peptide Two or more amino acids covalently joined by peptide bonds.

peptide bond A substituted amide linkage between the α-amino group of one amino acid and the α- carboxyl group of another, with the elimination of water.

pH The negative logarithm of hydrogen ion concentration of an aqueous solution.

phosphatases Enzymes that hydrolyse a phosphate ester or anhydride, releasing inorganic phosphate, P_i.

phosphodiester linkage A chemical grouping that contains two alcohols esterified to one molecule of phosphoric acid, which thus serves as a bridge between them.

phosphogluconate pathway (pentose phosphate pathway) An oxidative pathway beginning with glucose 6-

phosphate and leading, via 6-phosphogluconate, to pentose phosphates and yielding NADPH.

phospholipid A lipid containing one or more phosphate groups.

phosphorylation Formation of a phosphate derivative of a biomolecule, usually by enzymatic transfer of a phopshoryl group from ATP.

pK_a The negative logarithm of an equilibrium constant.

pleated sheet The side-by-side, hydrogen-bonded arrangement of polypeptide chains in the extended β-conformation.

polar Hydrophilic, or "water-loving"; molecules or groups that are soluble in water.

polymerase chain reaction (PCR) A repetitive procedure that results in a geometric amplification of a specific DNA sequence.

polynucleotide A covalently linked sequence of nucleotides in which the 3′ hydroxyl of the pentose of one nucleotide residue is joined by a phosphodiester bond to the 5′ hydroxyl of the pentose of the next residue.

polypeptide A long chain of amino acids linked by peptide bonds; the molecular weight is generally less than 10,000.

polysaccharide A linear or branched polymer of monosaccharides units linked by glycosidic bonds.

porphyrin Complex nitrogenous compound containing four substituted pyrroles covalently joined into a ring; often complexed with a central metal atom.

projection formula A method for representing molecules to show the configuration of groups around chiral centres; also known as Fischer projection formula.

prostaglandins A class of lipid-soluble, hormone-like regulatory molecules derived from arachidonate and other polyunsaturated fatty acids.

prosthetic group A metal ion or an organic compound (other than an amino acid) that is covalently bound to a protein and is essential for its activity.

protein A macromolecule composed of one or more polypeptide chains, each with a characteristic sequence of amino acids linked by peptide bonds.

protein kinase Enzymes that transfer the terminal phosphoryl group of ATP or another nucleoside triphosphate to a Ser, Thr, Tyr, Asp, or His side chain in a target protein, thereby regulating the activity or other properties of that protein.

proteoglycan A hybrid macromolecule consisting of an oligo or polysaccharide joined to a polypeptide.

purine A nitrogenous heterocyclic base found in nucleotides and nucleic acids; containing fused pyrimidine and imidazole rings.

pyridoxal phosphate A coenzyme containing the vitamin pyridoxine (vitaminB_6) and functions in reactions involving amino group transfer.

pyrimidine A nitrogenous heterocyclic base found in the nucleotides and nucleic acids.

quaternary structure The three-dimensional structure of a multisubunit protein; particularly the manner in which the subunits fit together.

reduction The gain of electrons by a compound.

regulatory enzyme An enzyme having a regulatory function through its capacity to undergo a change in catalytic activity by allosteric mechanisms or by covalent modification.

regulatory gene A gene that gives rise to a product involved in the regulation of expression of another gene; for example, a gene coding for a repressor protein.

renaturation Refolding of an unfolded (denatured) globular protein so as to restore native structure and protein function.

repression A decrease in the expression of the gene in response to a change in the activity of the regulatory protein.

repressor The protein that binds to the regulatory sequence or operator for a gene, blocking its transcription.

respiration-linked phosphorylation ATP formation from ADP and P_i, driven by electron flow through a series of membrane-bound carriers, with a proton gradient as the direct source of energy driving rotational catalysis by ATP synthase.

respiratory chain The electron transfer chain; a sequence of electron-carrying proteins that transfer electrons from substrates to molecular oxygen in aerobic cells.

ribonuclease A nuclease that catalyses the hydrolysis of certain internucleotide linkages of RNA.

ribonucleic acid See RNA.

ribosomal RNA (r RNA) A class of RNA molecules serving as components of ribosomes.

ribosome A supramolecular complex of rRNAs and proteins, approximately 18 to 22nm in diameter; the site of protein synthesis.

ribozymes Ribonucleic acid molecules with catalytic activities; RNA enzymes.

RNA (ribonucleic acid) A polyribonucleotide of a specific sequence linked by successive 3, 5 phospho diester bonds.

RNA polymerase An enzyme that catalyses the formation of RNA from ribonucleoside 5-triphosphates, using a strand of DNA or RNA as template.

saponification Alkaline hydrolysis of triacylglycerols to yield fatty acids as soaps.

saturated fatty acids These are fatty acids which do not contain double bonds.

second messenger An effector molecule synthesised within a cell in response to an external signal (first messenger) such as hormones.

secondary structure The residue-by-residue conformation of the backbone of a polymer.

simple protein A protein yielding only amino acids on hydrolysis.

specific activity The number of micromoles (μ mol) of a substrate transformed by an enzyme preparation per minute per milligram of protein at 25°C; a measure of enzyme purity.

specific heat The amount of energy (in joules or calories) needed to raise the temperature of 1 g of a pure substance by 1°C.

substrate The specific compound acted upon by an enzyme.

substrate level phosphorylation Phosphorylation of ADP or some other nucleoside 5 –diphosphate coupled to the dehydrogenation of an organic substrate; independent of the electron transfer chain.

suicide inhibitor A relatively inert molecule that is transformed by an enzyme, at its active site, into a reactive substance that irreversibly inactivates the enzyme.

synthases Enzymes that catalyse condensation reactions in which no nucleoside triphosphate is required as an energy source.

synthetases Enzymes that catalyse condensation reactions using ATP or another nucleoside triphosphate as an energy source.

tertiary structure The three dimensional-conformation of a polymer in its native folded state.

thioester An ester of a carboxylic acid with a thiol or mercaptan.

toxins Proteins produced by some organisms and toxic to certain other species.

transaminases See aminotransferases.

transamination Enzymatic transfer of an amino group from an α-keto acid.

transcription The enzymatic process whereby the genetic information contained in one strand of DNA is used to specify a complementary sequence of bases in an mRNA chain.

transfer RNA(tRNA) A class of RNA molecule each of which combines covalently with a specific amino acid as the first step in protein synthesis.

translocase (1) An enzyme that catalyses membrane transport, (2) An enzyme that causes a movement , such as the movement of a ribosome along mRNA.

triacylglycerol An ester of glycerol with three molecules of fatty acid; also called a triglyceride or neutral fat.

tricarboxylic acid cycle See citric acid cycle.

triose A simple sugar with a backbone containing three carbon atoms.

tRNA See transfer RNA.

tropic hormone (tropin) A peptide hormone that stimulates a specific target gland to secrete its hormone; for example, thyrotropin produced by the pituitary stimulates secretion of thyroxine by the thyroid.

uncompetitive inhibition The reversible inhibition pattern resulting when an inhibitor molecule can bind to the enzyme–substrate complex but not to the free enzyme.

unsaturated fatty acid A fatty acid containing one or more double bonds.

urea cycle A metabolic pathway in vertebrates for the synthesis of urea from amino groups and carbon dioxide; occurs in the liver.

ureotelic Excreting excess nitrogen in the form of urea.

uricotelic Excreting excess nitrogen in the form of urate (uric acid).

V_{max} The maximum velocity of an enzymatic reaction when the binding site is saturated with substrate.

vitamin An organic substance required in small quantities in the diet of some species; generally functions as a component of a coenzyme.

zwitter ion A dipolar ion, with spatially separated positive and negative charges.

zymogen An inactive precursor of an enzyme; e.g. pepsinogen, the precursor of pepsin.

REFERENCES

Adam, R.L.P., Knowler, J.T. and Leader, D.P., *The Biochemistry of Nucleic acids*. Tenth Edition. Chapman and Hall Publishers, London, New York. 1986

Harper, H.A., *Review of Physiological Chemistry*. Muruzen Asian Ed. 1973

Jain, J. L., *Fundamentals of Biochemistry*. Chand and Company (Pvt.) Ltd. 1988

Martin, D. W., Mayes, P.A. and Rodwell, W.W., *Harper's Review of Biochemistry*. Lange Medical Publications. 1983

Nelson, D.L. and Cox M.M., *Lehninger Principles of Biochemistry*. Third Edition. MacMillan, Worth Publishers. 2001

Stryer, L., *Biochemistry*. Wiley International. 1992

West, E.S., Todd, W.R., Mason, H.S. and Van brugen, J.T., *Text Book of Biochemistry*, Macmillan Co. 1963

White, A., Handler, P. and Smith, E.L., *Principles of Biochemistry*. Fifth Edition. McGraw-Hill Koagakusha Publishers. 1973

INDEX

www.ingramcontent.com/pod-product-compliance
Lightning Source LLC
Chambersburg PA
CBHW022051210326
41519CB00054B/308

* 9 7 8 8 1 8 0 9 4 0 0 4 0 *